Heating Boiler Operator's Manual

ABOUT THE AUTHOR

Mohammad A. Malek, Ph.D., P.E., is an internationally recognized expert in boiler and pressure vessel technology. He is a professional engineer registered in both the United States and Canada. He has more than 30 years of experience in design, construction, installation, operation, maintenance, inspection, and repair of boilers and pressure vessels. He has published numerous technical articles and authored chapters in a few books. Dr. Malek is a member of the American Society of Mechanical Engineers, National Society of Professional Engineers, American Society of Safety Engineers, Association of Energy Engineers, Association for Facilities Engineers, National Association of Power Engineers, Florida Engineering Society, and Society of Operations Engineers, U.K. He is a speaker for ASME Code Section I—Power Boilers. Currently he is chief boiler inspector for the State of Florida. Dr. Malek is an adjunct professor at the FAMU-FSU College of Engineering, Tallahassee, Florida. He has previously authored the following books: *Power Boiler Design, Inspection, and Repair* (2005) and *Pressure Relief Devices* (2006), both from McGraw-Hill.

Heating Boiler Operator's Manual

Maintenance, Operation, and Repair

Mohammad A. Malek, Ph.D., P.E.

McGraw-Hill

New York Chicago San Francisco Lisbon London Madrid
Mexico City Milan New Delhi San Juan Seoul
Singapore Sydney Toronto

The McGraw-Hill Companies

1 2 3 4 5 6 7 8 9 0 DOC/DOC 0 1 2 1 0 9 8 7 6

ISBN-13: 978-0-07-147522-8
ISBN-10: 0-07-147522-2

The sponsoring editor for this book was Kenneth P. McCombs and the production supervisor was Richard C. Ruzycka. It was set in Century Schoolbook by International Typesetting and Composition. The art director for the cover was Anthony Landi.

Printed and bound by RR Donnelley.

This book is printed on acid-free paper.

This book is dedicated to my beloved daughters Anjali and Emily.

Contents

Preface

This book covers complete technology starting from definitions to design, construction, installation, operation, maintenance, and inspection of all types of heating boilers. The book refers to the requirements of the American Society of Mechanical Engineer's (ASME) Boiler and Pressure Vessel Code applicable to heating boilers. The ASME Code Section IV describes rules for construction of heating boilers whereas ASME Section Code VI is recommended rules for the care and operation of heating boilers.

Heating boilers are low-pressure boilers used for steam heating, hot water heating, and domestic hot water supply. Approximate 80 percent of the total boilers in use are heating boilers and the remaining 20 percent are high-pressure boilers or power boilers. Perhaps every building you work or live in has a low-pressure boiler. Heating boilers are found in office buildings, residential buildings, apartment buildings, schools, universities, hospitals, nursing homes, day care centers, hotels, motels, and other locations where heating the building or hot water is required.

Low working pressure of heating boilers does not mean that it is not dangerous. Failure of any pressure parts or auxiliary systems of heating boilers can lead to severe explosion. In fact, heating boilers cause more number of accidents than caused by the high-pressure boilers.

I am thankful to my friends, manufacturers, suppliers, repairers, installers, inspectors, insurance companies, and jurisdictions for providing valuable information for the book. Also, I am thankful to many Web sites from where I collected information on the latest technology. The information presented in this book is to educate the reader on heating boiler technology. I will consider my endeavor successful if the reader can apply information from the book to make any heating boilers safe protecting human lives and property.

Mohammad A. Malek, Ph.D., P.E.

Chapter

1

Heating Boiler Basics

A boiler is a closed vessel where water or other liquid is heated, steam or vapor is generated, steam is superheated, or any combination of these functions is accomplished under pressure or vacuum for use external to itself by the direct application of energy from the combustion of fuels, or from electricity, or solar energy. The term *boiler* includes fired units, separate from processing systems and complete within themselves, for heating or vaporizing liquids other than water.

In 1914, the American Society of Mechanical Engineers (ASME) published the first boiler code with the title *Rules for the Construction of Stationary Boilers and for Allowable Working Pressures.* The contents of the rules were: Section I—Power Boilers; Section II—Heating Boilers, under the heading Part I—New Installations. The ASME has expanded the Code over a period of time and latest boiler and pressure vessel codes are:

- ASME Section I—Rules for Construction of Power Boilers
- ASME Section II—Materials
- ASME Section III—Nuclear Systems
- ASME Section IV—Rules for Construction of Heating Boilers
- ASME Section V—Nondestructive Examination
- ASME Section VI—Recommended Rules for the Care and Operation of Heating Boilers
- ASME Section VII—Recommended Guidelines for the Care of Power Boilers
- ASME Section VIII—Rules for Construction of Pressure Vessels
- ASME Section IX—Welding and Brazing Qualifications

- ASME Section X—Fiber-Reinforced Plastic Pressure Vessels
- ASME Section XI—Rules for In-service Inspection of Nuclear Power Plant Components
- ASME Section XII—Rules for Construction and Continued Service of Transport Tanks

Based on the construction, the boilers are classified as *power boilers* and *heating boilers*, according to the ASME Boiler and Pressure Vessel Code (Figure 1.1).

A power boiler is defined as a boiler in which steam or vapor is generated at a pressure more than 15 psi for use external to itself. A power boiler is constructed in accordance with the rules of ASME Section I—Rules for Construction of Power Boilers. These rules cover construction of power boilers, which include high-temperature water boilers, miniature boilers, electric boilers, and organic fluid vaporizers. A power boiler is considered a high-pressure boiler, and it may be designed for up to a maximum working pressure of 5000 psi.

A heating boiler is defined as a boiler in which steam or other vapor is generated at a pressure less than 15 psi, or a boiler in which hot water is generated at pressures not exceeding 160 psi and/or temperatures not exceeding 250°F. Like the power boiler, a heating boiler is constructed in accordance with the rules of ASME Section IV—Rules for Construction of Heating Boilers. These rules cover construction of steam

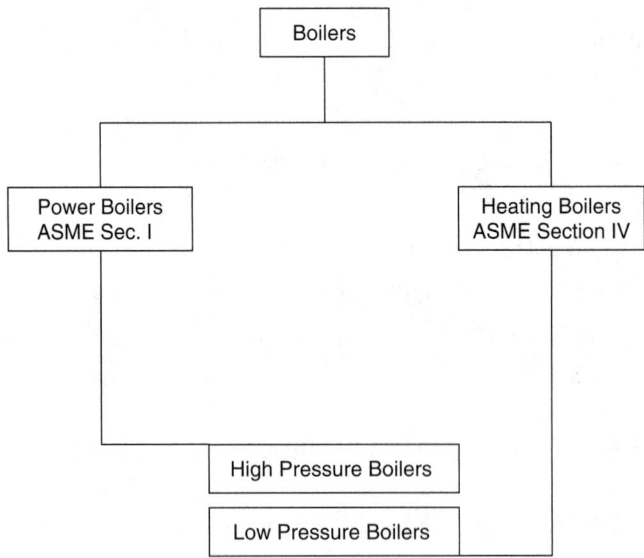

Figure 1.1 Classification of boilers according to ASME Code.

heating boilers, hot water heating boilers, hot water supply boilers, and water heaters. A heating boiler is considered a low-pressure boiler.

1.1 Hot Water and Steam

Hot water is generated by hot water boilers, which include hot water heating boilers, hot water supply boilers, and hot water heaters. Hot water is water to which heat energy has been added. The hot water becomes hotter when more heat is added. A system design engineer knows the temperatures of hot water needed for various applications through consultation with the users.

Steam is generated by steam boilers. The primary purpose of an effective steam generation system is to produce steam under the conditions—flow rates and pressures—required for the systems end-use requirements. A system design engineer knows how to maximize the effective use and heat content of the steam needed for various end-use applications.

1.1.1 Properties of hot water

Water is everywhere—it defines our planet. Water is composed of two hydrogen atoms and one oxygen atom. Water is unique as it is the only natural substance that is found in all three states—liquid, solid (ice), and gas (steam). Water freezes at 32°F and boils at 212°F at sea level.

Appendix M shows properties of water. Some of the other important properties of water are:

- Water has a high specific heat index.
- Water has a very high surface tension.
- Water has a neutral pH of 7, which is neither acidic nor basic.
- Water is a universal solvent.
- Water absorbs a large amount of heat.

A hot water boiler generates hot water for residential and commercial uses other than space heating. The maximum temperature of outlet water is 210°F (98.5°C). A system designer should know the coldest temperature in order to determine temperature rise.

In practice, a system designer establishes design temperatures for various applications. Figure 1.2 shows a water temperature guide for typical water heating system design temperatures.

1.1.2 Properties of saturated steam

Steam is water in a vapor condition. Change of water to steam takes place at a temperature that is dependent on the pressure inside the vessel. At atmospheric pressure (0 psi), water boils at 212°F. Water is releasing steam

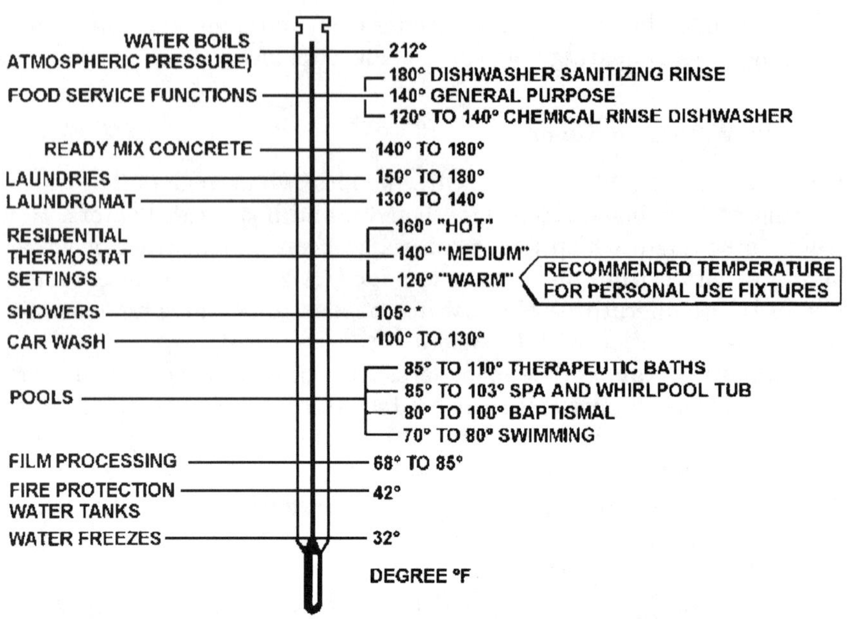

Figure 1.2 Temperature guide for system design.

at 212°F while the water is also at 212°F. When pressure is increased to 50 psi, the water boils at 297.7°F and the steam is also at 297.7°F.

The boiling point of water increases with increase in pressure until the critical pressure-temperature is reached. Critical pressure is 3206 psia and its corresponding temperature is 705°F. When this pressure-temperature point is reached, water is converted directly to steam.

Steam today is an essential part of modern society. Without steam, our food, textile, medical, power, heating, and transport industries could not exist. Steam provides controllable amounts of energy from a boiler room, where it is generated, to the point of use. Since most of the heat content of steam is stored as latent heat, large amounts of heat can be transferred efficiently at constant temperature for many heating applications.

Steam is one of the most widely used mediums for conveying heat energy. Steam has the following advantages for delivering energy:

- Economic to generate.
- Easy to distribute.
- Easy to control.
- Easy to transfer to the process.

- Steam is flexible.
- Steam plant is easy to manage.

There are two types of steam: saturated steam and superheated steam. When steam is removed from its liquid and heated to a higher temperature, it is superheated a given number of degrees. The number of degrees of superheat is the difference between saturated steam temperature at the given pressure and actual steam temperature.

Saturated steam is a vapor at a temperature that corresponds with its pressure. Any removal of heat from steam causes a portion of the steam to convert into a liquid state. For example, water is heated and starts to boil at 212°F and continues to boil until the pressure changes.

Superheated steam is a vapor at a temperature above its corresponding pressure. Heat is added to the steam in the super heater after it is removed from its liquid. Steam pressure leaving the super heater remains the same but its temperature increases.

A heating boiler produces saturated steam. The properties of saturated steam are shown in Appendix N.

1.2 History of Heating Boilers

The 1914 edition of the ASME Code—*Rules for the Construction of Stationary Boilers and Allowable Working Pressures*—was revised in 1918. The ASME Boiler Code Committee adopted the new Code edition on December 3, 1918. At that time, revision of Part I, Section II was incomplete.

The ASME appointed a fourteen-member Subcommittee on Heating Boilers on September 26, 1919, with S.F. Jeter as the chairperson, to revise the section of the Code that discussed heating boilers. The Subcommittee presented its final report to the Boiler Code Committee after holding a series of meetings and public hearings.

The 1923 edition of the Code for *Rules for the Construction of Low Pressure Heating Boilers*, known as the ASME Code Section IV, was approved by the ASME Council on May 23, 1923. It contained 31 pages and consisted of the following parts:

- Part I, Steel Plate Boilers (paragraphs H-1 to H-83)
- Part II, Cast Iron Boilers (paragraphs H-84 to H-120)

In Part I, paragraph H-1, the Code was limited to steam boilers with pressures of 15 psi and hot water boilers to 160 psi and 250°F. The method of computing the maximum allowable working pressure (MAWP) was stated in paragraph H-4, and the specifications of the materials to be used followed the rules for power boilers. Paragraphs H-15 to H-26 addressed the requirements for joints, braces, and stayed surfaces. Paragraphs H-27 to H-64 addressed the requirements for boiler openings, supports, settings, installations, and fittings and appliances including safety valves. Paragraphs H-65 to H-68 addressed hydrostatic tests with shop inspection, stamping, and marking. The remaining paragraphs, H-76 to H-83, addressed welded boilers.

In Part II, paragraphs H-84 to H-86, the pressures and temperatures to be used with cast iron boilers were specified. Paragraphs H-87 to H-89 contained the requirements for boiler openings, flanged connections, and threaded openings. Six paragraphs addressed boiler installation, and twelve paragraphs addressed both safety valves and water-relief valves. Paragraph H-120 specified the ASME marking requirements on the cast iron boilers.

Appendix A shows ten pages from the ASME Low Pressure Heating Boilers Code Section IV 1923 edition. These pages are: 1–6, 18–19, and 24–25.

The 1927 edition of Section IV (Low Pressure Heating Boilers) incorporated a number of revisions to the 1923 edition. The important changes were the following:

- The inclusion of specifications for flanged-quality steel plates used for forged welding
- The inclusion of specifications for base metal used in autogenous welding
- The basing of safety valve–relieving capacity on the grate area rather than the radiating surface
- The inclusion of formulas for copper-tube thickness
- The inclusion of an index of seven pages

The 1931 edition of Section IV was prepared from addenda to clarify statements of the 1927 edition. Minor changes were made to paragraphs H-40, H-55, H-56, H-64, H-93, H-108, H-117, and H-120, and the portion of the Code entitled "Autogenous Welded Boilers" was changed to "Fusion Welded Boilers."

The Section IV Code was published at irregular intervals from 1931 to 1952, in the following years: 1931, 1932, 1935, 1937, 1940, 1941, 1943, 1946, 1949, and 1952. Since 1953, the Code has been published every three years, and the latest edition always includes all previous addenda. It has been published in the following years: 1953, 1956, 1959, 1962, 1965, 1968, 1971, 1974, 1977, 1980, 1983, 1986, 1989, 1992, 1995, 1998, 2001, and 2004.

Although heating boilers are constructed safely under ASME Section IV Code, the rules of ASME Section VI are recommended for their care and operation. These rules are nonmandatory guidelines for the safe operation of steam heating boilers, hot water supply boilers, and hot water heating boilers after installation (however, they are not applicable to potable water heaters).

1.3 Types of Heating Boilers

The first 1923 ASME Code Section IV—Rules for Construction of Low Pressure Heating Boilers remained the same for 83 years. Due to technological development in the last century, the ASME has expanded the Code from its original 31 pages to 232 pages (2004 edition). The current name of the Section is ASME Code Section IV—Rules for Construction of Heating Boilers.

Heating boilers are mostly used for building heating and water supplies. The heating boilers are selected for low-pressure and low-temperature applications. Typical locations for heating boilers are: office buildings, schools, nursing homes, hospitals, hotels, motels, laundromats, churches, etc. It is estimated that 80% of the boilers used belong to the category of heating boilers.

There are many types of heating boilers used all over the world. General classification of the heating boilers based on the following criteria is shown in Figure 1.3.

- Based on use
- Based on material
- Based on fuel
- Based on tube type
- Miscellaneous

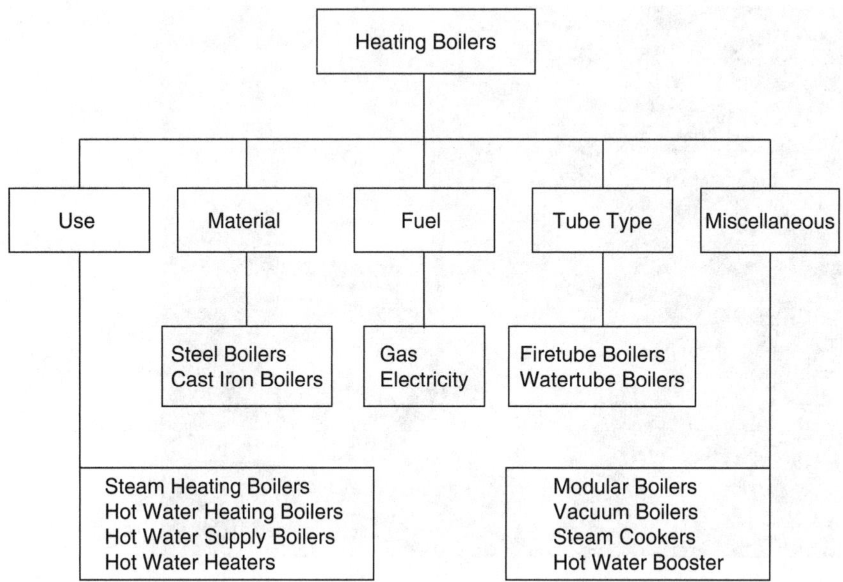

Figure 1.3 Classification of heating boilers.

1.3.1 Classification based on use

Steam heating boilers. A steam heating boiler is a boiler in which the operating pressure does not exceed 15 psi when directly fired by gas, oil, electricity, or other solid or liquid fuels. Steam heating boilers are used for generating heat for industrial use or for the heating systems found in buildings. Figure 1.4 shows a steam heating boiler.

Hot water heating boilers. A hot water heating boiler is a boiler in which the operating pressure does not exceed 160 psi and/or the temperature does not exceed 250°F. Such a boiler is used for heating buildings or for generating hot water for industrial processes.

Hot water supply boilers. A hot water supply boiler is a boiler in which the operating pressure does not exceed 160 psi and/or temperature does not exceed 250°F. Such a boiler is used for supplying hot water to buildings and for industrial use (Figure 1.5).

Hot water heaters. A hot water heater is a potable water boiler in which the operating pressure does not exceed 160 psi and/or temperature does not exceed 210°F. A hot water heater is used for supplying potable water in buildings. Figure 1.6 shows a hot water heater.

Figure 1.4 Steam heating boiler.

Figure 1.5 Hot water supply boiler. *(Courtesy: RBI Boiler, Canada)*

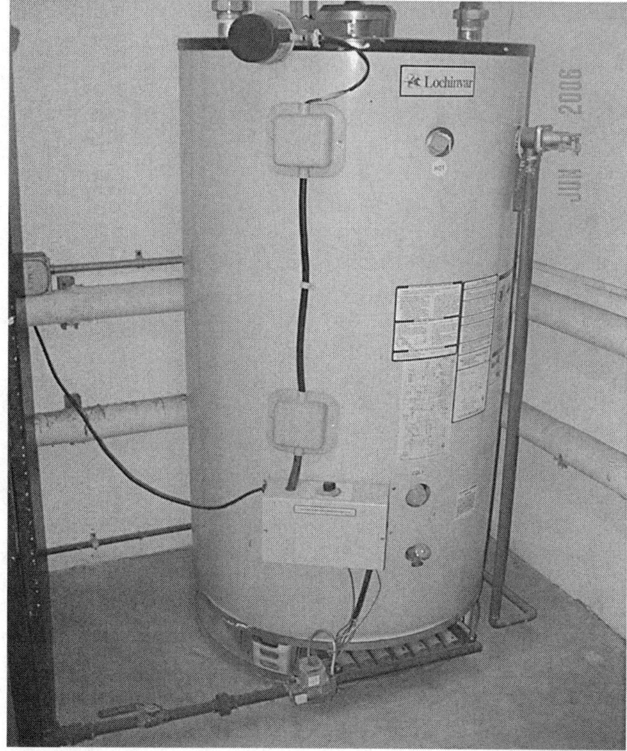

Figure 1.6 Hot water heater.

1.3.2 Classification based on material

Steel boilers. A steel boiler is a boiler in which the construction material for the main boiler and its pressure parts are made of steel.

Cast iron boilers. A cast iron boiler is a boiler in which the construction material for the main boiler and its pressure parts is cast iron. Such boilers are initially fabricated in sections, all of which are assembled as a complete boiler unit with push nipples or elastomer grommet-type seals. Figure 1.7 shows a cast iron boiler.

1.3.3 Classification based on fuel

Gas-fired boilers. As the name indicates, a boiler that is fired by gas is classified as a gas-fired boiler. The gas used for the boiler may be natural gas, propane gas, or any other liquid petroleum gas.

Oil-fired boilers. This type of boiler is fired by oil. Normally, heavy oil known as *Bunker C* is used because of low fuel cost. The no. 6 oil is referred to as Bunker C. Other fuel oils are: no. 1 oil, no. 2 oil, no. 3 oil, no. 4 oil, no. 5 oil, diesel, and kerosene.

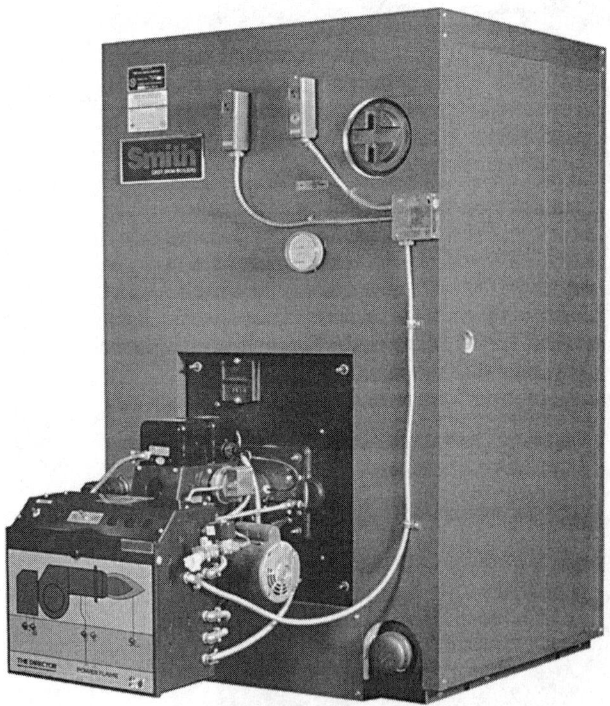

Figure 1.7 Cast iron boiler. *(Courtesy: H.B. Smith)*

Coal-fired boilers. Different varieties of coal are used for boilers, mostly for industrial and utility boilers. Coal is a very economical fuel for use in heating boilers, too.

Electric boilers. Electricity is used for this type of boiler instead of conventional fuels. Electricity is a clean fuel and popular for boilers used in research laboratories, food industries, and other industries where the nuisance of smoke is a problem.

1.3.4 Classification based on tube type

Firetube boilers. A firetube boiler is a boiler in which the products of combustion pass through the tubes, which are surrounded by water. The body of the boiler is called the *shell* and it contains water. Firetube boilers are used for applications ranging from 15 to 3000 hp. Firetube boilers are classified according to the:

- Position of the furnace
- Layout position
- Flue gas travel
- Type of firebox

According to position of furnace. Firetube boilers may be internally or externally fired. In an internally fired boiler, the grate and combustion chamber are enclosed within the boiler shell. In an externally fired boiler, on the other hand, the grate and combustion chamber are located outside the boiler shell.

According to layout position. Firetube boilers may be horizontal or vertical depending on the layout position.

The following horizontal firetube boilers are popular: Horizontal-Return-Tubular (HRT) boiler and Scotch Marine boiler.

1. *Horizontal-Return Tubular (HRT) boiler.* This boiler has a long cylindrical shell supported by the furnace sidewalls and set on saddles to permit the expansion and contraction of the boiler (Figure 1.8).
2. *Scotch Marine boiler.* This is a firetube-return tubular boiler consisting of a cylindrical shell that contains either the combustion chamber or the furnace and tubes (Figure 1.9).

The vertical firetube boiler consists of a cylindrical shell with an enclosed firebox. There are two types of vertical firetube boilers: vertical-exposed tube boiler and vertical-submerged tube boiler.

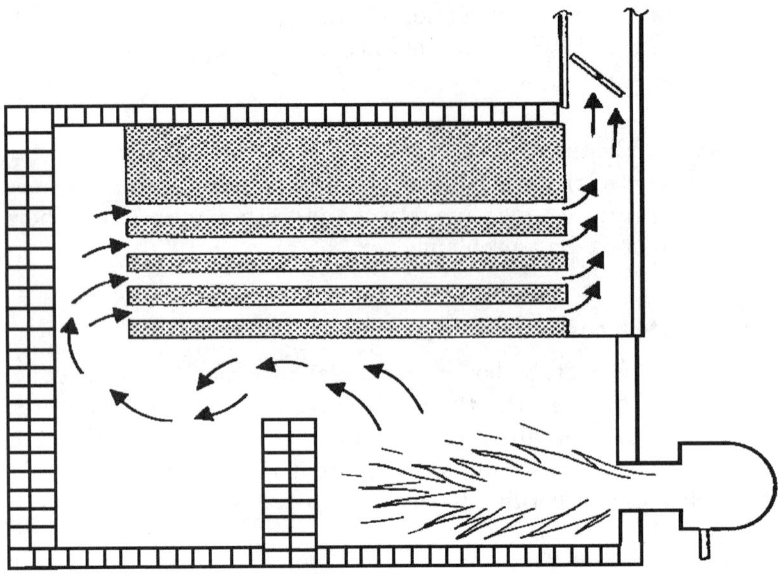

Figure 1.8 Horizontal-Return Tubular (HRT) boiler. *(Courtesy: ASME Section VI)*

1. *Vertical-exposed tube boiler.* In this boiler, the upper tube sheet and tube ends are above the normal water level, extending to the steam space (Figure 1.10).
2. *Vertical-submerged tube boiler.* In this boiler, the tubes are rolled into the upper tube sheet, which is below the water level.

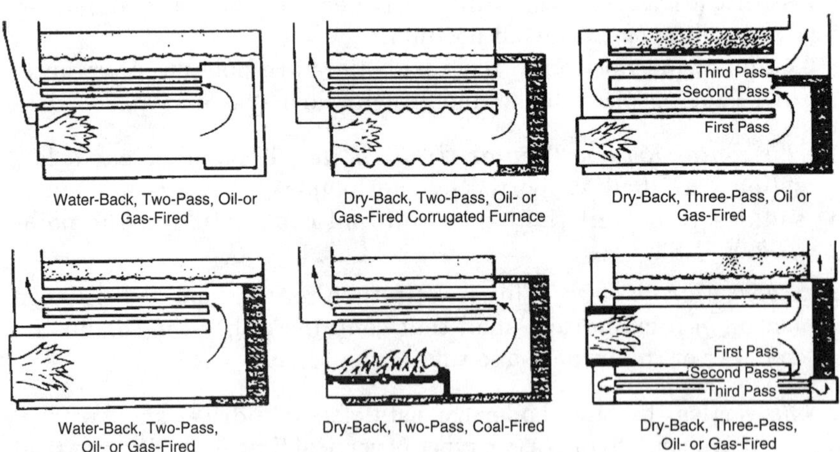

Figure 1.9 Scotch Marine-type boiler. *(Courtesy: ASME Section VI)*

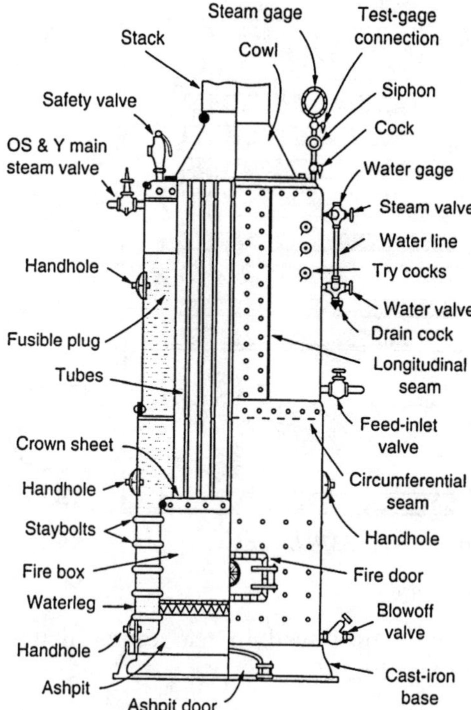

Figure 1.10 Vertical-exposed tube boiler.

According to flue-gas travel. Firetubes boilers may be classified as *single-pass, two-pass, three-pass, or four-pass*, depending on the number of flue-gas passes.

According to type of firebox. Firebox boilers have the firebox integral with the boiler. Different types of firebox boilers are: locomotive type and multipass type.

1. *Locomotive type.* A firetube boiler used for locomotive engines (Figure 1.11).
2. *Multipass type.* A firetube boiler with multiple passes of flue-gas travel with an integral furnace, usually having a refractory or dry bottom.

Watertube boilers. A watertube boiler is a boiler in which the water passes through the tubes and products of combustion surround the tubes. The tubes are encaged in a furnace that the burners fire into.

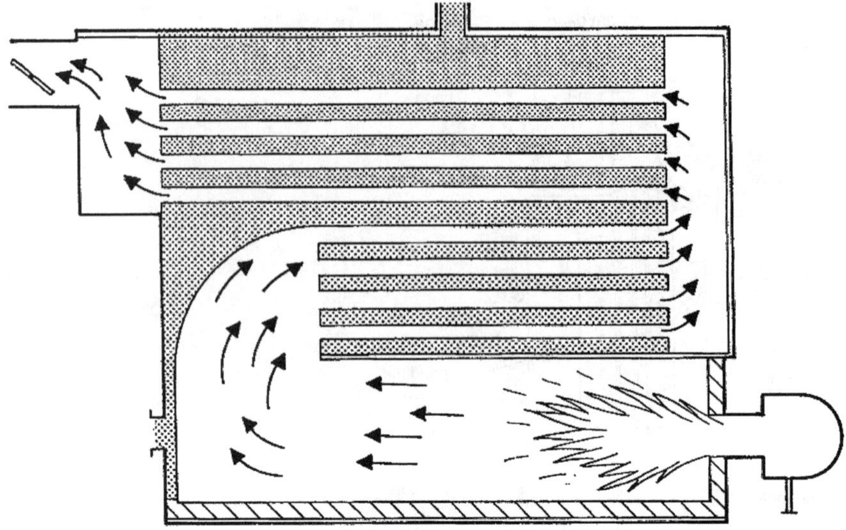

Figure 1.11 Locomotive firebox boiler. *(Courtesy: ASME Section VI)*

The water is heated and steam is produced in the upper drum. Watertube boilers are classified according to:

- Tube and drum configuration
- Water circulation

According to tube and drum configuration. Watertube boilers are typically produced as: Flex or Serpentine tube boilers, straight or inclined tube boilers, and straight copper-finned tube boilers.

1. *Flex or Serpentine tube boiler.* The bent tubes are installed in two drums, located one above the other.
2. *Straight or inclined tube boiler.* The tubes run between two header boxes. Figure 1.12 shows a typical packaged boiler.
3. *Straight copper-finned tube boiler.* The tubes run either vertically or horizontally between header boxes.

According to water circulation. There are two kinds of water circulation–type boilers: natural circulation and forced circulation.

1. *Natural circulation–type boiler.* This uses movement of heated and cooled water for circulation. As heated water rises, cooler water drops down to take its place; then steam bubbles form and move to the steam drum, which increases the water circulation.

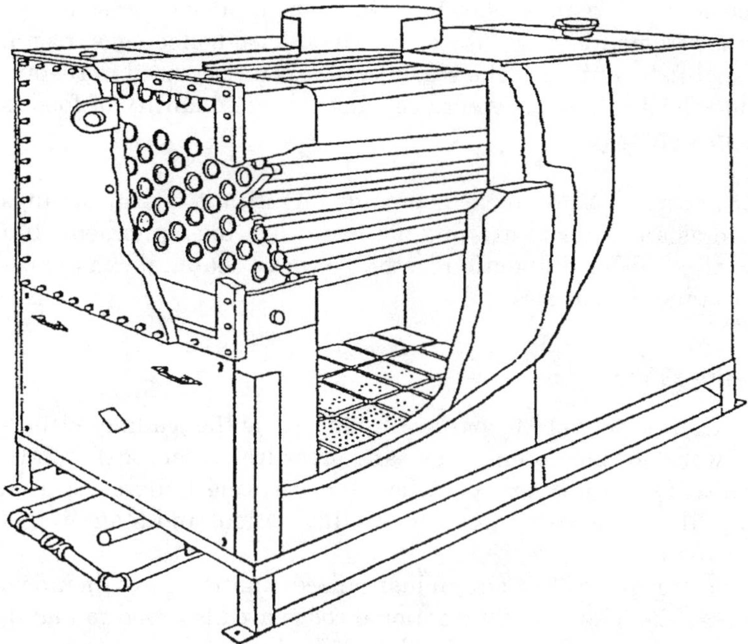

Figure 1.12 Packaged inclined watertube boiler with atmospheric burner.

2. *Forced circulation–type boiler.* This uses a high-pressure pump to force water continuously through the coils that form the boiler.

1.3.2 Miscellaneous boilers

Modular boilers. A modular boiler is a group of small boilers, made of either steel or cast iron, that collectively act as a larger boiler with modulated input. These small boilers, known as modules, are joined together in the field.

Because of the smaller size modules, start-up of the modular boiler is faster than a conventional system, with less loss of heat, providing a better response to system needs. Modular boilers can be used for steam generation, space heating, or domestic hot water.

Vacuum boilers. A sealed steam boiler that operates below atmospheric pressure is known as a vacuum boiler. The sealing of this boiler is done in the manufacturer's factory. A vacuum boiler is designed and constructed according to the rules of mandatory Appendix 5 of ASME Code Section IV.

Steam cookers. A steam cooker generates steam at low pressure for cooking food items like vegetables, rice, fish, hard-boiled eggs, pasta, poultry, shellfish, and ribs. The steam cooker is mostly used in restaurants and school cafeterias where cooking a large quantity of food is required in a short time.

Hot water booster. A booster can provide significant energy savings when used as a booster to existing hot water heating equipment. If a booster is used with a conventional storage water heater, it can almost eliminate standby heat loss.

1.4 Heating Boiler Code

The American Society of Mechanical Engineers is the leading authority in the world for publishing boiler and pressure vessel codes. ASME Code Section IV—Rules for Construction of Heating Boilers—is used internationally for design, construction, inspection, and stamping of heating boilers.

In addition to the ASME Code, industrialized countries have their own boiler codes. The following international codes are also used to manufacture heating boilers around the globe:

1. Canadian standards
   CSA B51—Boiler, pressure vessel, and pressure piping code
2. British standards
   BS 1113—Specifications for design and manufacture of watertube boilers
   BS 2790—Specifications for design and manufacture of shell boilers of welded construction
3. European (EU) standards
   prEN 12952—Standard for watertube boilers
   ptEN 12953—Standard for shell boilers
4. German standards
   TRD—Technical rules for steam boilers
   DIN 2918—Stationary shell boilers of welded construction (other than watertube boilers)
5. French standards
   NFE 31-001—Boilers operating with solid, liquid, or gas fuels
   NFE 32-100—National Standard for boilers
6. Japanese standards
   JIS B8201—Construction of steel boilers for land use

7. Indian standards
   IBR 1950—Indian boiler regulations
8. International Standard Organization (ISO) standards
   ISO 831—Rules for construction of stationary boilers
   ISO 5730—Stationary shell boilers of welded construction (other than watertube boilers)

1.5 Adoption of Boiler Code

The ASME *Boiler and Pressure Vessel Code* Section IV is composed of a set of rules for the construction of heating boilers. The objective of the Code is to establish safety rules for providing reasonable protection to life and property. The Code encompasses requirements, prohibitions, and guidance for the design, fabrication, inspection, testing, and certification of heating boilers.

Similarly, ASME Code Section VI is a set of guidelines for the care and operation of heating boilers. The objective of the Code is to establish safety rules for ensuring reasonable protection of life and property.

The Code rules are nonmandatory unless they are adopted into the laws of a governmental jurisdiction—a state, province, county, or city. A jurisdiction may adopt these rules either partly or completely. Once adopted, the jurisdiction enforces these nonmandatory rules by using authorized inspectors under the supervision of a chief boiler inspector. Currently, most of the United States, and provinces of Canada have adopted ASME Code Section IV. Though few jurisdictions have adopted ASME Code Section VI, many jurisdictions use this Code as a guideline for heating boiler operation.

Chapter 2

Steam Heating Boilers

A steam boiler, which operates at less than 15 psig, is mostly used for heating purposes. The word *heating* used in this book refers to the heating for buildings or spaces. Besides heating, a low-pressure steam boiler is used for industrial process applications and producing hot water by mixing steam with water.

2.1 Steam Boilers

When a steam boiler is used for heating purposes, it is classified as a steam heating boiler. The steam produced by the steam heating boiler is used for heating buildings or completing processes. A steam boiler is simple, efficient, and reliable. But despite this simplicity, with any steam boiler, problems can arise if it is not operated and maintained properly.

The process of steam generation is very simple. In a steam boiler, input is fuel and feedwater, whereas the output is products of combustion and steam. Feedwater is heated by a burner in the furnace and steam is generated within the boiler. The pressure in the boiler forces the steam to discharge through the main steam pipe. A gas-fired steam boiler is shown in Figure 2.1.

Steam for heating purposes can also be generated from an electric steam boiler. An electric steam-heating boiler is shown in Figure 2.2.

2.2 Types of Steam Heating Boilers

The type of steam heating boiler depends on the steam requirements, purpose of heating, and location. The following steam heating boilers are most commonly used:

Figure 2.1 Steam heating boiler—gas fired.

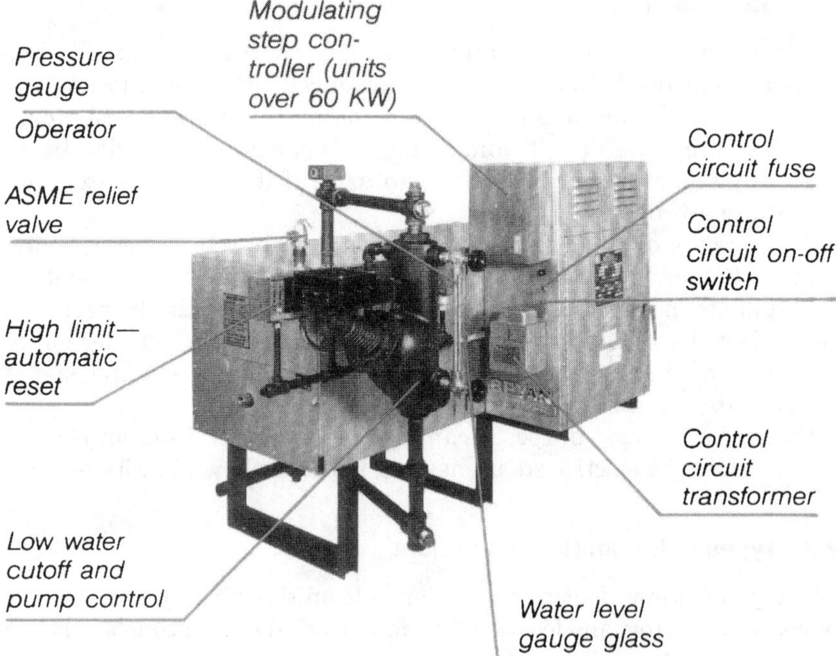

Figure 2.2 Electric steam-heating boiler. *(Courtesy: Bryan Steam, LLC)*

- Steam boiler for space heating
- Sterilizer for sterilizing operation
- Steam cooker for cooking food

2.2.1 Space-heating steam boilers

A steam boiler used exclusively for heating of building space is known as a space-heating steam boiler. Steam boilers look quite similar to hot water boilers. The major differences between them are noticed when you look at the steam or hot water heating system beyond the boiler.

In a hot water heating system, the hot water boiler and distribution piping are completely filled with water all the time. In a steam heating system, the steam boiler is approximately 75 percent full of water. The remaining portion of the steam boiler, the distribution piping, and the radiators are filled with air. They get filled with steam when the system is under operation. A steam boiler for a space heating system is shown in Figure 2.3.

Any type of steam boiler can be used for space heating (e.g., firetube, watertube, steel, copper, or cast iron).

2.2.2 Steam cooker

A steam cooker, also known as steamer, is used in cafeterias and restaurants to cook food by steam. Steam cooking uses moisture for

Figure 2.3 Steam heating boiler. *(Courtesy: Rite Boilers)*

maintaining flavor and nutrients. The steamer cooks vegetables, rice, fish, hard-boiled eggs, pasta, poultry, meat, stews, shellfish, and ribs. A steamer can be used for faster cooking of large volume food items. Heavy load capacity and preset controls save labor and time. A steam cooker is shown in Figure 2.4. The two upper chambers are for the storage of trays with food. The boiler is installed in the lowest chamber (Figure 2.5).

The steam cooker is basically a low-pressure boiler, which produces steam at a maximum pressure of 15 psig. The steamer is fired either by gas or electricity. The steam is supplied to the compartments where food items are placed on trays. Cooking time varies depending on the food item.

Typical features of a steamer are:

- Stainless steel exterior, frame and flanged feet
- Stainless steel cooking compartments with coved interior corners
- Separate manual control for each compartment with constant steam feature
- Electric ignition

Figure 2.4 Gas-fired steamer.

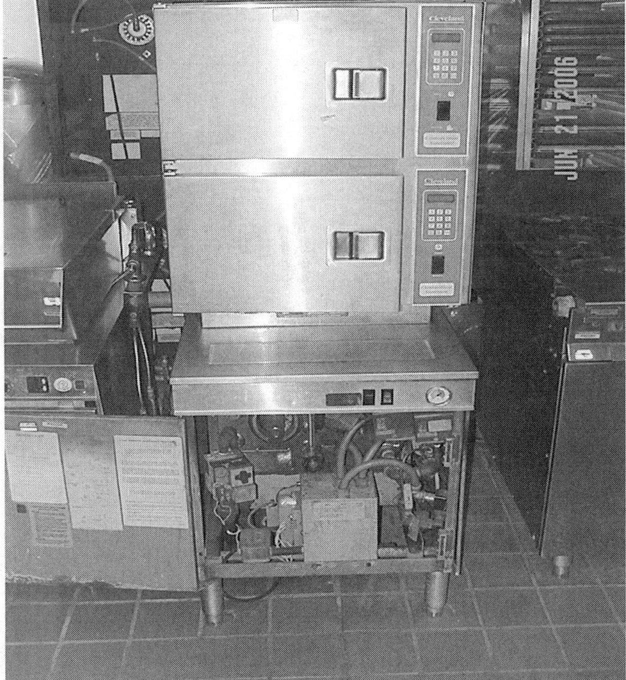

Figure 2.5 Lowest section shows steam boiler.

- Split water line connection, single drain connection
- Input 270,000 Btu/hour
- Output 5.5 BHP
- Options and accessories

2.2.3 Steam sterilizers

A steam sterilizer is used for sterilizing a wide variety of materials. The steam sterilizer, also called an autoclave, is the most popular sterilizing equipment in the world (Figure 2.6). Steam sterilizers are mostly used by health care facilities and laboratories for sterilization of instruments.

A boiler generates steam, which is distributed to a chamber containing trays. The materials to be sterilized are placed on the trays. The sterilization is achieved by controlling temperature, pressure, and time.

A typical steam sterilizer has the following features:

- Chamber dimensions 6 inches dia. × 14 inches deep, upto 16 inches dia. × 26 inches deep
- Tray size is 12 inches × 20 inches × 4 inches
- Electricity is used as the source of fuel for the boiler

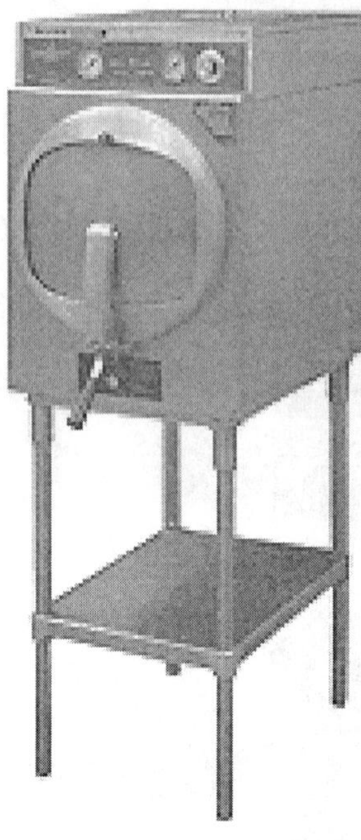

Figure 2.6 Steam sterilizer. *(Courtesy: Market Forge)*

- Standard equipment includes an automatic temperature regulator, thermometer, safety valve, steam gauge 0–30 psi, 0–60 minute timer, low water cut-off, thermostatic steam trap, and signal light
- Normal operating temperature is 250°F; adjustable temperature is 230–250°F
- Operating pressure is 0–15 psi

2.3 Steam Heating System

A steam heating system provides excellent heating and efficient operation if the system is designed, installed, and maintained properly. Although there have been changes in the design and control of space heating systems, the basic fundamentals still apply to the modern systems.

A steam heating system has the following advantages:

- Steam gives up a large amount of heat by condensation so that only a small amount needs to be circulated.
- High temperatures and better heat transfer permits smaller heating surfaces.
- Positive circulation is insured by the pressure of the steam itself.

- System is flexible and less expensive to install.
- Less installation is required, which reduces cost.
- Does not require any special equipment, such as pumps,—to move the steam. It flows under its own pressure.

Disadvantages of a steam system are:

- High cost of maintenance (cleaning traps, checking valves, regulators, and so on).
- Maintaining system tightness—due to slow leakage in system.
- Water-treatment chemicals are costly.

Steam heating systems are classified by:

1. Piping arrangements
2. Direction of steam flow
3. Pressure or vacuum conditions maintained during operation
4. Method of returning condensate to the boiler

2.3.1 Piping arrangements

Piping arrangements for steam systems can be classified as one-pipe systems and two-pipe systems.

One-pipe system. This system was designed many years ago to meet the needs of that time. The one-pipe system is a single pipe that serves the dual purpose of supplying steam to radiators and providing a path to return the condensate to the boiler.

One-pipe system with gravity return. This system operates on gravity circulation. All heating units and the end of the supply main are above the boiler water line so that condensate can flow back to the boiler by gravity. The one-pipe steam-heating system is shown in Figure 2.7.

The system is simple but heat emissions in radiators are hard to control. Modulating heat causes the heating elements to become partly filled with air. The system can work properly if the heat can be modulated directly at the boiler. Air valves are necessary for evacuating air during start-up.

Advantages of the system are:

- Simplicity.
- Low installation cost.
- Suitable for small installations.

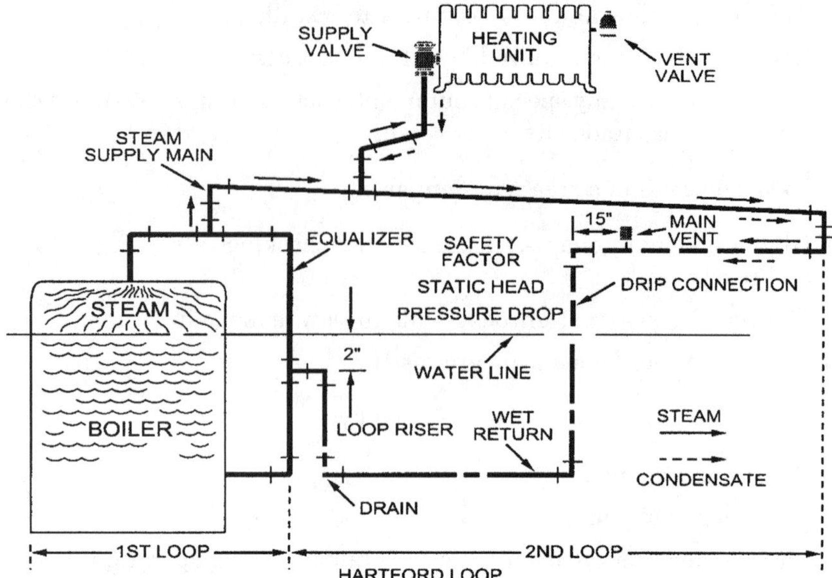

Figure 2.7 One-pipe system with gravity return. *(Courtesy: ITT Industries)*

Disadvantages of the system are:

- Lack of modulating quality.
- Not suitable for large installations.
- The system is subjected to water hammer as steam and condensate flow in the same pipe but in the opposite direction.
- Water enters the boiler above the waterline, interfering with steam release.

One-pipe system with dry return. In this system, steam rises to a point from where the flow in the steam and return mains is always in the same direction. This is a better design because the steam and condensate are separated in different pipes. This arrangement helps to prevent water hammer. A single-pipe system with dry return is shown in Figure 2.8.

In the risers to the radiator the same pipe is used for steam and condensate. There is very little condensate piping below the waterline.

One-pipe system with waterline wet return. This system is a further improvement on the one-pipe system with dry return. The waterline return can protect the boiler for a considerable length of time in case of failure in the return line below the waterline. A one-pipe system with waterline wet return is shown in Figure 2.9.

Each boiler that has a stop valve in the steam outlet should have either a check valve or a waterline return connection in the return to

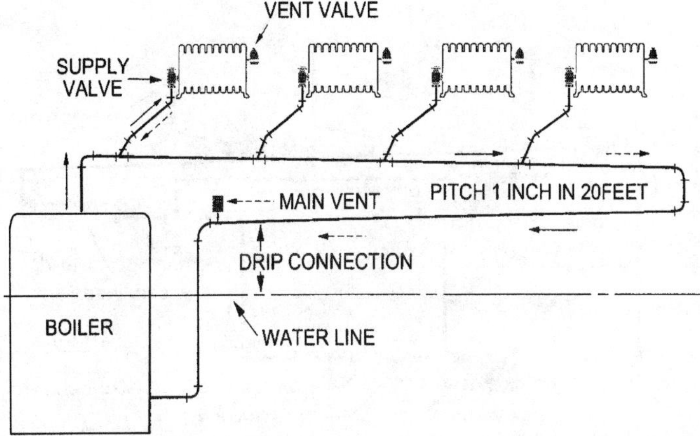

Figure 2.8 One-pipe system with dry return. *(Courtesy: ITT Industries)*

the boiler. The waterline return connection prevents the loss of water below the waterline of the boiler.

Two-pipe system. A steam heating system is known as a *two-pipe system* when steam and condensate flow in separate mains. In a two-pipe system, each heating unit is provided with two piping connections.

Two-pipe system with waterline returns. A two-pipe system with waterline return is shown in Figure 2.10. In this case, a return piping below the waterline has been installed so that a leak or break in the return line below the waterline will not result in the immediate draining of the boiler.

Direction of steam flow. Heating systems can also be classified based on the direction of steam flow in the risers. The steam leaves the boiler

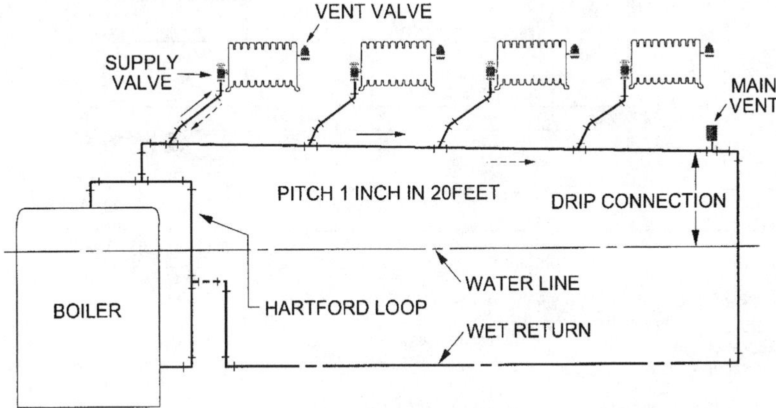

Figure 2.9 One-pipe system with waterline wet return. *(Courtesy: ITT Industries)*

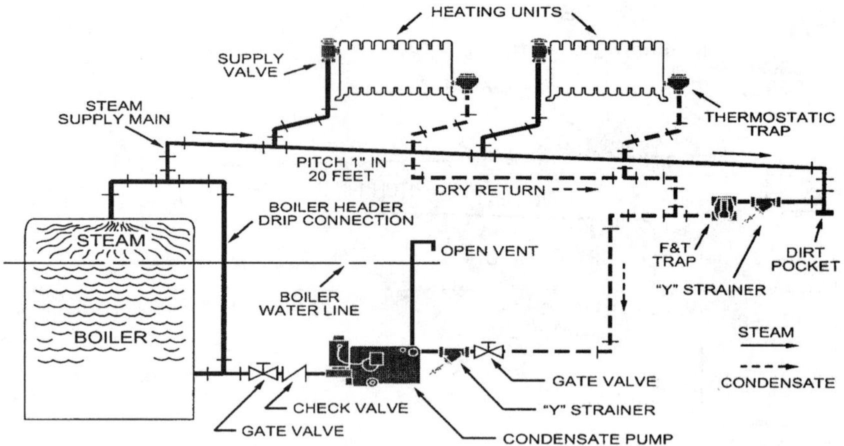

Figure 2.10 Two-pipe steam system with waterline wet returns. (*Courtesy: ITT Industries*)

through distributed pipes and enters into the radiators. Based on the flow of steam, the classifications are: *up-flow* and *down-flow*.

Up-flow system. In an up-flow system the steam flows directly from the boiler to the radiators utilizing the steam. An up-flow system is shown in Figure 2.11.

Down-flow system. In a down-flow system, steam is taken to the top of the building and then is distributed through down-comers to the radiators.

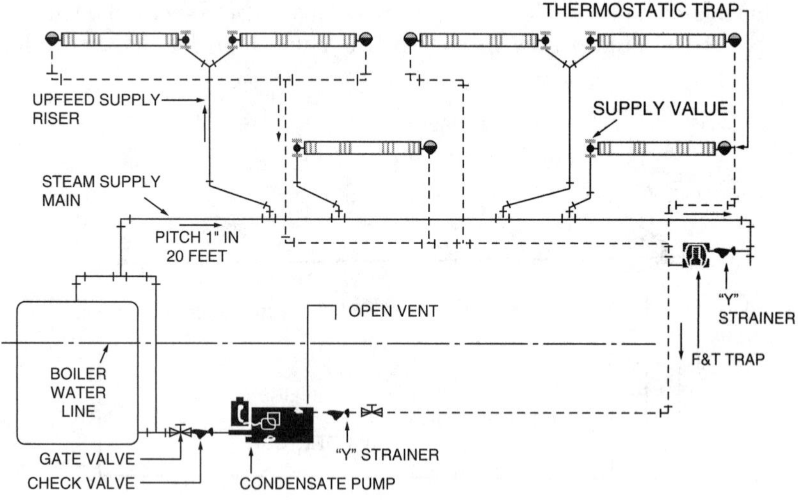

Figure 2.11 Up-flow system. (*Courtesy: ITT Industries*)

Pressure conditions. Depending on the pressure conditions under which the system is designed to operate, steam heating systems may be classified as: high-pressure, low-pressure, vapor, or vacuum.

High-pressure system. A high-pressure system operates at pressures above 15 psi. A high-pressure boiler generating steam at more than 15 psi is incorporated in the system. A high-pressure steam boiler can be used for both heating and process steam. When a part of the steam is used for process work, it cannot be returned to the boiler in the form of condensate.

Low-pressure system. A low-pressure system operates at 0 to 15 psi. A low-pressure boiler generating steam at maximum pressure of 15 psi is incorporated in the system. A low-pressure steam boiler can be used for both heating and process steam. When part of the steam is used for process work, it cannot return to the boiler in the form of condensate.

Vapor system. A vapor system operates under very low pressure or slight vacuum conditions without the use of a vacuum pump.

Vacuum system. A vacuum system operates under vacuum and low-pressure conditions with the use of a vacuum pump (Figure 2.12).

This system is a two-pipe system with a vacuum pump return, which is used to maintain subatmospheric pressures in the system. The pump withdraws the air and water from the system, separates air from the water and expels it to the atmosphere, and pumps it back to the boiler.

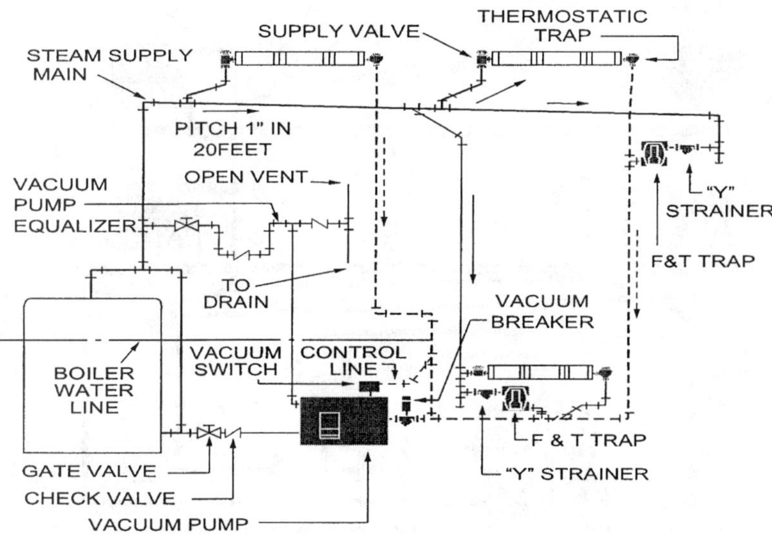

Figure 2.12 Vacuum system. (*Courtesy: ITT Industries*)

No connection should be made from the supply side to the return side of the system except through a suitable trap.

Condensate return. In a closed heating system, 100 percent of the condensate is returned to the boiler. There are two methods of returning condensate to the boiler:

- Gravity system
- Mechanical return system

Gravity system. In this system, all radiators or heating units should be elevated sufficiently above the waterline of the boiler to permit condensate to flow freely back to the boiler.

Mechanical return system. If it is necessary to return the condensate to the boiler by some means other than gravity, a mechanical system is used. The mechanical systems used are: receiver or trap, and condensate pump.

1. *Two-pipe system with return trap.* A return trap is used where pressure conditions vary between those of a gravity-return and a forced-return system. The receiver or trap is used to assist in the return of the condensate to the boiler. Figure 2.13 shows a trap being used where pressure conditions make it necessary to assist the return of the condensate.
2. *Two-pipe system with condensate return pump.* In this system, the condensate is returned to the boiler under pressure of the atmosphere

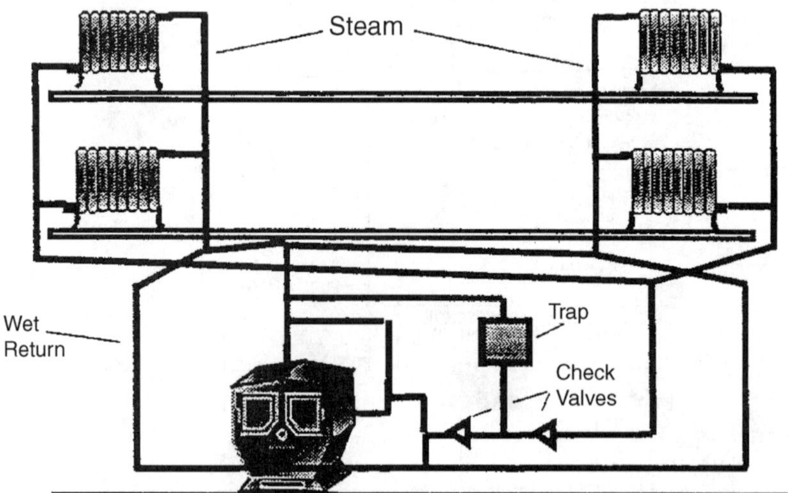

Figure 2.13 Two-pipe system with return trap.

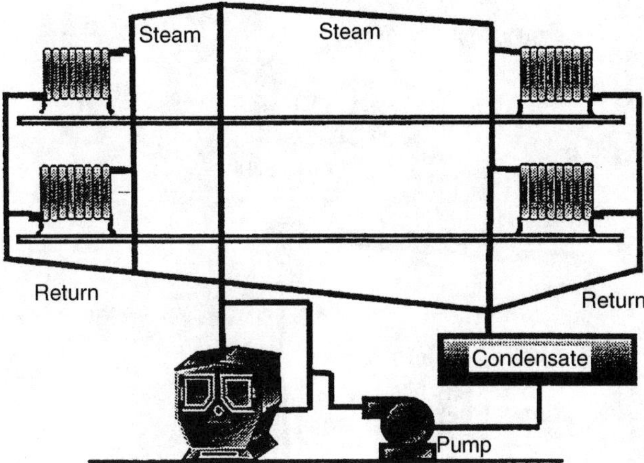

Figure 2.14 Two-pipe system with condensate pump.

or above, by means of a pump. The steam leaves the boiler through the distribution pipes. Air in the pipes and heating units should be vented to allow steam traps the air to discharge into the return line and to a vented condensate. As the steam gives up its heat, it condenses and the condensate drains through the steam trap to a condensate pump. A two-pipe system with condensate return pump is shown in Figure 2.14.

The advantage with this system is better individual modulation of radiators. The disadvantages are more equipment needed and thus higher costs.

2.4 Steam Heating System Components

A steam heating system uses steam to move heat from where it is produced to where it is needed. Steam is delivered via pipes to the exact locations where it is used. A steam heating system is shown in Figure 2.15. Various components of a steam heating system are given below:

Boiler. The boiler heats water and produces steam that feeds the system.

Condensate pump. The pump circulates the condensate through the distribution system.

Distribution system. This is, in fact, the piping system that distributes steam to individual units.

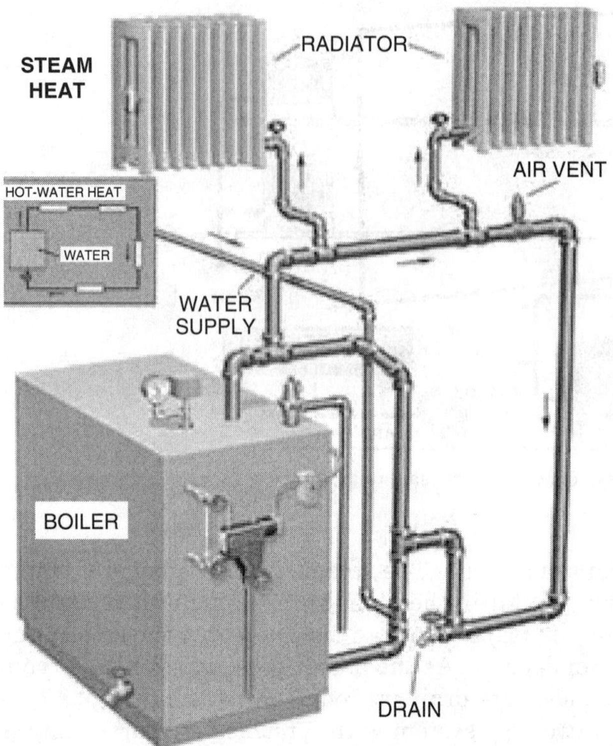

Figure 2.15 Steam heating system. *(Courtesy: Popular Mechanics)*

Controls. The devices that control the temperature of the hot water, or that switch the pump and boiler on and off include:

- Aquastat—operates boiler at constant temperature between 180–220°F regardless of heating load.
- Outdoor reset—operates boiler at varying water temperatures based on outdoor temperature.
- Outdoor cutout—turns off boiler system when outdoor temperature indicates that no heating is required.
- High limit—turns off boiler system if water temperatures exceed a safe operating limit.

Fin tubes. The equipment in the radiator through which steam, from the distribution system and boiler, circulates.

Zone valves. Valves on baseboard radiators.

Figure 2.16 Steam Radiator. *(Courtesy: Steam Radiators)*

Radiator. Device that distributes heat in individual units (Figure 2.16).

Thermostat. Regulates the temperature of the radiator.

Steam traps. An automatic device to stop the flow of steam so that the heat energy can be transferred, and the condensate and air can be discharged as required.

Combustion air intakes. Vents that draw air in and out of the building.

2.5 ASME Code Requirements

A heating boiler may be fabricated in a boiler shop either by welding or brazing. The rules in Part HF, Subpart HW in ASME Section IV are applicable for steam heating boilers, hot water heating boilers and hot water supply boilers, and their parts fabricated by welding. As the construction method of the boiler is welding, all the applicable rules of welding in ASME Code are applied while fabricating the boiler.

Manufacturers or contractors are held responsible for the welding done on the boiler. Manufacturers will have welding procedures in place in accordance with the requirements of ASME Section IX. Manufacturers will qualify the welding procedures required for construction of boilers under ASME Section IV. They are also responsible for the performance tests of welders and welding operators who will apply welding procedures.

2.5.1 Materials requirements

Materials used for pressure parts are required to conform to the specifications given in ASME Section II. The stress values of such materials should conform to the values given in Appendices B and C. Materials for welding are grouped as P-numbers and the assigned P-numbers may be found in ASME Section IX. The P-numbers may also be found in Appendices B and C.

Carbon content in carbon or alloy steel should not be more than 0.35%. Such plates should not be shaped by oxygen cutting or any thermal cutting process. Stud materials should not contain more than 0.27% carbon with minimum tensile strength of 60,000 psi (410 Mpa).

Different specifications. Requirements of ASME Section IX are applicable when welding two materials for construction. Two different materials of different specifications are welded in accordance with the requirements of paragraph QW-251.2 of ASME Section IX.

Small parts. Small parts such as pipe fittings, welding caps, and so on when formed by casting, forging or die-forming may be used as standard pressure parts. Such parts should be of good weldable quality when used for construction.

2.5.2 Welding requirements

The welding processes as described in ASME Section IX are used for construction of boilers. But not all the welding processes are approved for use in boiler fabrication.

The following processes may be used for welding of pressure and non-pressure parts:

1. Arc or gas-welding processes: The processes are limited to shielded metal arc, submerged arc, gas metal arc, gas tungsten arc, plasma arc, atomic hydrogen metal arc, and oxyfuel gas-welding.
2. Pressure welding processes: The processes are limited to flash, induction, resistance, pressure thermite, pressure gas, and inertia and continuous-drive friction welding.

Welding qualifications. The welding on pressure parts, and joining non-pressure parts to pressure parts is done by qualified procedures, welders, and the welding operators. The requirements of ASME Section IX are applied to qualify the welding process, the welders, and welding operators.

Production work. Production work will be undertaken if the welding procedures, the welders, and the welding operators have been qualified in accordance with the requirements of ASME Section IX. Welding may be performed within the first 3 ft (0.9 m) of the first production weld if welders are qualified as per QW-304 and welding operators are qualified as per paragraph QW-305 of ASME Section IX.

Interchanging qualifying tests. A manufacturer is responsible for performance qualification tests of the welders and welding operators employed by his organization. The performance tests conducted by one manufacturer or contractor will not qualify a welder or welding operator to work for any other manufacturer or contractor.

Maintenance record. A complete record of qualifications and identifying marks will be maintained by a manufacturer or contractor. The record includes welding procedures, welders, and welding operators employed by the manufacturer, date and results of the test, and the identification marks assigned to each welder. The records will be certified by the manufacturer and be available to the inspector. The welder or welding operator will stamp his identification mark on or adjacent to all welding joints. Alternatively, the manufacturer may keep records of all the welded joints, and the welders and welding operators who performed these joints.

2.5.3 Design of weldments

All welds will be made to ensure satisfactory penetration and fusion into the base metal to the root of the weld. All members should be properly prepared, fitted, aligned, and retained in the position in accordance with the specifications of the welding procedures.

Butt joints. A butt joint, longitudinal, circumferential, or other types will be used for welding plates of a drum, shell, or other pressure parts. The butt joint should be double welded, or filler metal may be added from one side to make the weld penetration complete with reinforcement on both sides of the joint.

Lap joints. A lap joint may be used for longitudinal or circumferential joints for welding plates of a shell if the welding joint is not in direct contact with products of combustion. The lap joint should be a full fillet weld inside and outside, and the throat should be a minimum of 0.7 times the thickness of the thinner plate. The length of the overlap should be a minimum of four times the thickness of the thinner plate.

Corner or tee joints. A corner or tee joint with a single full-fillet weld may be used for a boiler designed for a maximum of 30 psi (207 kPa). The throat of the fillet weld should be a minimum of 0.7 times the thickness of the thinner plate. A corner or tee joint with a double full-fillet weld, or full penetration, should be used for a hot water boiler designed for more than 30 psi (207 kPa).

Joint efficiency. The joint efficiency E is used in the formulas to calculate the minimum thickness and design pressure of a boiler part under internal pressure. Joint efficiency factor E is not required when the boiler part is designed only under external pressure.

The following joint efficiencies, E, are used for the welding joints performed by a gas welding process:

E = 85% for full penetration double-welded butt joints
E = 80% for full penetration single-welded butt joints with backing strips
E = 60% for single-welded butt joints without the use of backing strips
E = 65% for double full-fillet lap joints (boiler designed for more than 30 psi)
E = 49% for double full-fillet lap joints (boiler designed for a maximum 30 psi)

Horizontal stays. The horizontal stays are inserted through holes and the clearance between the holes and the stays should be a maximum of 1/16 inch (1.6 mm). The ends of the stays should be clearly visible and should not project more than 3/8 inch (10 mm) beyond the surfaces.

Diagonal stays. Diagonal stays are attached to the inner surfaces of the shell by fillet welds.

Flanged heads or tubesheets. An outwardly or inwardly flanged head or tubesheet may be attached to the shell by fillet welding if the head or tubesheet is supported by tubes or braces, or both. The welding joint attaching an outwardly flanged head or tubesheet should be completely within the shell. The inwardly flanged heads or tubesheets should be full-fillet welded inside and outside. The throats of the full-fillet welds should be a minimum of 0.7 times the thickness of the head or tubesheet.

Unflanged heads or tubesheets. Unflanged heads or tubesheets may be attached to the shell by welding if the head or tubesheet is supported by tubes, or braces, or both. The weld for boilers designed for more than 30 psi (207 kPa) should be a full-penetration weld or a double-full fillet weld.

2.5.4 Attachment welding

Furnace attachment for boiler pressure at maximum 30 psi. A furnace or crown sheet may be attached to a head or tubesheet with a full fillet weld if the furnace will not extend beyond the outside face of the head or tubesheet for a distance more than the thickness of the head. The throat of the full fillet weld should be a minimum of 0.7 times the thickness of the head or tubesheet.

Furnace attachment for boiler pressure more than 30 psi. A furnace or crown sheet in a hot water boiler may be attached to a head or tubesheet by a full penetration weld, with the furnace or crown sheet extending, at a minimum, through the full thickness of the head or tubesheet. When exposed to primary gases, furnace or crown sheet projections will not extend beyond the face of the plate by more than 3/8 inch (10 mm).

Tube attachment. The tubes may be attached to the tube sheet by welding. The projection of the tubes beyond the tubesheet will not exceed a distance equal to the tube thickness. The maximum projection of 1/2 inch (13 mm) is allowed for water tubes.

Head to shell attachments. The form heads, ellipsoidal, torispherical or hemispherical, concave or convex to the pressure side, may be welded to the shell in accordance with paragraph HW-715 of ASME Section IV. A skirt is not required when head thickness does not exceed 1-1/4 times the shell thickness. If head thickness exceeds 1-1/4 times the shell thickness, a skirt is required to be provided having a length not less than 3 times the head thickness, or 1-1/2 inches (38 mm), whichever is less. An integral skirt is not required for formed heads of full hemispherical shape, concave to pressure. Flanged ellipsoidal or torispherical heads convex to pressure may be welded to the shell with a full fillet weld with the throat not less than 0.7 times the head thickness.

2.5.5 Inspection

The inspector will designate the stages of inspection during fabrication of the boiler and other pressure parts. The inspector's inspection point is known as the *hold point* and is recorded in a document called a Traveler or a Process Sheet. The manufacturer will offer the boiler and pressure part for inspection at the hold point. Fabrication work cannot proceed without the release of the hold point by the inspector.

Welding procedure qualifications. The inspector should be satisfied that the welding procedures used in fabrication have been qualified in accordance with ASME Section IX. The manufacturer will submit documents to the inspector showing that he has complied with the requirements of the Code. The inspector may call for and witness the test welding and testing at any time if he or she desires to do so.

Performance qualifications. The inspector should be satisfied that welding performed by welders or welding operators is qualified in accordance with ASME Section IX. The manufacturer will make available to the inspector the record of performance qualification tests as required by the Code. The inspector may call for and witness the test welding and testing at any time if he desires to do so.

2.6 Specifying a Steam Heating Boiler (Combination Gas/No. 2 Oil-Fired, 200 HP, MAWP 15 psi)

Boiler capacity. Furnish and install one 200 HP watertube steam boiler in accordance with plans and specifications. The boiler shall be designed for a maximum allowable working pressure of 15 psi and maximum operating pressure of 15 psig. The boiler shall have an output of 6,900 lb/hr at the steam nozzle when fired with a no. 2 oil and/or natural gas at 1,000 Btu/ft^3 with a minimum pressure of 10 psi. Electrical power available will be 115 V, 60 HZ, and single phase.

Construction. The boiler shall be designed and constructed in accordance with ASME Code Section IV—Rules for Construction of Heating Boilers. It shall be of the inclined water tube design with 2 inch non-proprietary straight steel tubes (SA 178 Grade A, 13 Gauge) rolled between two drums. Headers shall be free to expand and contract in order to reduce stresses caused by thermal shock.

Both headers (steam drums) shall incorporate bolted and gasketed removable head-plates that will completely expose all waterside surfaces for inspection and cleaning when opened. Header flanges shall have drilled and tapped smooth surfaces for easy gasket clean-up and flange maintenance.

The boiler shall be completely factory-preassembled on a structural steel base and be ready for the attachment of steam, fuel, blowdown, electrical and vent connections. The manufacturer shall assume responsibility for the complete boiler unit.

Burner. The boiler burner package shall be UL listed. The burner shall be factory-mounted on a hinged door to swing out for easy inspection and shall include the following:

- Forced draft combustion air blower
- Control panel—National Electric Manufacturer's Association 1 (NEMA 1) with combustion control, motor starter, control switches, transformer and indicator lights.
- Gas trim shall include main and pilot fuel valves, regulators and high and low gas pressure switches.
- Oil trim shall include oil pump with capacity of twice the firing rate, oil pump coupled to a separate motor, pressure-atomized burner, oil-metering valve, dual-oil valves, strainer, gauge, and check valve.
- Ignition transformer with electrode.
- Flame monitoring ultraviolet scanner.
- Air pressure proving switch.
- An observation port shall be installed through the front door to inspect the flame.
- Burners and controls conform to the Factory Mutual (FM) standards.

Steam trim. Standard steam trim shall include the following components:

- Pop safety valve set for 15 psi set pressure.
- Low water cutoff water column, try cocks, gauge glass set with drain cock, feedwater pump control, and blow down valve.
- Secondary low water cutoff with manual reset and blow down valve.
- Steam pressure gauge.
- High limit pressure control with manual reset.

Efficiency guarantee. The boiler must be guaranteed to operate at minimum fuel-to-steam efficiency of 82 percent from 30–100 percent of rating.

Shop tests. The completed fully assembled unit shall be factory test-fired to check for proper operation of burner, controls, and safety devices.

Start-up services. A factory authorized representative of the manufacturer shall provide start-up service and training of the operator.

Warranty. The manufacturer shall warrant all furnished parts for 18 months from the date of boiler shipment. The pressure vessel shall carry a 5-year warranty against thermal shock damage.

Documents. The manufacturer shall provide the following documents:

- Three copies of perfect bound owner's manuals
- Three copies of the manufacturer's data reports
- Three copies of the certified prints

Chapter

3

Hot Water Heating Boilers

A hot water heating boiler, which operates at pressures not exceeding 160 psi (1100 kPa) and/or for temperatures not exceeding 250°F (121°C), is mostly used for heating purposes. The word *heating* refers to heating for buildings or spaces. Besides heating, a hot water heating boiler is used for industrial process applications.

A hot water heating boiler is simple, efficient, and reliable. But despite its simplicity, as in any hot water heating boiler, problems can arise if it is not operated and maintained properly.

3.1 Hot Water Heating Boilers

When a hot water boiler is used for heating purposes, it is classified as a hot water heating boiler. The hot water produced by a hot water heating boiler is used for heating buildings or applied to industrial processes. A hot water heating boiler is shown in Figure 3.1.

The process of hot water generation is very simple and is similar to the steam generation process, the difference being the boiler generates hot water instead of steam. In a hot water boiler, input is fuel and feedwater, and output is the products of combustion and hot water. Feedwater is heated by a burner in the furnace and hot water is generated within the boiler. The pressure in the boiler forces the hot water to discharge through the main pipe.

3.2 Types of Hot Water Heating Boilers

Hot water heating boilers are classified according to their use and purposes. The following are the common hot water heating boilers:

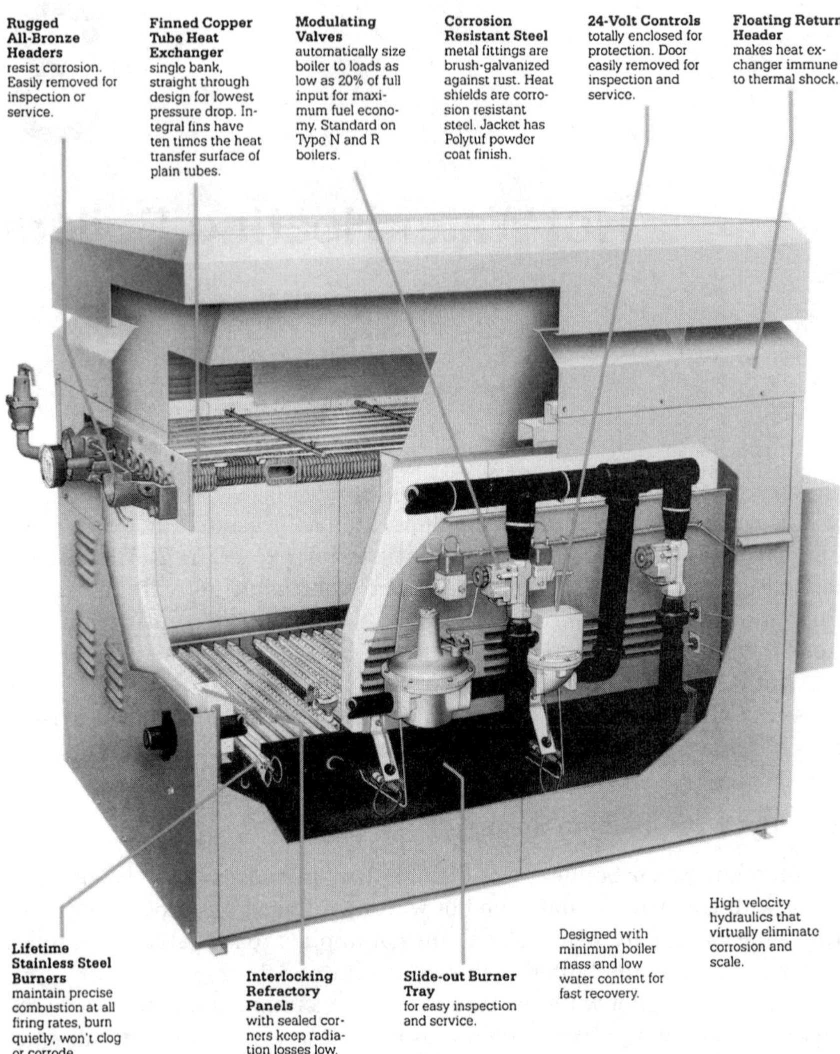

Figure 3.1 Hot water heating boiler. *(Courtesy: Raypak, Inc.)*

- Hot water heating boilers for space heating
- Boosters for commercial dish washers

3.2.1 Hot water heating boilers for space heating

A hot water heating boiler may be used for heating building space. Most small- and medium-sized commercial buildings and business centers use

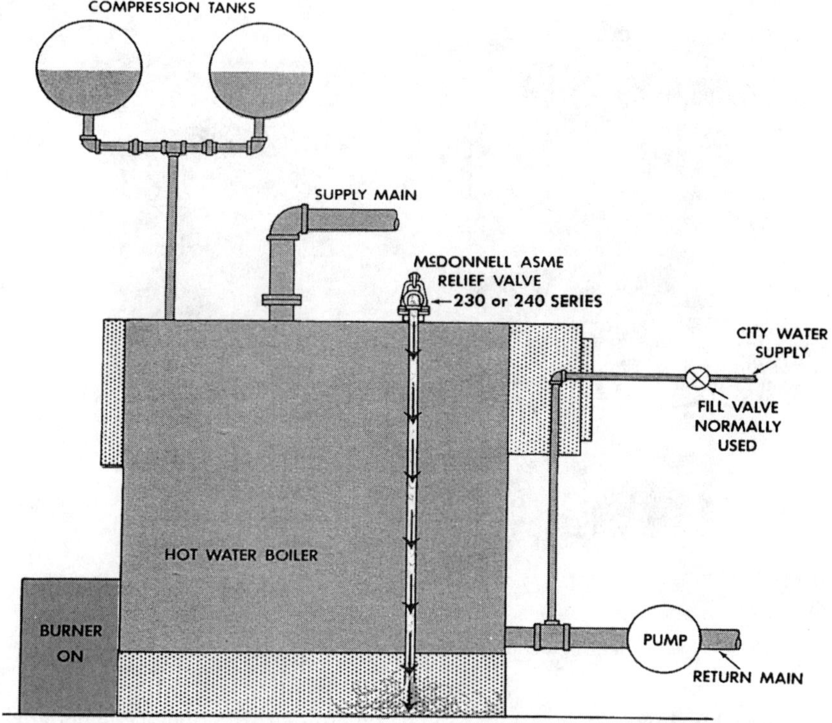

Figure 3.2 Hot water heating boiler. *(Courtesy: McDonnell & Miller, Inc.)*

hot water heating boilers for space heating. Typical hot water heating boilers are shown in Figures 3.2 and 3.3.

Water is heated to dispersing temperatures, typically 140–180°F, and normally returned about 20°F lower. The system is called a *closed loop,* with minimum fresh water makeup. Hot water heating boilers are preferred as they generally do not require boiler operators. Also, perfect water treatment is not required because of the closed loop. Since a hot water heating boiler operates at low temperatures, the boiler can operate at higher fuel conversion efficiency than a steam boiler.

3.2.2 Boosters for commercial dishwashers

A booster is an instantaneous fin tube-type heater designed to boost the temperature of regular water from 110–150°F to 180°F. Water at 180°F may be used as sanitizing rinse water in commercial dishwashers.

Figure 3.4 shows a booster connected near the commercial dishwashing machines. The booster, which is designed to fit under the dish table, near the dishwasher, minimizes heat loss that could occur if it were installed in a remote location.

Figure 3.3 Hot water heating boiler. *(Courtesy: Ajax Boiler, Inc)*

Figure 3.5 shows the sectional view of a single tank, door-type booster. The booster has a heat input capacity of 58,000 Btu/hr with a storage capacity of 3.2 gallons. The output equivalent capacity is 13.5 kW. The unit may safely be vented directly into the room. This booster is capable of increasing the temperature of 135 gal/hr by 40°F.

A booster heater can use either gas or electricity as fuel. An electric booster heater is shown in Figure 3.6.

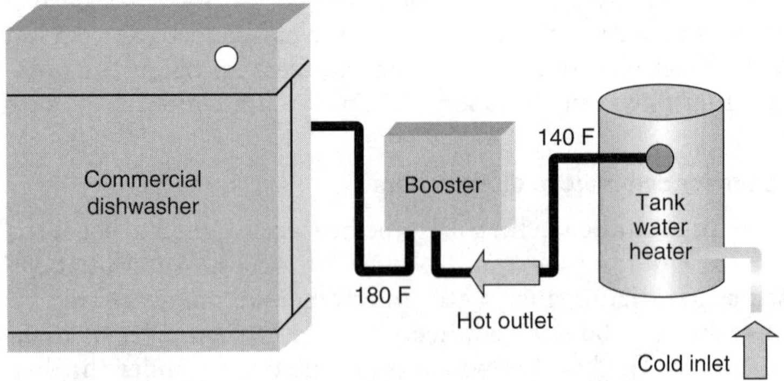

Figure 3.4 Booster installed near the dishwasher. *(Courtesy: tanklessheaterguide.com)*

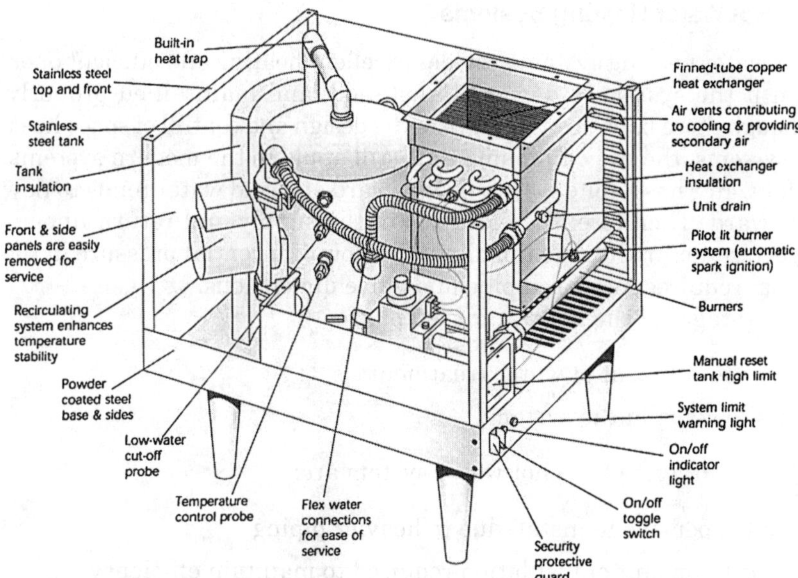

Figure 3.5 Sectional view of a booster heater. *(Courtesy: Hatco Corporation)*

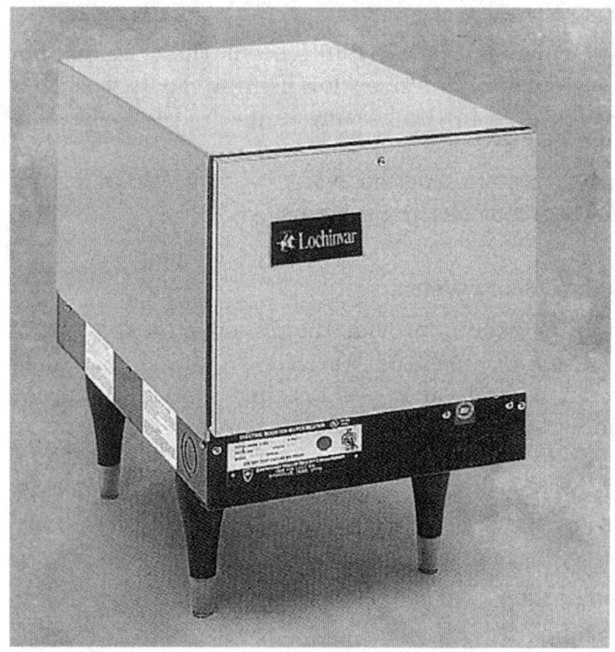

Figure 3.6 Electric booster heater. *(Courtesy: Lochinvar Corporation)*

3.3 Hot Water Heating Systems

A hot water heating system provides excellent heating and efficient operation if the system is designed, installed, and maintained properly. Although there have been changes in the design and control of space heating systems, the basic fundamentals still apply to the modern systems.

Hot water has a tendency to flow upward and cold water tends to flow downward. Therefore, the elevation of the supply and return lines is important for proper functionality. Also, low differential pressure in the system requires increased pipe and valve dimensions.

Advantages of the hot water system are:

- Less requirement of water treatment
- Less leakage in the system

Disadvantages of the hot water system are:

- More expensive to install due to heavier piping
- Greater amount of insulation required to maintain efficiency

Hot water heating systems may be classified into three types: gravity systems, forced circulation systems, and zone-controlled systems.

3.3.1 Gravity heating system

In this system, the flow depends on the difference in the weight of hot and cold water. The system has a relatively low heating capacity because of low medium temperatures in the heating units. A gravity heating system is shown in Figure 3.7.

The gravity system is a simple, older hot water heating system. Its use is limited to smaller hot water heating systems.

3.3.2 Forced circulation heating system

In this system, a pump is used to provide the pressure for the flow of water through the system (Figure 3.8). Water is circulated regardless of temperature between hot and cold water. The pipes, valves, radiators, and the like can be downsized because of higher flow and higher mean temperatures.

The forced circulation heating system is the type of system generally installed today. This is the only practical alternative for bigger systems.

3.3.3 Zone-controlled heating system

In this system, the temperature in different zones is controlled by flow control valves, which control the amount of hot water flowing in the radiators and thus control heat emission from them (Figure 3.9). Multiple

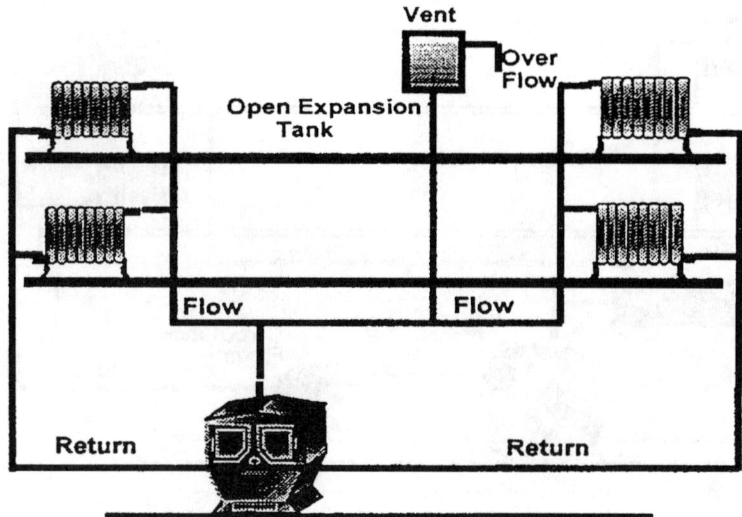

Figure 3.7 Gravity heating system.

zones can be controlled by a single boiler. The system can also be designed to control the flow to each zone by separate circulating pumps.

In some of these systems, hot water is circulated continuously, and the flow to the individual zones or apartments is controlled by flow-control valves. In other systems, a number of pipes are connected to a common header and water is taken from that header. Water is circulated through the individual system, and returned to the boiler by a circulating pump controlled by the thermostat at a particular location.

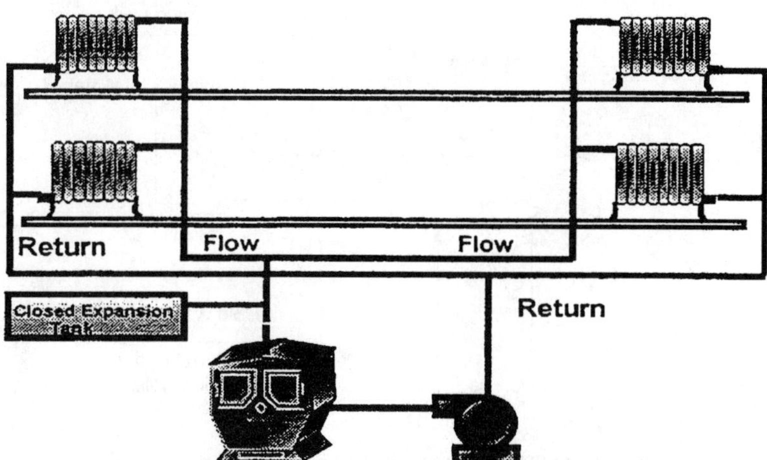

Figure 3.8 Forced circulation heating system.

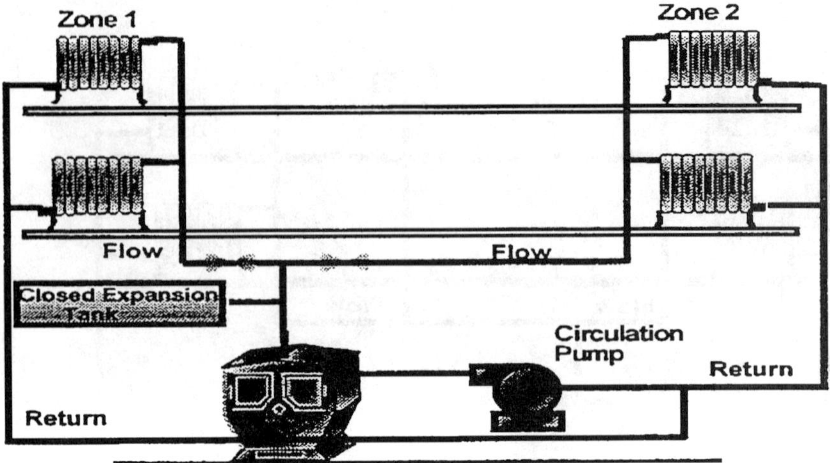

Figure 3.9 Zone-controlled hot water system.

In all hot water systems, the piping should be installed in such a way that the entire system can be drained. Each radiator and high points of a hot water heating system should be equipped with vent valves so that air may be vented, thus keeping the radiators full of water for maximum heat transfer.

3.3.4 Solar system application

Modern installations that utilize solar energy to supplement the operation of boilers and air-conditioning systems are also interconnected with basic hot water heating systems (Figure 3.10).

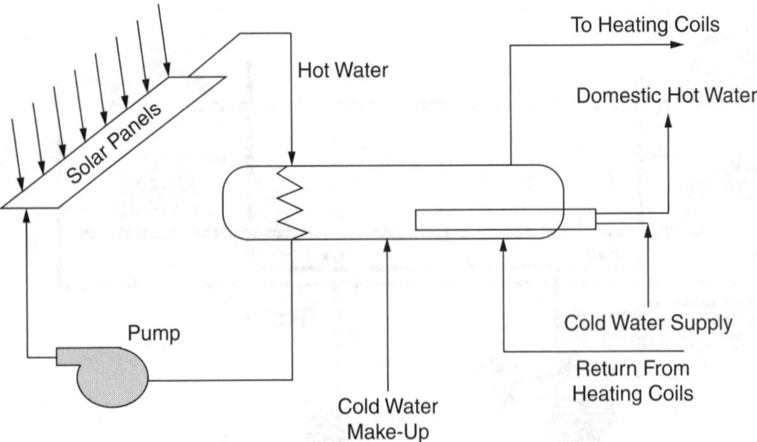

Figure 3.10 Solar system application.

3.4 Hydronic Heating Systems

A hydronic heating system uses hot water to move heat from where it is produced to where it is needed. Hot water is delivered via a pipe to the exact location where it is used. The system is relatively easy to zone, which is an advantage for comfort and economy. A hydronic heating system is shown in Figure 3.11.

The advantages of a hydronic system are:

- Comfortable heating
- Design flexibility
- Easy installation
- Clean operation
- Quiet operation
- Flexible zoning
- Energy efficiency

3.4.1 Components of hydronic heating systems

Boiler. The boiler heats water and produces the hot water that feeds the system.

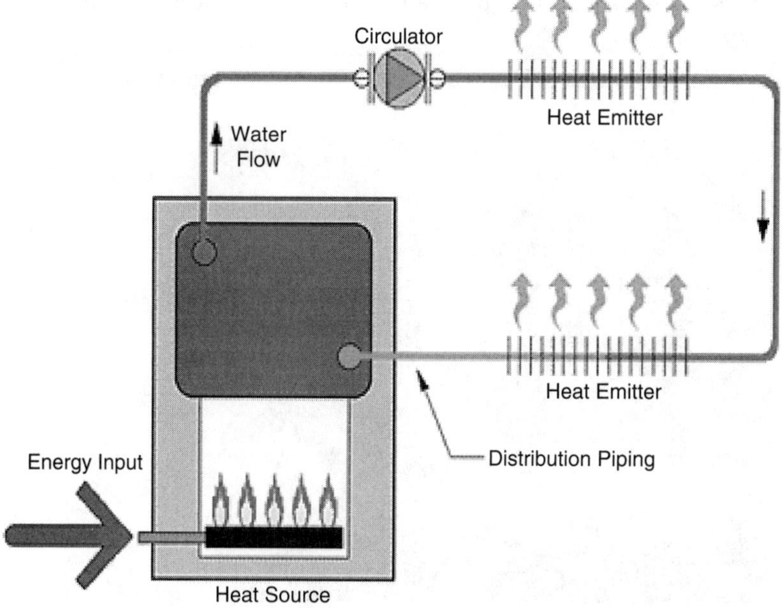

Figure 3.11 Hydronic heating system. *(Courtesy: Hydronicpros.com)*

Pump. The pump circulates water through the distribution system.

Distribution system. This is the piping system that distributes hot water to individual units.

Expansion tank. A tank, partially filled with air and water, which allows for expansion and contraction of hot water as it heats up and cools down.

Controls. These are the devices that control the temperature of hot water, or that switch pumps and boilers on and off include:

- Aquastat—operates boiler at constant temperature between 180–220°F regardless of heating load.
- Outdoor reset—operates boiler at varying water temperatures based on outdoor temperature.
- Outdoor cutout—turns off boiler system when outdoor temperature indicates that no heating is required.
- High limit—turns off boiler system if water temperature exceeds a safe operating limit.

Fin tubes. A feature in the radiator through which water from the distribution system and boiler circulates.

Zone valves. Valves on baseboard radiators.

Radiator. A device that distributes heat or cool air in individual units (Figure 3.12).

Thermostat. An automatic device that regulates the temperature of the radiator.

Combustion air intakes. Vents that draw air in and out of the building.

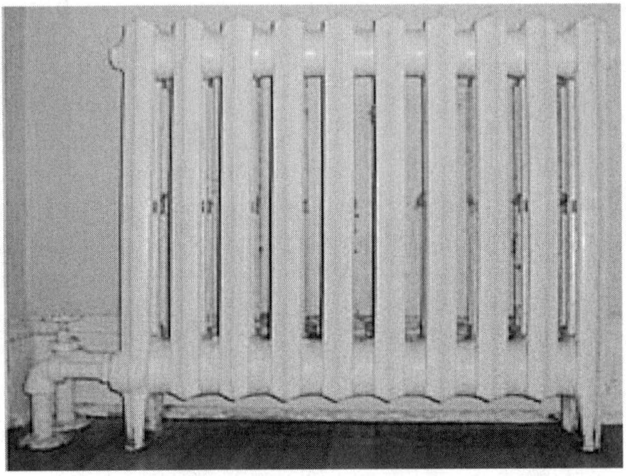

Figure 3.12 Hydronic radiator. *(Courtesy: Steam Radiators)*

3.5 Specifying a Hot Water Heating Boiler (Gas-Fired, Capacity 500 MBtu/hr, MAWP 150 psi)

The hot water heating boiler shall have a capacity of 500 MBtu/hr input and 425 MBtu/hr output, and a 504 gal/hr recovery rate at a 100°F temperature rise across the heat exchanger. The unit shall be designed for maximum allowable working pressure of 160 psi and inspected in accordance with ASME Code Section IV—Rules for Construction of Heating Boilers.

Construction. The watertube heat exchanger shall be a straight tube design with no blind pockets, of 7/8 inch inside diameter integral finned copper tubes, 0.33 inch minimum fin height. The tubes shall be rolled directly into ASME headers rated for 160 psi. Headers shall have covers permitting visual inspection and cleaning of all internal surfaces. The piping-side header shall have removable flanges to facilitate maintenance and permit vertical removal of a complete heat exchanger for service or replacement.

A flow-sensing device shall be factory-mounted as an integral part of the heat exchanger assembly to stop the flow of gas to the burners whenever water flow is inadequate or interrupted.

The unit is for indoor installation and shall have an integral draft diverter for a gas-tight connection. The outer jacket shall be a unitized shell, finished with acrylic thermoset paint baked at not less than 325°F. The frame shall be constructed of galvanized steel for strength and protection.

Burner. The burner shall be the atmospheric type and constructed of stainless steel. The combustion chamber shall be lined with a cast refractory of at least 2 inches in thickness to retain heat, and approved for service temperatures of not less than 2000°F. All gas manifolds shall be outside the combustion chamber and all primary combustion air shall be drawn directly from outside the heater.

The firing mode shall be a 4-stage control, which shall be electronically operated, and shall continually monitor the water temperature—staging the heater for the most efficient operation. A gas train shall be provided in accordance with the requirements of ASME CSD-1 Code—Controls and Safety Devices for Automatically Fired Boilers.

The ignition safeguard system shall be intermittent electronic ignition with electronic flame supervision. Responds should be in less than 0.8 second upon flame failure, 115 V control circuit.

Controls. The standard control system operates on 24 VAC power from a class 2 transformer. All the controls shall meet the requirements of ANSI Standard Z21.13 and ASME CSD-1 Code. The controls include:

- Ignition safeguard
- High-temperature limit
- Operating temperature control
- Gas pressure regulator
- Redundant electric gas valve
- Water flow sensing
- Manual shut-off gas valve

Start-up services. A factory authorized representative of the manufacturer shall provide start-up service and training of the operator.

Warranty. The manufacturer shall warrant all-furnished parts for 18 months from the date of boiler shipment. The pressure vessel parts shall carry a 5-year warranty against thermal shock damage.

Documents. The manufacturer shall provide the following documents:

- Three copies of perfect bound owner's manuals
- Three copies of the manufacturer's data reports
- Three copies of the certified prints

Chapter

4

Hot Water Supply Boilers

A hot water supply boiler, which operates at pressures not exceeding 160 psi (1100 kPa) and/or for temperatures not exceeding 250°F (121°C), is mostly used for supplying domestic hot water. Besides hot water supply, a hot water supply boiler is used for industrial process applications.

4.1 Hot Water Supply Boilers

When a hot water boiler is used for domestic water supply, it is classified as a hot water supply boiler. The hot water produced is used for domestic purposes.

The process of hot water generation by a hot water supply boiler is similar to the generation of hot water by a hot water heating boiler, as shown in Figure 4.1. In a hot water supply boiler, input is fuel and feedwater, and output are the products of combustion and domestic water. Feedwater is heated by a burner in the furnace and domestic water is generated within the watertube. The pressure in the watertube discharges hot water through the main pipe.

4.2 Types of Hot Water Supply Boilers

Hot water supply boilers are classified according to their uses and purposes. The following are the most common hot water supply boilers:

- Hot water supply boilers without tanks
- Hot water supply boilers with tanks
- Pool heaters

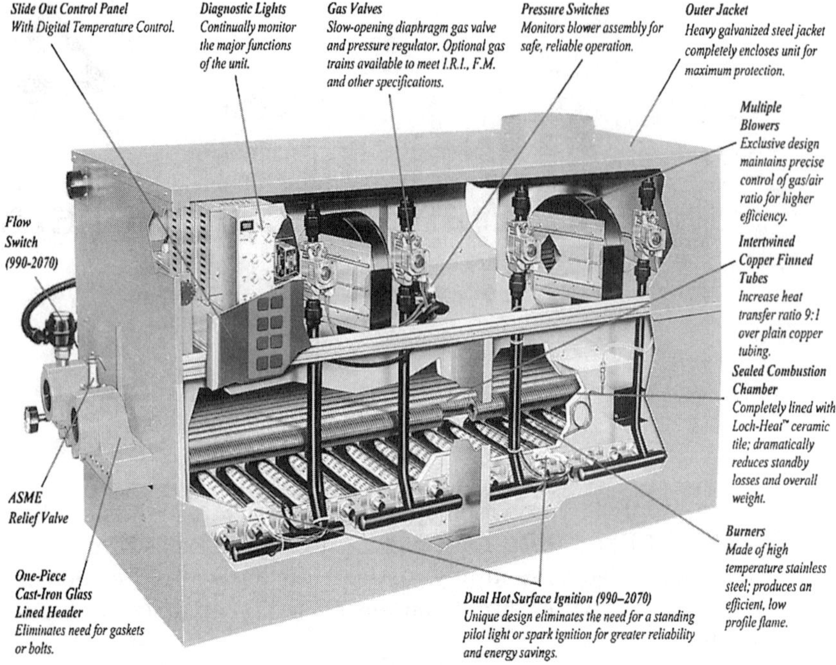

Figure 4.1 Hot water supply boiler. *(Courtesy: Lochinvar Corporation)*

4.2.1 Hot water supply boilers without tanks

In this system, hot water is supplied directly from the hot water supply boiler to the endpoint by the piping distribution network. The hot water supply boiler, if designed and operated properly, can give long service life. Typical efficiency for such a boiler is more than 80 percent. A hot water supply boiler with sectional views is shown in Figure 4.2.

A hot water supply boiler has the following features:

- Efficient and compact design
- Stainless steel burner tubes
- Solid bronze headers
- Slide-out heat exchanger
- Wide range of capacities from 300 to 2500 MBH

4.2.2 Hot water supply boilers with tanks

In this system, a tank is attached close to the hot water supply boiler to meet the demand of hot water supply. The system also includes a

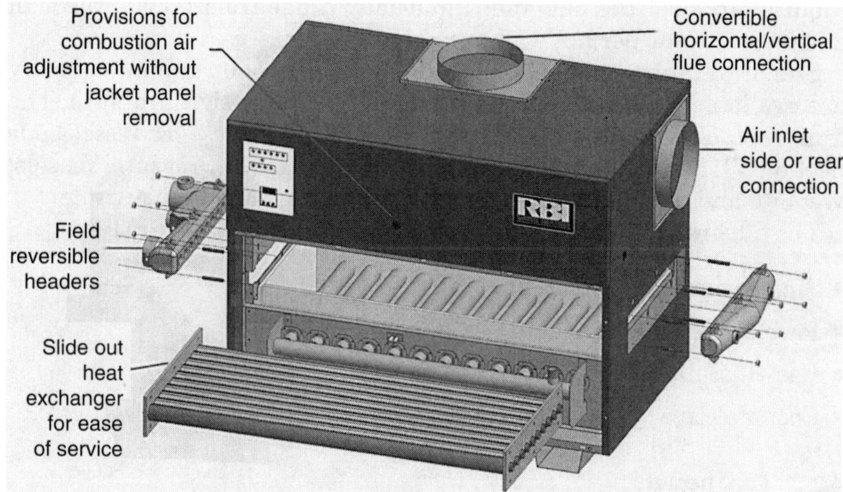

Figure 4.2 Hot water boiler without tank. *(Courtesy: RBI Water Heaters, Canada)*

water pump attached to the hot water supply boiler, generally prepiped and prewired. A package system of a hot water supply boiler with a storage tank is shown in Figure 4.3.

Designing a system like this depends on the total capacity required for end use. The designer may add as many water supply boilers as

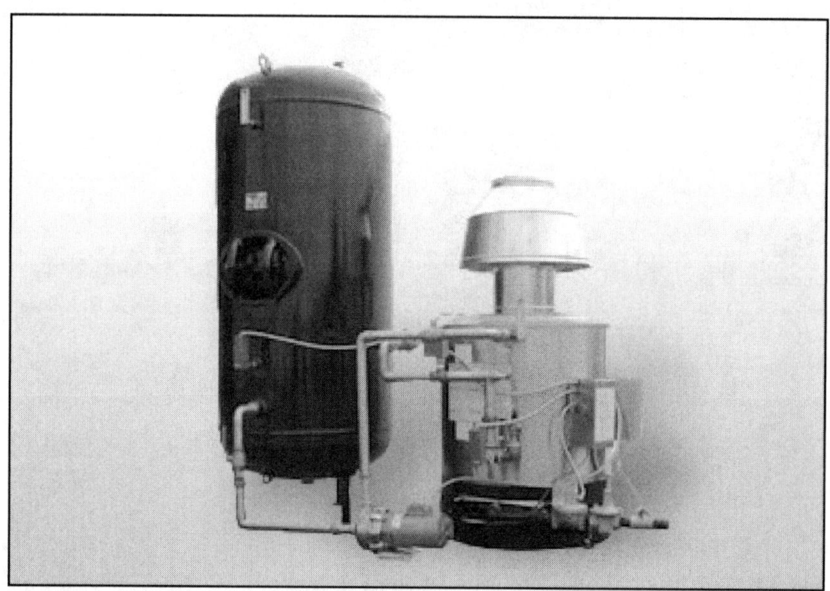

Figure 4.3 Hot water supply boiler with tank. *(Courtesy: Weben Jarco)*

required to meet the demand. Input may range from 50,000 through 2,500,000 Btu/hr per hot water supply boiler.

Standard tanks up to 72 inches diameter and 3000 gallons capacity are available in the market for 125 and 150 psi design pressure. The tank should be designed under ASME *Boiler and Pressure Vessel Code* Section VIII—Division 1, Rules for Construction of Pressure Vessels. Webnite and polykad linings are also available for the tank inside.

The advantages of a hot water supply boiler with tank are:

- Single source responsibility
- Preplumbed, prewired, and skid mounted
- Easy installation
- Cost effective

4.2.3 Pool heaters

A pool heater is used to supply hot water to the swimming pool. The heater allows setting the desired water temperature like a thermostat in home. When the water temperature in the pool drops below the set temperature, the heater starts and supplies hot water. When the water temperature in the pool reaches the set temperature, the heater shuts off. A pool heater is shown in Figure 4.4.

Figure 4.4 Pool heater for outdoor installation. *(Courtesy: Raypak, Inc.)*

Pool heaters are designed for either indoor or outdoor installation. Most heaters are installed at a location higher than the pool water level. If the heater is installed outside, it should be placed on a concrete, brick, or stone pad. The heater for outdoor use should have a built-in hood at the top. Proper venting is required if the pool heater is installed in a pump house or boiler room.

4.3 Hot Water Supply System

The purpose of a hot water supply system is to provide the building occupants with enough hot and cold water. Old buildings have hot water supply systems with gravity storage tanks on the top floor of the buildings. Modern systems have pressurized tanks connected to the supply pumps.

Hot water supply systems can be classified as: hot water supply system with gravity tank, and hot water supply system with pressure tank.

4.3.1 Hot water supply system with gravity tank

This system uses a gravity tank at least 30 ft (10 m) above the highest outlet. It may be necessary to install pressure reducing valves on the lowest floor of a tall building. The volume of the tank should be designed to compensate for the limited capacity of the supply lines. A hot water supply system with a gravity tank is shown in Figure 4.5.

Disadvantages of the system are:

- Open gravity tanks have the potential danger of freezing during winter conditions.
- Huge tanks may affect the construction of the building.

4.3.2 Hot water supply with a pressurized tank

This system uses a pressurized tank, which is partly filled with air behind a membrane. The air compensates for pressure variations during consumption and when the supply pump starts and stops. A pressurized tank has a limited compensating capacity for storage in main supply lines.

A hot water supply system with a pressurized tank is shown in Figure 4.6.

4.4 Specifying a Hot Water Supply Boiler (Gas-Fired, Capacity 2,450,000 Btu/hr, MAWP 160 psi)

Domestic hot water supply boilers shall have a capacity of 2,450,000 Btu/hr input and 2,000,000 Btu/hr output, and a 2411 gal/hr recovery

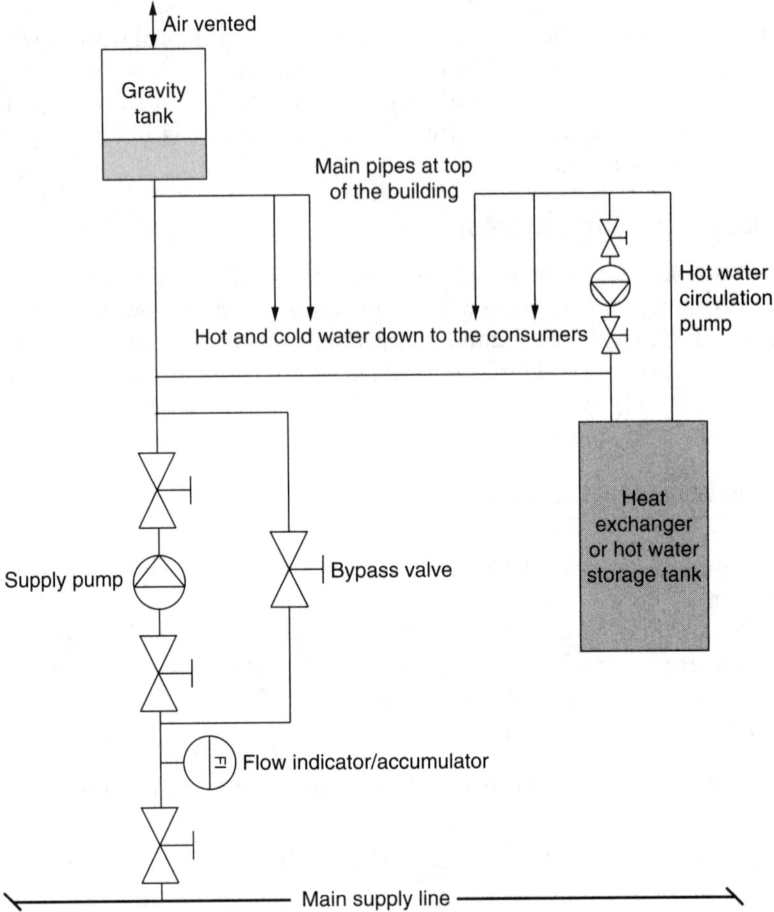

Figure 4.5 Hot water supply system with a gravity tank.

rate at a 100°F temperature rise. The unit shall be designed for maximum allowable working pressure of 160 psi and inspected in accordance with ASME Code Section IV—Rules for Construction of Heating boilers.

Construction. A watertube heat exchanger shall be a straight tube design with no blind pockets, 7/8 inch inside diameter, integral finned copper tube with a 0.33 inch minimum fin height. E tubes shall be rolled directly into ASME headers rated for 160 psi. Headers shall have covers permitting visual inspection and cleaning of all internal surfaces. The piping-side header shall have removable flanges to facilitate maintenance and permit vertical removal of a complete heat exchanger for service or replacement.

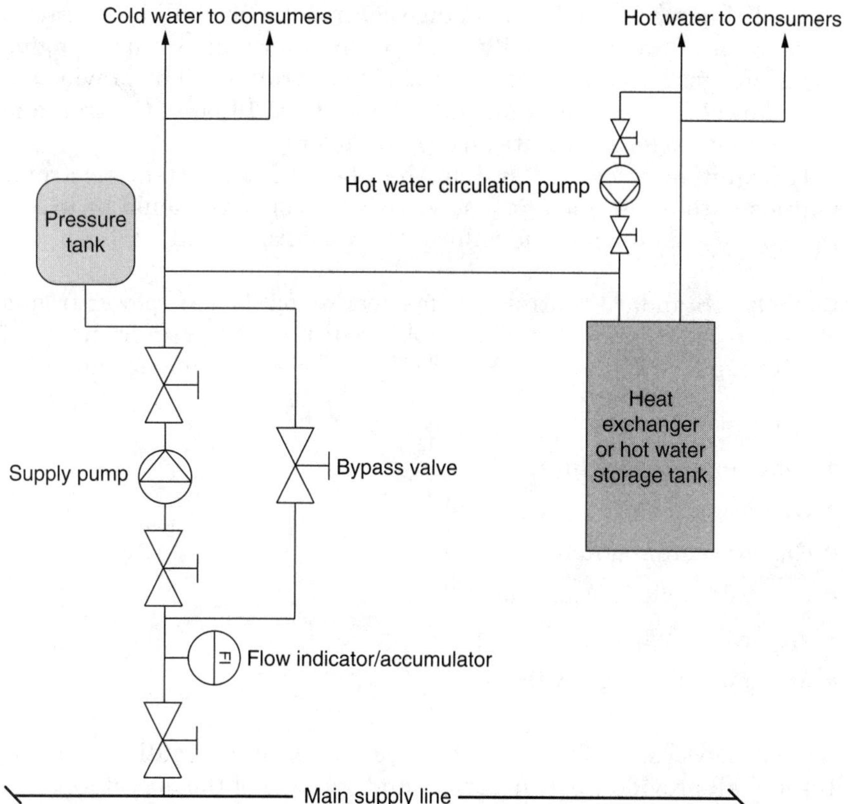

Figure 4.6 Hot water supply system with a pressurized tank.

A flow-sensing device shall be factory-mounted as an integral part of the heat exchanger assembly to stop the flow of gas to the burners whenever water flow is inadequate or interrupted.

The unit is for indoor installation and shall have an integral draft diverter for gas-tight connection. The outer jacket shall be a unitized shell, finished with acrylic thermoset paint baked at not less than 325°F. The frame shall be constructed of galvanized steel for strength and protection.

Burner. The burner shall be the atmospheric type, constructed of stainless steel. The combustion chamber shall be lined with a cast refractory of at least two inches in thickness, to retain heat, and approved for service temperatures of not less than 2000°F. All gas manifolds shall be outside the combustion chamber and all primary combustion air shall be drawn directly from outside the heater.

The firing mode shall be a 4-stage control, which shall be electronically operated and continually monitor the water temperature, staging the heater for most efficient operation. A gas train shall be provided in accordance with the requirements of ASME CSD-1 Code—Controls and Safety Devices for Automatically Fired Boilers.

The ignition safeguard system shall be an intermittent electronic ignition with electronic flame supervision. Responds should be in less than 0.8 second upon flame failure, 115 V control circuit.

Controls. Standard control systems operate on 24 VAC power from a class 2 transformer. All the controls shall meet the requirements of ANSI Standard Z21.13 and ASME CSD-1 Code. The controls include:

- Ignition safeguard
- High-temperature limit
- Operating temperature control
- Gas pressure regulator
- Redundant electric gas valve
- Water-flow sensor
- Manual shut-off gas valve

Start-up services. A factory authorized representative of the manufacturer shall provide start-up service and training of the operator.

Warranty. The manufacturer shall warrant all furnished parts for 18 months from the date of boiler shipment. The pressure vessel parts shall carry a 5-year warranty against thermal shock damage.

Documents. The manufacturer shall provide the following documents:

- Three copies of perfect bound owner's manuals
- Three copies of the manufacturer's data reports
- Three copies of the certified prints

Chapter

5

Hot Water Heaters

A hot water heater supplies hot water at pressures not exceeding 160 psig (1100 kPa) and temperature not exceeding 210°F (99°C). A hot water heater is similar to a hot water supply boiler, and is used to supply potable water to any building. The difference in application between the hot water supply boiler and the water heater is as follows:

- The water temperature of a water heater is limited to 210°F (99°C).
- A corrosion resistant lining or material is used in the construction of a water heater.
- A water heater is used for supplying potable water, with makeup water from a potable supply system.

Water heaters, in general, are classified as residential water heaters and commercial water heaters. Residential water heaters are used for supplying potable water for houses, apartments, condominiums and other residential buildings. Commercial water heaters are used for supplying potable water for commercial buildings, industries, office buildings, and other commercial locations.

5.1 Working Principle

A water heater works on the simple principle of water heating. In a gas-fired water heater, gas is fed through a control valve and a thermostat switch. The burner is usually installed under the tank. The exhaust gases are vented either through a hollow core at the center of the tank or around the tank sides. The burner heats the tank, which in turn heats the water. A gas-fired water heater is shown in Figure 5.1.

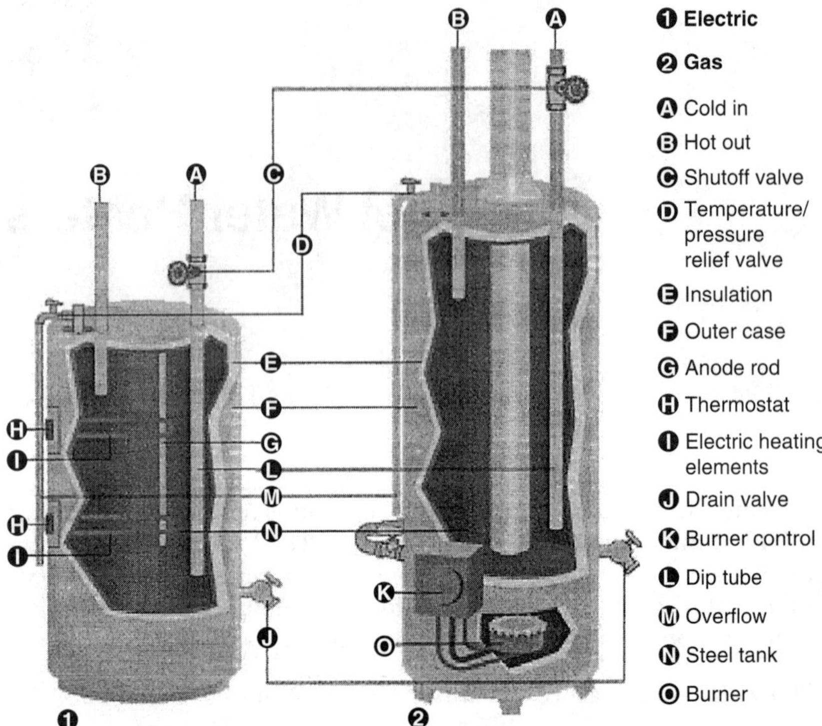

Figure 5.1 Water heaters—electric and gas fired. *(Courtesy: howstuffworks.com)*

The water heater heats and stores water. The speed at which a water heater heats water is called recovery rate. If water is drawn faster than it is heated, the temperature drops. The demands of hot water supplies determines the capacity and recovery rate of the water heater. Typically, a water heater with a low recovery rate has a high-capacity tank. On the other hand, a water heater with a high recovery rate does not require a large tank.

5.2 Types of Hot Water Heaters

Water heaters are classified according to the following:

- *Use*—residential, commercial
- *Holding tank*—tank, tank with storage tank, tankless
- *Fuel type*—gas, oil, electricity, or solar
- *Miscellaneous types*—heat pump heater, condensing heater

5.2.1 Use type

Water heaters can be used for residential or commercial applications. The classifications based on the use are:

- Residential
- Commercial

Residential water heaters. A water heater for residential use ranges from 5 to 120 gallons nominal tank capacity.

Commercial water heaters. A water heater for commercial application may range from 40 to 5000 gallons. These heaters are available in vertical or horizontal models. A commercial water heater is shown in Figure 5.2.

5.2.2 Tank type

Most of the water heaters used today have heavy inner steel holding tanks. Typically the tank holds 30–500 gallons water. The tank is designed for 150 psi maximum allowable working pressure. The steel tank normally has a bonded glass liner to prevent rusting from water.

Figure 5.2 Commercial water heater.

Tank-type water heater. This is the most common water heater used for commercial and residential purposes. A heavy steel tank holds the hot water, which is generated within the tank. A tank-type water heater is shown in Figure 5.3.

The essential features of the tank-type heater are:

- A glass liner protects the inside of the tank.
- Insulation is installed surrounding the tank.
- A dip tube is installed to let cold water into tank.
- A pipe is installed to let hot water out of tank.
- A thermostat is required to control the temperature of water inside the tank.
- A drain valve allows draining the tank.
- A T&P relief valve is installed to keep the tank from exceeding maximum allowable working pressure.

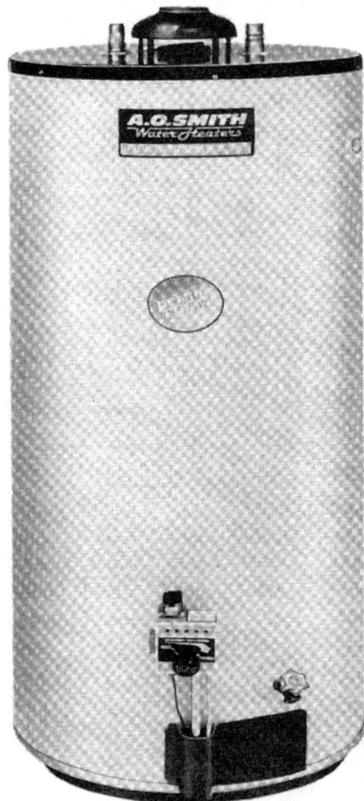

Figure 5.3 Tank-type water heater. *(Courtesy: A.O. Smith Corporation)*

Tank-type water heater with storage tank. Many tank-type water heaters use storage tanks for temporary storage of water. A water heater with a storage tank is shown in Figure 5.4.

Tankless heaters. In a tankless water heater, cold water travels through a pipe inside the unit, which is turned on only when water is required. This type of water heater is also called a demand water heater or instantaneous water heater as it operates on demand, and reduces the standby loss.

An electric tankless water heater is shown in Figure 5.5. When a hot water tap is turned on, cold water enters through a pipe into the unit. In the electric tankless heater, an electric element heats the water. In a gas-fired tankless heater, a gas burner heats the water. As a result, a tankless water heater delivers a constant supply of water.

5.2.3 Water heaters based on fuels

Various fuels are used for the heat source of water heaters. The most common fuels are:

- Gas
- Oil
- Electricity
- Solar energy

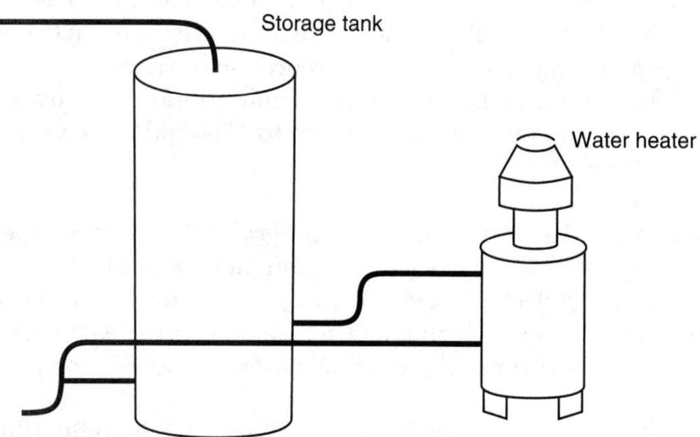

Figure 5.4 Water heater with a storage tank.

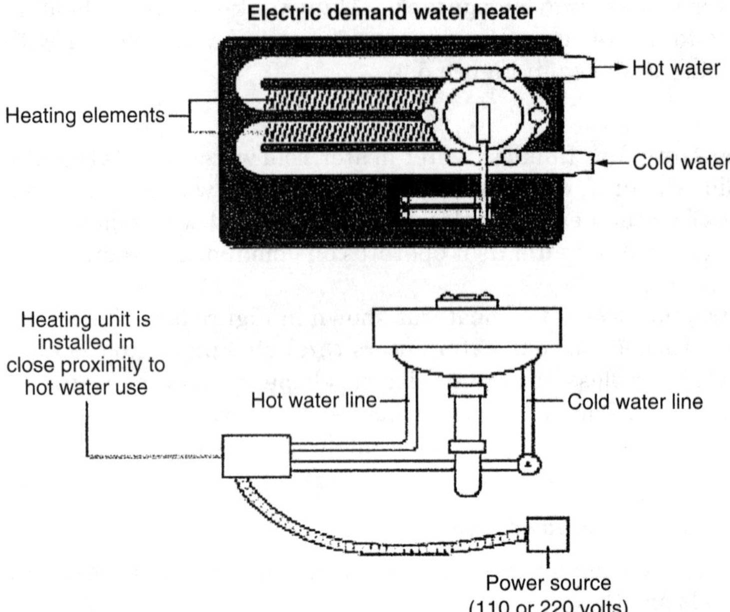

Figure 5.5 Electric tankless water heater. *(Courtesy: tankslesswaterheater-guide.com)*

Gas-fired water heaters. Common fuel for water heaters is natural gas. Gas-fired water heaters are used for both residential and commercial application. A gas-fired water heater is shown in Figure 5.6.

Oil-fired water heaters. Oil-fired heaters are generally the storage type, direct-fired. These heaters have high recovery rates. The fuel used is Garden No. 1 or No. 2. Direct oil-fired water heaters are manufactured with pressure aromatizing, aspirating, or vaporizing burners.

An oil-fired storage water heater ranges from 40 gal/hr recovery (100°F rise) with a 20-gallon storage tank, up to 1300 gal/hr recovery with a 200-gallon tank.

Electric water heaters. An electric heater is generally the storage type, consisting of a tank with one or more immersion metal sheathed tubular heating elements or electrode sets. Wattages are selected to meet recovery requirements. The tank and heating elements are designed to withstand the water conditions. An electric water heater is shown in Figure 5.7.

Residential and commercial electric water heaters are rated for 120, 208, 240, and 480 volts. Tanks for electric heaters are tested at 300 psi

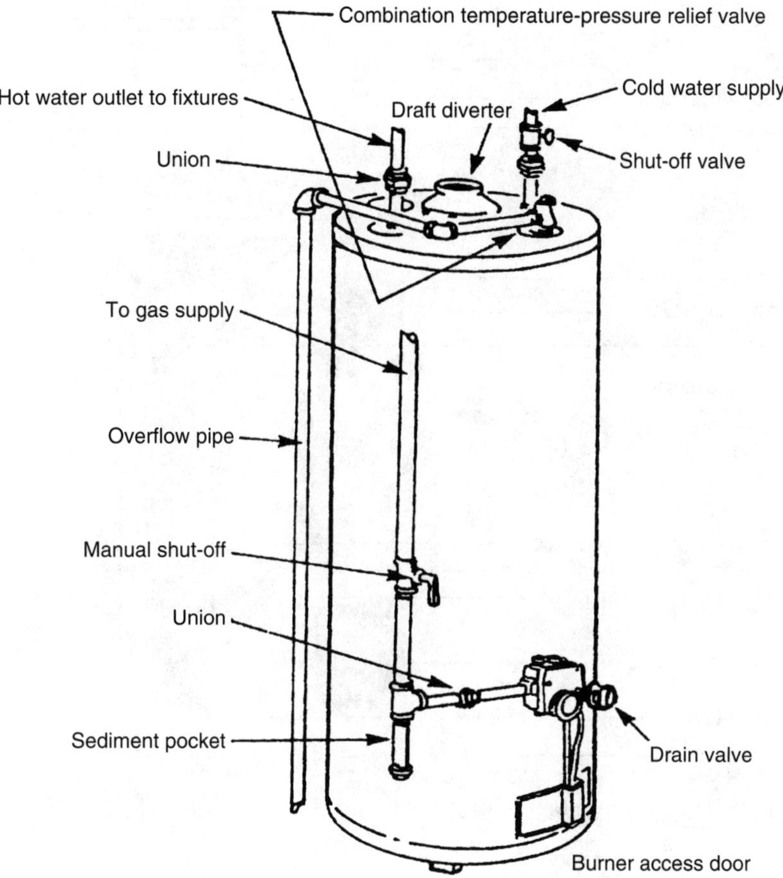

Figure 5.6 Gas-fired water heater.

hydrostatic test pressure, and are rated for a safe working pressure of 122–155 psig.

Solar heaters. The basic components of a solar water heater are solar collectors, storage tank, associated piping, and controls. Water heaters may be of the natural convection or forced circulation type. A solar water heater system is shown in Figure 5.8.

5.2.4 Miscellaneous water heaters

Heat-pump water heater. A heat-pump water heater operates the same way that HVAC-system heat pump does. The heat pump water heater extracts heat from surrounding air and transfers it to the water in a tank.

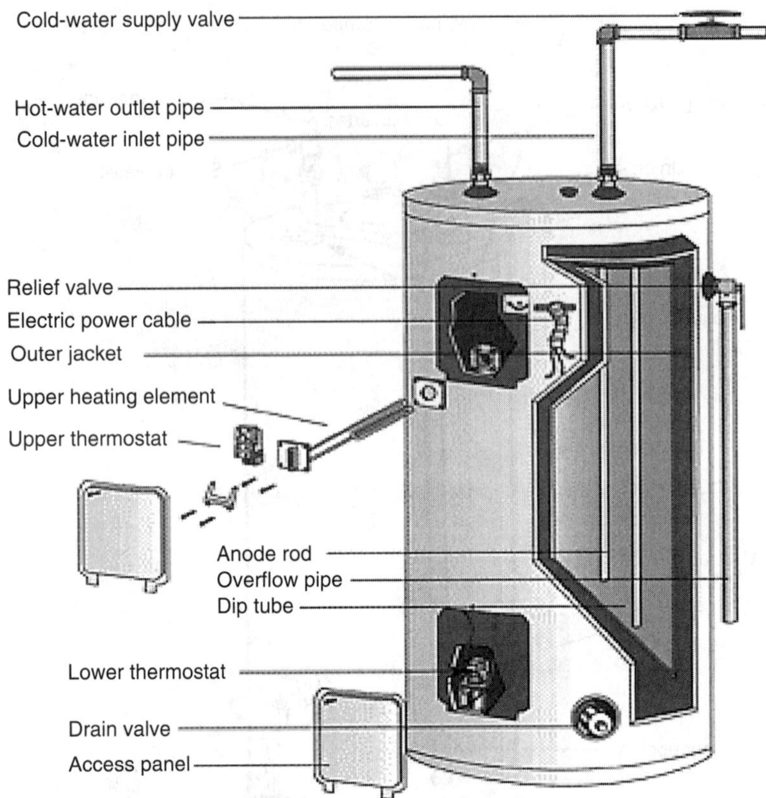

Figure 5.7 Electric water heater.

Heat-pump water heaters use electricity to move heat from one place to another instead of generating heat directly. To operate most efficiently, these heaters should be placed in areas having excess heat, such as furnace rooms.

Condensing water heater. A conventional gas-fired heater has a peak operating efficiency of 78–82 percent. The heater burns fuel in a combustion chamber, and the products of combustion pass through a heat exchanger, which transfers heat from the gases to water. Much of the heat loss occurs due to water, a by-product of combustion vapor, which carries out part of the energy lost in the flue gases.

A condensing water heater recovers much of this energy by using a larger heat exchanger or a second heat exchanger that reduces the flue gas temperature to the point where this water vapor condenses, releasing more energy. Condensing water heaters offer a peak operating efficiency of 88–90 percent. Figure 5.9 shows a comparison between a forced draft heater and a condensing water heater.

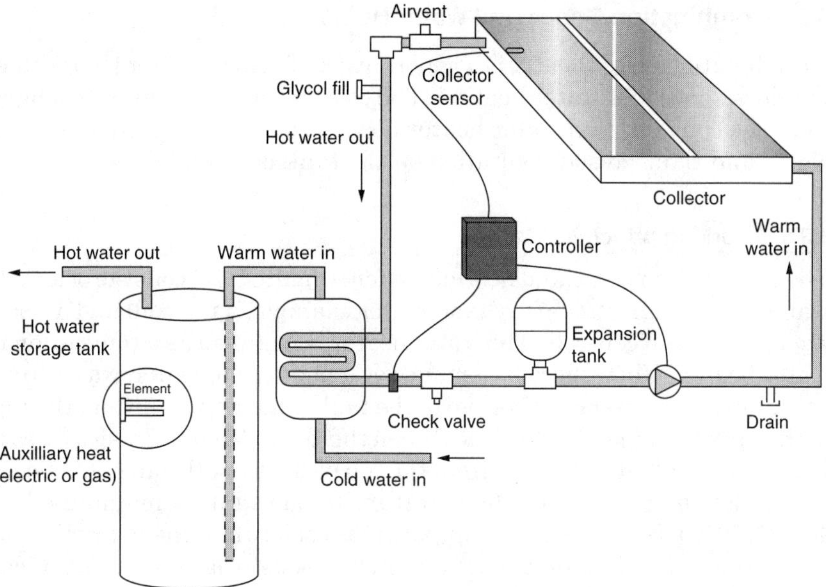

Figure 5.8 Solar water heating system.

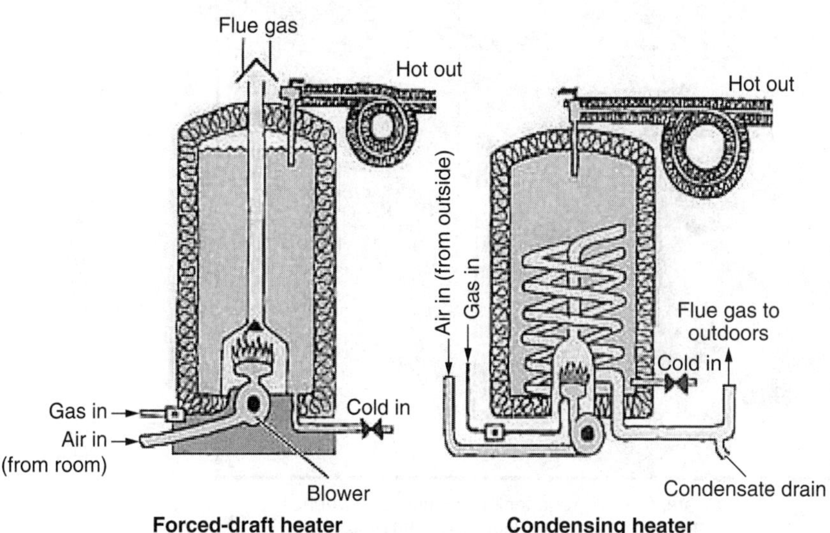

Figure 5.9 Condensing heater vs. forced draft heater. *(Courtesy: Santa Monica Green Building Program)*

5.3 Combination Space and Water Heaters

A combination water heater is a generally gas-fired hot water heater that provides space and water heating. The combination system lasts longer than the conventional water heater because of increased circulation of water and reduced sedimentation on the tank bottom.

5.3.1 Working principle

The combination space and heating system (Figure 5.10) consists of a tank water heater, a heat delivery system (for example, a fan coil) and associated pumps and controls. The water heater is installed as a conventional water heater. When there is demand for domestic hot water, water from the mains pipe enters the bottom of the tank, and hot water from the top of the tank is supplied to the load. When there is a demand for space heating, a pump circulates water from the top of the tank through a fan coil.

The domestic hot water temperature in the tank is maintained at 140°F (60°C). Because this temperature is cooler than the hydronic systems, the space heater delivery system needs a larger than typical heater. The water heater can be either a conventional tank-type water heater or a condensing gas boiler.

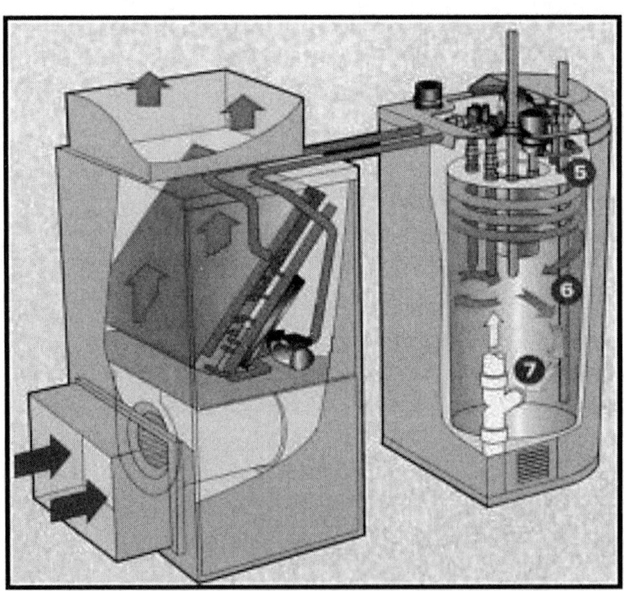

Hot water from storage tank is circulated through an air handler to provide space heating

Figure 5.10 Combination space and water heater. *(Courtesy: Advanced Buildings Technologies and Practices)*

The advantages of the combination space and water heaters are:

- Reduced floor space
- Lowest capital cost
- Improved energy efficiency
- Increased tank life

The disadvantages of the combination space and water heaters are:

- Available only in small sizes
- Not feasible for large loads

5.3.2 Application

All space heating piping has to be designed for potable water. Ferrous metals or lead-based solder cannot be used. A combination space and water heater is suitable for use in office buildings, apartments, retail food service, institutional arenas, and single family residential buildings.

5.4 Hot Water Heater System

A hot water system, as shown in Figure 5.11, is composed of water heaters, water supply connections, bottom drain valves, thermometers, piping, and expansion tank. The recirculation system may be gravity or pump-actuated.

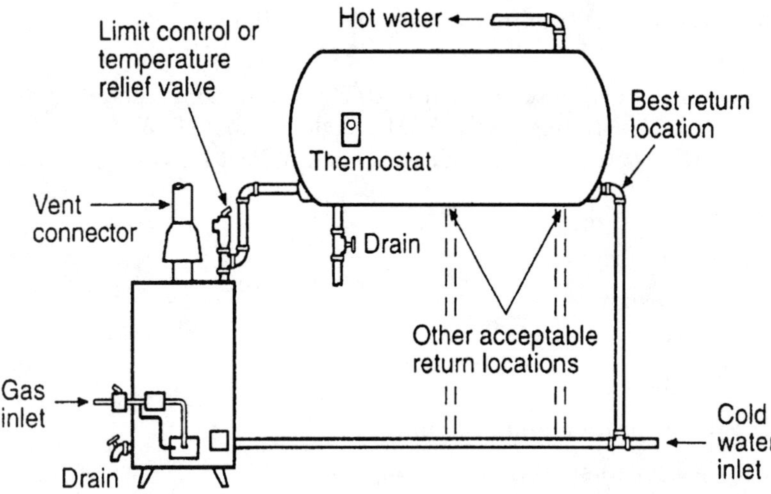

Figure 5.11 Hot water heater system with gravity circulation. *(Courtesy: NFPA 54)*

5.5 ASME Requirements

ASME Section IV, Part HLW applies to water heaters in commercial or industrial sizes for supplying potable water for commercial purposes at pressures not exceeding 160 psig (1100 kPa) and temperatures not exceeding 210°F (99°C). This Part is not applicable to residential-size water heaters.

5.5.1 General outline

The scope of Part HLW covers water heaters and water storage tanks for supplying potable hot water. The water heaters and water storage tanks are constructed with corrosion resistant materials. Part HLW is not applicable for hot water heating boilers. The requirements of linings for lined water heaters are specified in paragraph HLW-200. The materials used for construction of unlined water heaters are listed in Appendix E. Water heaters and tanks may be provided with cathodic protection.

Service restriction. The requirements are applicable to water heaters and water storage tanks for working pressures not more than 160 psig (1100 kPa) and water temperatures not more than 210°F (99°C).

Exception. Water heaters are exempted from the requirements of Part HLW if none of the following limitations is exceeded:

1. Heat input of 200,000 Btu/hr (58.6 kW)
2. Water temperature of 210°F (99°C)
3. Nominal water-containing capacity of 120 gallons

Stamping. A water heater or water storage tank constructed under Part HLW may be stamped with ASME Code Symbol HLW.

A brief outline of the contents of each Article of Part HLW is as follows:

Article 1—General

Article 2—Materials

Article 3—Design

Article 4—Weldments

Article 5—Tests

Article 6—Inspection and Stamping

Article 7—Controls

Article 8—Installation

5.5.2 Material requirements

Lining materials. A water heater or water storage tank is provided with a corrosion-resistant material lining for supplying potable water. The specifications of lining materials should conform to the requirements of this chapter. The following types of lining are used for construction of potable water heaters and water storage tanks:

- *Glass lining.* The surface of water heater or tank that is exposed to hot water is glass-lined and the minimum average thickness of the lining should be 0.005 inch (0.13 mm). Glass lining may be applied to the surfaces before assembly.
- *Galvanizing.* The galvanized coating may be used on the water heater vessel as a corrosion-resistant material. The amount of coating should be minimum 1 oz of zinc per square foot of surface based on a coating thickness of 0.007 inch (0.043 mm). The zinc used for coating should conform to ASTM B6, specifications for zinc (slab zinc) and should be equal to the grade "Prime Western."
- *Cement lining.* A cement lining may be applied to resist corrosion. The cement should be applied to provide a minimum thickness of 3/16 inch (4.8 mm). The lining should be completely cured and cover the interior of the vessel.
- *Copper lining.* Copper of weldable or brazeable quality may be applied as a corrosion-resistant material. The minimum thickness of such lining should be 0.005 inch (0.13 mm). Welding or brazing for lining attachments to steel backing should conform to the requirements of ASME Section IX.
- *Fluorocarbon polymer lining.* Fluorocarbon polymer lining is corrosion resistant and suitable for hot water service. The minimum thickness of such lining should be 0.003 inch (0.076 mm).The surface should be cleaned and free from scale, oxidation, oil, and the like before application of the lining. Lining should be cured at a proper temperature and duration to ensure continuity of lining.
- *Amine or polyamine epoxy lining.* The amine or polyamine epoxy linings may be used for electric water heaters with immersion-type elements, storage tanks, and surfaces of water heaters that are not directly heated by the products of combustion. The minimum thickness should be 0.003 inch (0.076 mm). Lining should be cured at a proper temperature and duration to assure continuity of lining.
- *Thermally sprayed metallic lining.* The lining material is any type of copper or copper alloy suitable for spraying. The inner surfaces should be cleaned by grit blasting before spraying. Minimum lining thickness should be 0.005 inch (0.13 mm). The spraying process

should be controlled so the surface temperature does not exceed 650°F (343°C).

- *Polymeric, flexible lining.* The material listed by the National Sanitation Foundation International (NSFI) should be used. Lining material should also meet the requirements of ANSI/NSF 14-1900 for potable water service at minimum temperature of 210°F (99°C). The minimum thickness of lining should be 0.020 inch (0.51 mm). Inner surfaces of the vessel should be free from discontinuities that exceed one-half times the thickness of the liner.
- *Autocatalytic (electroless) nickel-phosphorus lining.* The composition of nickel-phosphorus should conform to ASTM B 733-90 SC3, Type 1, Class 1. Minimum lining thickness should be 0.0003 inch (0.0076 mm). The phosphorus content should be a minimum of 9 percent and maximum of 13 percent.

Pressure parts. The materials for primary pressure parts such as shells, heads, flues, headers, or tubes should conform to the specifications of ASME Section II. Only those materials listed in Appendices D and E will be used. A material test report is required for plate material to verify its chemical and mechanical properties. The mill of origin shall provide this mill test report. Materials used for product forms will require identification marking in accordance with specifications of ASME Section II. Painted color identification or any other method of marking acceptable to the authorized inspector may be used. A welded assembly may be supplied as a part of the completed water heater. In that case, the parts manufacturer will submit a Manufacturer's Partial Data Report Form HLW-7.

Unidentified materials. When it cannot be identified, each piece of material is required to show that the chemical composition and mechanical properties of ASME Section II have been met. The materials will be tested as per the requirements of ASME Section II, and the manufacturer of the completed vessel is responsible for verifying all the requirements.

5.5.3 Design

Water heaters are fired with gas, oil, or electricity. Water heaters are designed in accordance with the requirements of this part. The following are the design parameters of water heaters:

Maximum allowable working pressure	160 psi (1100 kPa)
Minimum allowable working pressure	100 psi (690 kPa)
Maximum water temperature	210°F (99°C)

All the pressure parts of water heaters should be designed to prevent flame impingement. The design temperature should not be less than the mean metal temperature expected during operation. Metal temperature may be determined by calculations or by actual measurement of temperature under equivalent operating conditions. The proof test described may be used if the maximum allowable working pressure of the water heater is not known.

The minimum thickness of plate material used for construction of any lined or unlined water heater should be a minimum of 1/8 inch (3.2 mm).

Shells. Pressure parts such as shells, headers, and pipes of water heaters are subjected to internal pressure. The maximum allowable working pressure and thickness of the pressure parts under internal pressure are calculated as per the following formulas:

$$t = \frac{PR}{SE - 0.6P}$$

$$P = \frac{SEt}{R + 0.6t}$$

where P = maximum allowable working pressure, psi (but not less than 100 psi)

S = maximum allowable stress values from Appendices D and E, psi

t = required wall thickness, exclusive of liner, inch

R = inside radius of cylinder, inch

E = efficiency of longitudinal joint or ligament between tube holes, whichever is less. For seamless shells, $E = 1$.

Dished heads. Blank unstayed dished heads such as ellipsoidal heads, torispherical heads may be used for pressure either on the concave side or the convex side. The following symbols are used in the formulas to calculate the maximum working pressure and thickness of a blank unstayed dished head:

t = required wall thickness, exclusive of liner, inch

P = maximum allowable working pressure, psi [but not less than 100 psi (690 kPa)]

D = inside diameter of the head skirt; or inside length of the major axis of an ellipsoidal head, inch

S = maximum allowable stress values from Appendices D and E, psi

L = inside spherical or crown radius, inch

E = lowest efficiency of any joint in the head. For seamless heads, $E = 1$.

Ellipsoidal heads. The maximum allowable working pressure and thickness of a dished head of semiellipsoidal form under pressure on the concave side is calculated by the following formulas:

$$t = \frac{PD}{2SE - 0.2P}$$

$$P = \frac{2SEt}{D + 0.2t}$$

Torispherical heads. The maximum allowable working pressure of a torispherical head under pressure on the concave side is calculated by the following formulas:

$$t = \frac{0.885PL}{SE - 0.1P}$$

$$P = \frac{SEt}{0.885L + 0.1t}$$

Crown radius. The inside crown radius of an unstayed formed head should not be more than the outside diameter of the skirt of the head. The inside knuckle radius of a torispherical head should be a minimum 6 percent of the outside diameter of the skirt but not less than three times the head thickness.

Hemispherical heads. Hemispherical heads are not allowed for construction of potable water heaters, as joint design is complicated.

Pressure on convex side. Unstayed dished heads with pressure on the convex side will have maximum allowable working pressure of 60 percent of that for the same heads with pressure on the concave side.

Tubes. The thickness of tubes, seamless or welded, subjected to internal pressure should be calculated by using the formula of paragraph HG-3.01. If the tubes are subjected to external pressure, the formula of paragraph HG-3.12 will be used to calculate thickness.

Openings. All openings except single openings will be reinforced in accordance with paragraph HG-3.21. Reinforcement is not required for single openings if welded connections are attached according to the code and connections are not more than:

1. NPS 3 (DN 75) in shells or heads 3/8 inch (10 mm) or less; NPS 2 (DN 50) in shells or heads over 3/8 inch (10 mm).
2. Threaded, studded, or expanded connection is not more than NPS 2 (DN 50).

Tubes attachments. Tubes may be attached to the tubesheet by rolling. The minimum thickness of any tubesheet for installation of tubes should be 3/16 inch (4.8 mm). Tube attachment is accomplished by rolling the tubes into tube holes in the tubesheet. The tube holes may be formed by one of the following methods:

1. Using a method to the full size diameter that will not produce irregularities.
2. Using a method to a lesser diameter, then increase to full diameter by a secondary operation that will not produce irregularities.

5.5.4 Inspection and stamping

Water heaters are designed, constructed, inspected, and stamped in accordance with Part HLW of ASME Section IV. A manufacturer is responsible for complying with all the requirements of Part HLW. An authorized inspector is responsible for certifying that the water heater has been constructed in accordance with the code.

Authorized inspector's responsibility. The authorized inspectors are qualified by a written examination under the rules of any state of the United States or province of Canada. The authorized inspector is employed by an authorized inspection agency (AIA) accredited by the ASME. The authorized inspection agency may be the inspection organization of a state or municipality of the United States or a province of Canada. An authorized inspector is charged with the responsibility to certify that the water heaters have been constructed in accordance with the code. An authorized inspector has the following responsibilities:

1. Conducts necessary inspections to make sure that the manufacturer has complied with all the requirements of this part.
2. Make sure that the welding procedures used in construction have been qualified under ASME Section IX. The authorized inspector will review such evidence.
3. Verify that all welding done by welders or welding operators have been qualified under ASME Section IX. The authorized inspector will review evidence of performance qualification tests of each welder

and welding operator. He may call for and witness the test welding and testing if he has desires.

4. Make sure that the inspection and quality control procedure, if there is any, for fabrication of multiple duplicate water heaters and storage tanks has been followed.

Manufacturer's responsibilities. A manufacturer is overall responsible for complying with all the requirements of this part, if the water heater is to be marked with ASME Code Symbol HLW. The manufacturer has the following responsibilities under the code:

1. Complying with the code and make sure that part manufacturers will also comply with the requirements of the code.
2. Provide necessary information to the authorized inspector so that he can perform his duties under the code.
3. Make sure that quality control system has been complied with.
4. Submit or make available the following documents to the authorized inspector for his review:
   a. A valid certificate of authorization for use of ASME Symbol HLW
   b. Design calculations or certified proof test results
   c. Identification of materials
   d. Records of qualification of welding and/or brazing procedures
   e. Evidence of qualifications of each welder, welding operator or brazer
   f. Manufacturers' partial data reports, if required
   g. Records of traceability to the material identification, thickness, and acceptability of the materials as required by the code
   h. Offer the water heater or part for inspection at different stages as designated by the authorized inspector

Manufacturer's data reports. Data reports are the documents on design, construction, fabrication, inspection, and stamping on the water heaters. A manufacturer is required to complete a manufacturer's data report on the following ASME form:

- HLW-6—Manufacturer's data report for water heaters or storage tanks (Appendix K)

A single water heater or multiple identical water heaters with serial numbers in sequence and completed in a continuous 8-hour period may be included in this data report. The manufacturer will distribute the manufacturer's data reports among the user, the authorized inspection agency, the purchaser, and the jurisdiction. A copy of the manufacturer's data reports will be kept in the manufacturer's file for a minimum of 5 years.

Partial data reports. A parts manufacturer is required to complete a partial data report for each part requiring inspection under the code. All parts fabricated by a manufacturer other than the manufacturer of the complete water heater may be included in this data report. The parts manufacturer will forward the data reports in duplicate to the manufacturer of the complete water heater. The authorized inspector will witness the application of the Code Symbol stamp on the completed water heater based on the partial data report together with his own inspection. The partial data report shall be recorded on the following ASME form:

- HLW-7—Manufacturer's partial data reports for water heaters and storage tanks

Supplementary reports. When space on manufacturer's data report form HLW-6 is not sufficient to record data, additional data shall be recorded on the following ASME form. Form H-6 will be attached to the manufacturer's data report form.

- H-6—Manufacturer's data report supplementary form

Proof test reports. Tests to establish the maximum allowable working pressure of complete vessel or parts should be witnessed by the manufacturer's representative. These tests shall be witnessed and accepted by the authorized inspector. This proof test shall be recorded on the following ASME form:

- HLW-8—Manufacturer's master data proof test report for water heaters or storage tanks

Stamping. A manufacturer, who possesses a valid Code Symbol stamp and a certificate of authorization (Figure 5.12), will apply the Code Symbol stamp to water heaters constructed in accordance with the code. The water heater will be marked or stamped with the Code Symbol stamp as shown in Figure 5.13 and the form of stamping as shown in Figure 5.14. The following data required to be marked or stamped on the water heater:

1. The manufacturer's name, preceded by the words "Certified by"
2. Maximum allowable working pressure
3. Maximum allowable input in Btu/hr; electric heaters may use kW or Btu/hr
4. Manufacturer's serial number
5. Year built

CERTIFICATE OF AUTHORIZATION

This certificate accredits the named company as authorized to use the indicated symbol of the American Society of Mechanical Engineers (ASME) for the scope of activity shown below in accordance with the applicable rules of the ASME Boiler and Pressure Vessel Code. The use of the Code symbol and the authority granted by this Certificate of Authorization are subject to the provisions of the agreement set forth in the application. Any construction stamped with this symbol shall have been built strictly in accordance with the provisions of the ASME Boiler and Pressure Vessel Code.

COMPANY:

RAYPAK, INC.
2151 EASTMAN AVE.
OXNARD, CALIFORNIA 93030

SCOPE:

POTABLE WATER HEATERS AT THE ABOVE LOCATION ONLY (THIS AUTHORIZATION DOES NOT COVER WELDING OR BRAZING)

AUTHORIZED: **DECEMBER 30, 2003**
EXPIRES: **DECEMBER 16, 2006**
CERTIFICATE NUMBER: **26,041**

Chairman of The Boiler
And Pressure Vessel Committee

Director, Accreditation and Certification

The American Society of Mechanical Engineers

Figure 5.12 Certificate of authorization for symbol stamp HLW. *(Courtesy: Raypak, Inc.)*

Figure 5.13 ASME symbol stamp for water heaters. *(Courtesy: ASME Section VI)*

HLW

Certified by

____

(Name of manufacturer)

Maximum Allowable W. P.______psi

Maximum Allowable Input ______ Btu/hr ______ kW

Manufacturer's Serial No. ____________________

Year Built __________________________

Figure 5.14 Form of stamping on a completed water heater. *(Courtesy: ASME Section IV)*

Stamping of proof-tested vessel. A proof-tested vessel may be marked or stamped with the Code Symbol stamp if:

1. The proof test was stopped before any visible yielding
2. Welding was qualified as required by the code
3. MAWP was calculated by the formula based on minimum tensile strength[5]
4. Interior of the vessel was inspected for damage
5. Complete vessel was hydrostatically tested

Dimensions of stamping. The Code Symbol and marking are required to be stamped on the water heater with letters and figures a minimum of 5/16 inch (8 mm) high. The stamping may be done on a plate a minimum of 3/64 inch (1.2 mm) thick and the plate may be permanently fastened to the water heater. One of the following methods should be provided in case the stamping is to be covered by insulation:

1. An opening with a removable cover for view stamping
2. A nameplate with duplicated Code Symbol and data, and located in a conspicuous place

The plate dimension should be a minimum of 3 inches × 4 inches (76 mm × 102 mm) and marked with letters and numbers minimum 1/8 inch (3.2 mm) high.

5.5.5 Controls

Temperature control. An automatically fired water heater is fitted with the operating control for normal operation. A separate high-temperature limit actuated combustion control is required in addition to the operating control. The high-temperature limit will automatically cut off the fuel supply when water temperature reaches 210°F (99°C). When actuated, the high-temperature limit control will do the following functions for different types of water heaters:

1. *Gas-fired water heaters:* The high-temperature limit control will shut off the fuel supply with a shutoff means other than the operating control valve.
2. *Electrically heated boiler*: The high-temperature control limit will cut off all power to the operating controls.
3. *Oil-fired water heaters*: The high-temperature limit control will cut off all current flow to the burner system.
4. *Indirect water heater*: The high-temperature limit control will cut off the source of heat.

Limit controls. Limit controls, if used with electric circuit, will break the hot or line sides of the control circuit.

Safety controls. Safety controls such as primary safety controls, safety limit switches, burners, and electric elements are used for all water heaters. These controls are required by nationally recognized standards such as the following standards:

UL 732—Standards for Safety, Oil-Fired Heaters

UL 795—Standards for Safety, Commercial-Industrial Gas Heating Equipment

UL 1435—Standards for Safety, Electric Booster and Commercial Storage Tank Water Heaters

ANSI Z21.10.3—Standards for Gas Water Heaters, Volume III

A certifying organization will test the equipment under a national standard recognized by the authorities having jurisdiction. The monogram or symbol of the certifying organization will be affixed on the equipment, which will indicate that the controls and heat generating apparatus have been manufactured in accordance with that standard.

Electrical requirements. Water heaters are provided with electrical apparatus and controls either in the shop or in the field. All factory-mounted and wired controls, heat generating apparatus, and other appurtenances are installed in accordance with nationally recognized standards as listed in paragraph HLW-7.03. Provisions of the National Electric Code and/or electrical code enforced by local jurisdiction are applicable for installation of all field wiring for controls, heat generating apparatus, and other appurtenances. The National Electric Code will be applicable if there is no local code.

5.6 Installation

5.6.1 Safety relief valve requirements

At least one temperature and pressure relief valve or safety relief valve is required for each water heater. The valves should be ASME rated and be marked with Code Symbol V or HV. Minimum size of safety relief valve is NPS 3/4 (DN 20).

Set pressure. The pressure setting of the safety relief valves should be less than or equal to the maximum allowable working pressure. Set pressure of additional valves may not exceed 10 percent over the set pressure of the first valve. There may be different components of hot water system such as expansion tanks, storage tanks, pumps, valves, or piping with different working pressure than the hot water heater. In that case, set pressure of the safety relief valve should be the lowest maximum allowable working pressure of any component.

Safety relief valve capacity. The relieving capacity of the safety relief valve shall not be less than the maximum allowable input of the water heater. Rated burner input capacity may be used to determine relieving capacity of the safety relieve valve. The unit for relieving capacity is Btu/hr for gas or oil-fired water heaters. Relieving capacity for electric water heaters may be expressed in Btu/hr by multiplying kW input with 3500 Btu/hr. At maximum capacity, pressure cannot exceed more than 10 percent of the maximum allowable working pressure. Capacity of the safety valve has to be recalculated if operating conditions are changed or additional heating surface is installed.

Permissible mountings. Safety relief valves may be installed in the shop by the manufacturer or in the field by an installer. The safety relief valves will be installed properly before a water heater is placed in operation.

Safety relief valves are installed on the top of water heaters or directly to a tapped or flanged opening in the water heater. The spindles of the safety relief valves should be in upright and vertical position, except when the valve is mounted horizontally directly on the water heater shell and the outlet is pointed down. Piping or fitting used to mount the safety relief valve should not be of nominal pipe size less than the valve inlet size. The center line of the valve connection should not be at a point lower than 4 inches (102 mm) from the top of the water heater shell.

Common connection for multiple valves. The cross-section of a common connection should not be less than the combined areas of inlet connections of all the safety relief valves. When two or more valves are used on a water heater, they may be single, directly mounted, or mounted on a Y-base. When a Y-base is installed, the inlet area should not be less than the combined outlet areas.

Threaded connections. Threaded connections may be used for attaching safety relief valves to the water heaters.

Shutoff valves. Shutoff valves of any type should not be installed between the safety relief valve and the water heater. No shutoff valve will be placed between the discharge pipe and the atmosphere.

Discharge piping. Internal cross-sectional area of the safety relief valve discharge pipe should not be less than the full area of the valve inlet or of the total of the valve outlets. The discharge piping should be short and straight. If an elbow is used on the discharge pipe, it should be located close to the valve outlet. Safety valve discharge should be piped away from the water heater to a point where there is no danger for people. The discharge should be drained properly without any threat to public safety.

5.6.2 Water supply

Feed water is introduced into a water heater through an independent connection and not through any opening for cleaning, drain, pressure gage, or temperature gage. Stop valves are installed in the supply and discharge pipe connections to allow draining the water heater. A pressure reducing valve is required if pressure on the water supply line exceeds 75 percent of the set pressure of the safety relief valve.

5.6.3 Storage tanks

A storage tank if required should be designed and constructed in accordance with ASME Section VIII, Division 1 or ASME Section X. If a tank is constructed under ASME Section X, the maximum allowable temperature should be equal to or more than 210°F (99°C).

5.6.4 Expansion tank

An expansion tank is required if the hot water system is equipped with a check valve or pressure reducing valve in the cold waterline. The safety relief valve may lift periodically due to thermal expansion of water if an expansion tank is not provided. An expansion tank should be designed and constructed in accordance with ASME Section VIII, Division 1 or ASME Section X.

5.6.5 Piping expansion

Hot water mains connected to water heaters are subjected to expansion and contraction. Provisions should be made for the expansion and contraction of hot water mains connected to water heaters by installing substantial anchorage at suitable points and by providing swing joints when water heaters are installed in batteries. Figures 5.15 and 5.16 show typical acceptable schematic arrangements of piping incorporating strain absorbing joints.

5.6.6 Bottom drain valve

A bottom drain pipe connection fitted with a valve is provided for each water heater. The pipe should be connected to the lowest possible water

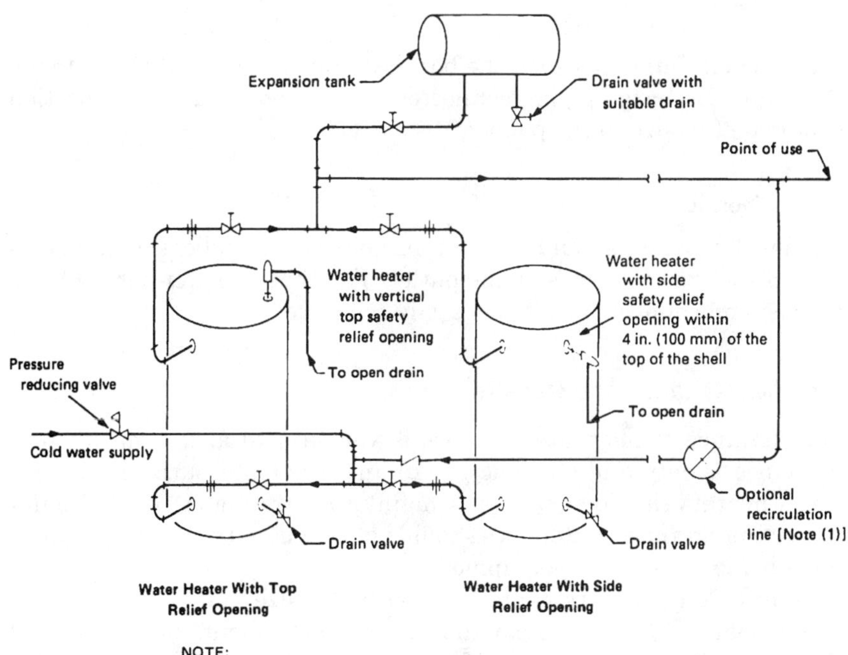

Figure 5.15 Acceptable piping installation of water heaters in battery. *(Courtesy: ASME Code Section IV)*

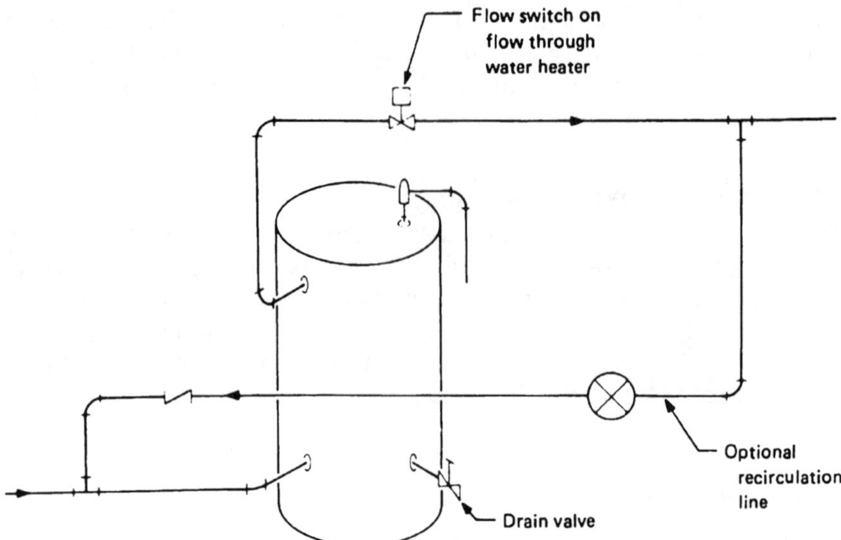

Figure 5.16 Acceptable piping installation for water heater. *(Courtesy: ASME Code Section IV)*

space. The minimum size of the bottom drain valve should be 3/4 inch (19 mm). Discharge piping connected to the bottom drain connection should be full size to the point of discharge.

5.6.7 Thermometer

Each water heater should have a thermometer for indicating temperature of the water at or near the outlet. The thermometer should be so located and connected that it is easily readable.

5.7 Maintenance and Repairs

Maintenance of hot water heaters is very essential for smooth performance, safety and efficiency. Like any other pressurized systems, water heaters should be regularly maintained through a regular maintenance program. Maintenance should not be ignored even though water heaters look like simple equipment.

Maintenance requirements vary with the size and type of the hot water heaters. The area near the water heater should be kept free of flammable liquids such as gasoline, adhesives, and other combustible materials. For proper combustion and ventilation, make sure that the flow of air to the water heater is not restricted.

Monthly maintenance. Technicians should carry out the following monthly maintenance:

- Visually check the burner while firing and pilot burner flame with main burner off. Rectify any unusual conditions.
- Clean the heater tank to remove solid suspended deposits.
- Check all hot water outlets for proper operation and leaks.
- Adjust water supply temperature whenever required, and check the entire system for leaks.

Annual maintenance. The following routine maintenance should be performed annually:

- The water heater's internal flue should be checked and cleaned.
- Check the burner and clean it by removing it from the water heater.
- Test the temperature and pressure (T&P) safety relief valve by lifting its lever handle and making certain that the valve operates freely.
- Check the anode rod and replace if necessary.
- Check the tank for scale and flush it to remove sediment.
- Check manifold pressure and controls.

Repairs. Welding repair to water heaters is generally costly and not economically feasible. It is recommended to replace the water heater rather than making major welding repairs.

5.8 Inspection

A water heater is required to be inspected at regular intervals. A certificate of operation is required in some jurisdictions to use commercial water heaters. The certificate is issued only after inspection of the water heater by an authorized inspector.

The owner should cooperate with the authorized inspector during inspection. Area around the water heater should be clean and free from all combustible materials. The water heater room should not be used as storage. Generally, the authorized inspector performs the following inspection:

T&P safety relief valve. This valve protects the water heater from excessive pressures and temperatures. The T&P valve should meet the requirements of ANSI Z21.22 Relief Valves and Automatic Gas Shutoff Devices for Hot Water Supply System. The capacity rating of the valve shall be equal to or more than the input capacity Btu/hr of the water heater. The valve should be marked with a maximum set pressure not to exceed the MAWP of the water heater. The discharge opening should not be blocked or reduced in size. No valve of any description should be placed between the tank and the T&P valve. The discharge pipe should not be smaller in size than the outlet pipe size of the valve.

The T&P valve should be tested at least once a year for proper operation. The valve will be tested by lifting the handle. Water should flow freely when the handle is lifted and should shut off completely when the handle is released. If the valve fails to reset completely, secure the water heater and replace the valve with a new one.

Controls. Operation of the controls, such as temperature control and high limit control, should be observed. The water heater should be taken out of service if any control is found defective, inoperative, or malfunctioning. The heater should not be operated until the deficiency is corrected.

Thermometer. Thermometer is installed to indicate the temperature of the water in the water heater. A defective thermometer should be replaced immediately.

Burner. Visual inspection of the main burner and pilot burner should be made at least once every 6 months. Burner surfaces should be checked for damage and soot deposits.

Outer casing. The outer casing is inspected for corrosion or deterioration of metals. Any hole in the casing is an indication of problems underneath the casing from external corrosion.

Accessories. An expansion tank, if installed, should be constructed according to ASME Code Section VIII, Division 1. A drain should be provided at or near the expansion tank. A stop valve should be installed between the water heater and the expansion tank. Operating stop valves should be installed on both the inlet and outlet sides of the water heater. The bottom drain valve should be tested to ensure proper operation and be discharged at a safe location. A pressure reducing valve is required if water pressure to the heater exceeds 75 percent of the maximum allowable working pressure of the unit.

Piping and connections. All piping and connections to the water heater should be checked for leakage and other possible defects. Control lines, safety valve connections, thermometer connections, drains, and blowoffs should be inspected thoroughly.

Venting system. A visual inspection of the venting system should be made once every 6 months. Obstruction could cause improper venting. Any causes of venting or leakage of combustion products should be checked.

5.9 Specifying a Hot Water Heater (Gas-Fired, 125 Gallons, MAWP 160 psi)

Furnish and install a water heater with a capacity of 125 gallons in accordance with plans and specifications, and ASME Code Section IV—Rules for Construction of Heating Boilers Part HLW. The water heater shall have an input rating of 720,000 Btu/hr and a recovery rate of 698 gal/hr at temperature rise of 100°F, with a minimum actual storage capacity of 125 gallons. The heater shall be equipped with 11/4 NTP at each water inlet and outlet opening, (2) 4-inch handhole cleanouts, and it shall have a working pressure of 160 psi.

The water heater shall be of glass-lined design and include a powered gas burner with electronic flame safeguard, intermittent ignition, main and pilot automatic gas valves, redundant solenoid gas valve, gas pressure regulator, diaphragm air switch for proof of blower operation, and flame inspection port. Controls shall include:

- High-temperature manual limit control (manual reset)
- Upper and lower thermostats
- Combination temperature and pressure gauge
- Low water cutoff
- ASME rated temperature and pressure (T&P) relief valve
- Draft regulator

The water heater shall be equipped with multiple anodes or cathodic protection. The heater shall be insulated with a vermin-proof glass fiber insulation or equal. The outer jacket shall have a baked enamel finish over a bonderized undercoating.

All internal surfaces of the heater exposed to water shall be glass-lined with an alkaline borosilicate, nickelous oxide composition that has been fused to the steel by firing a temperature range of 1400 to 1600°F. The glass lining shall be fused to steel by firing at a temperature range of 1400 to 1600°F. Heater tank shall have a 5-year limited warranty against corrosion.

Professional service for commissioning and start-up should be included. The manufacturer shall provide the following documents:

- Three copies of perfect bound owner's manuals
- Three copies of the manufacturer's data reports
- Three copies of certified prints

Chapter

6

Cast Iron Boilers

A cast iron boiler is a low-pressure boiler manufactured by casting the pressure components in sections from iron, bronze, or brass. This is basically a watertube boiler because the water is inside the cast sections and the products of combustion are on the outside. Cast iron boilers are considered a special type of boiler.

Cast iron boilers have been in use for the last century. The use of cast iron boilers is limited to low-pressure steam, hot water heating, and process applications. A cast iron boiler for steam heating is shown in Figure 6.1 and a cast iron boiler for hot water heating is shown in Figure 6.2.

The cast iron boiler assembly is manufactured and tested in accordance with ASME *Boiler and Pressure Vessel Code Section IV*—Rules for Construction of Heating Boilers. The boiler is certified by the American Gas Association (AGA). Annual Fuel Utilization Efficiencies (AFUE) are based on the U.S. Department of Energy's test procedures and FTC labeling regulations. AFUE and I=B=R ratings are certified in accordance with standards set by the Hydronics Institute Division of the General Appliance Manufacturers Association (GAMA).

Advantages of cast iron boilers are:

- Low initial installation cost
- Ease of operation and maintainability

Disadvantages of cast iron boilers are:

- High maintenance cost
- Limited use

Vent Damper
Automatically closes when unit turns off preventing heat loss.

Integral Draft Diverter
provides safety from back drafts and accommodates low ceiling heights. Transition directly to flue pipe from top of boiler.

Cast Iron Quality
Provides effective heat transfer, reliability and strength.

Electronic or Standing Pilot Ignition
Optional electronic ignition automatically lights the pilot only when needed, eliminating fuel waste.

Titanium Burners
Dunkirk's exclusive burners provide greater resistance to corrosion and oxidation.

Pressuretrol
This operating control prevents the boiler from building pressure levels beyond recommended limits. Will shut down the boiler if necessary.

Transformer
This receives the 120 volt power from the building's electrical service and steps it down to 24 volt power.

Water Sight Glass
Allows for easy viewing of the boiler's water level.

Low Water Cut-Off
Electronically monitors the water level in the heat exchanger shutting the unit off if the water level is too low.

Baked Enamel Steel Jacket
Factory installed insulation keeps off-cycle heat losses to a minimum in an attractive and compact package.

Figure 6.1 Cast iron steam boiler. *(Courtesy: Dunkirk Boiler)*

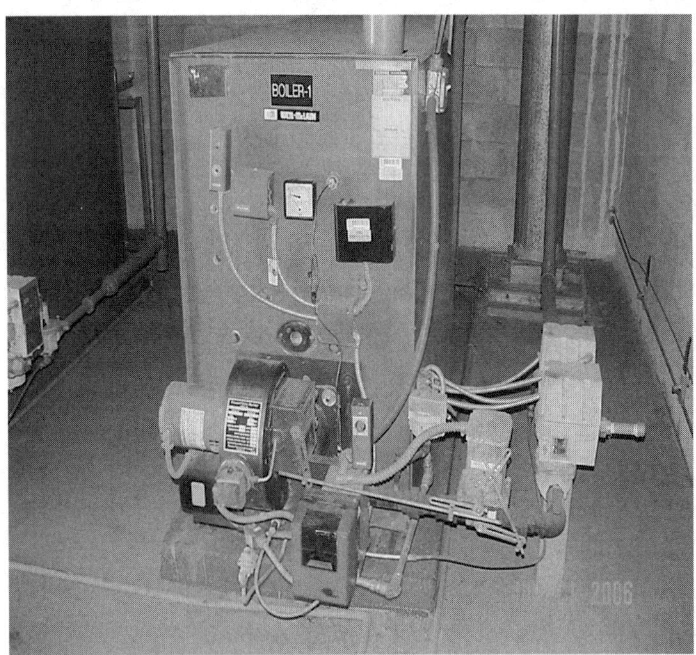

Figure 6.2 Cast iron hot water boiler.

6.1 General Features

A cast iron boiler is like any other boiler where steam or hot water is generated. All units are completely factory-assembled with controls and wiring. Prior to shipment, each unit is fire tested at the factory to ensure dependable performance.

6.1.1 Construction

A cast iron boiler consists of hollow cast iron sections, which are gasketed and bolted together. The cast iron boiler functions like a watertube boiler.

There are no tubes in a cast iron boiler, but the cast iron sections serve as tubes similar to watertubes in a watertube boiler. The cast iron sections form a vessel and contain water around the furnace. Water is heated in the sections and converted into steam in a steam boiler or hot water in a hot water heating boiler.

The material used for the major parts of the boilers is cast iron. Generally, gray cast iron is used in manufacturing cast iron boilers. The iron-carbon alloy is cast in a mold to make a particular shape. When the casting is cooled slowly in the mold, part of the carbon separates out as graphite. This makes gray cast iron less brittle and easier to machine.

6.1.2 Major components

Like any other boiler, a cast iron boiler has various components. Brief details of each major component are described below:

Cast iron boiler assembly. All the sections and push nipples are constructed of durable cast iron. When the boiler is heated, sections and push nipples expand and contract, providing a positive water seal.

A cast iron section is shown in Figure 6.3. Parts are thoroughly machined and joined with high-temperature flexible seals for assembly (Figure 6.4). Parts of a heat exchanger are shown in Figure 6.5. An assembled cast iron boiler before application of insulation is shown in Figure 6.6.

Cabinet. The cabinet is constructed of heavy-gauge steel with a baked-on enamel paint finish. The cabinet is insulated with fiberglass insulation, which keeps the surface temperature low.

6.2 Classification of Cast Iron Boilers

There are three types of cast iron boilers: horizontal-sectional, vertical-sectional, and one piece.

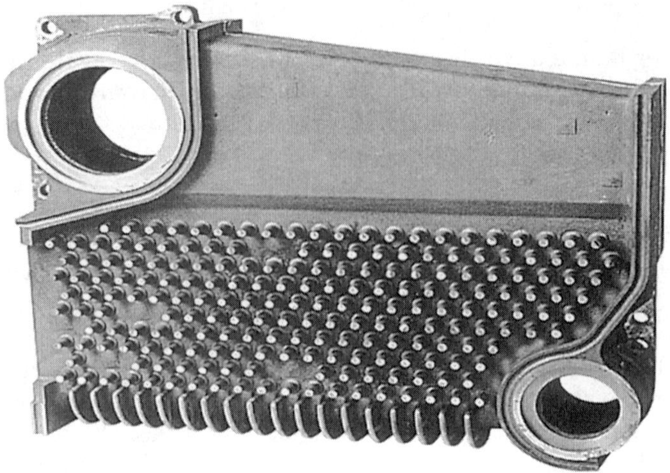

Figure 6.3 A cast iron section.

6.2.1 Horizontal-sectional cast iron boiler

This type of cast iron boiler is made up of sections stacked one above the other and is assembled with push nipples (See Figure 6.7).

6.2.2 Vertical-sectional cast iron boiler

This type of cast iron boiler is made up of sections standing vertically, as shown in Figure 6.8.

Figure 6.4 Parts are joined with flexible seals.

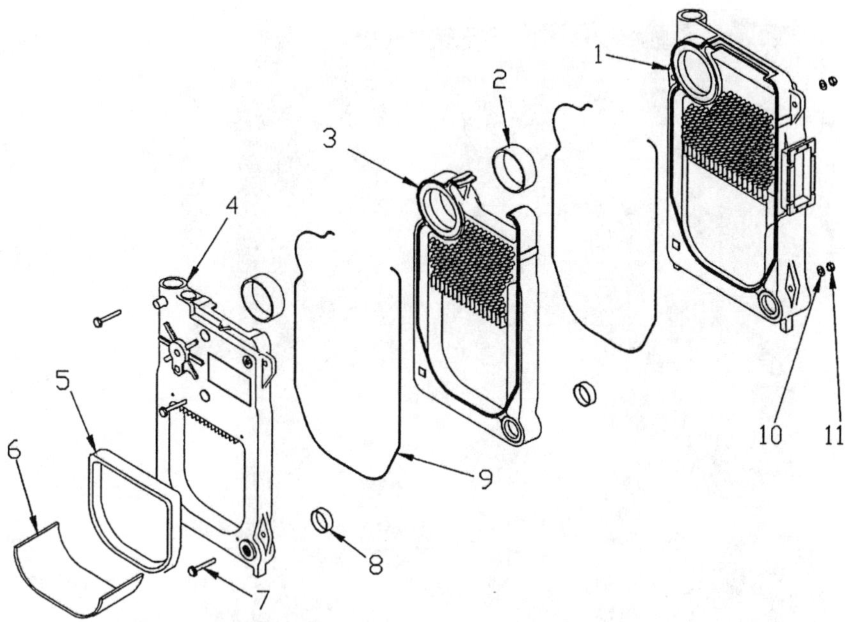

Parts of Heat Exchanger

Item	Description
1	Rear Section
2	Push Nipple
3	Center Section
4	Front Section Castover
5	Kit – Target Wall & Insulation Blanket
6	Insulation Blanket
7	Tie Rod
8	Push Nipple
9	Rope
10	Washer
11	Nut

Figure 6.5 Parts of heat exchanger. *(Courtesy: Weil McLain)*

6.2.3 One-piece cast iron boiler

In this type of cast iron boiler, the pressure vessel is made as a one-piece casting.

6.3 ASME Code Requirements

Cast iron boilers are widely used because their initial installation cost is low, they are easy to operate, and the maintenance is easy. Cast iron boilers are primarily used for steam heating and hot water heating.

Figure 6.6 Cast iron boiler assembly.

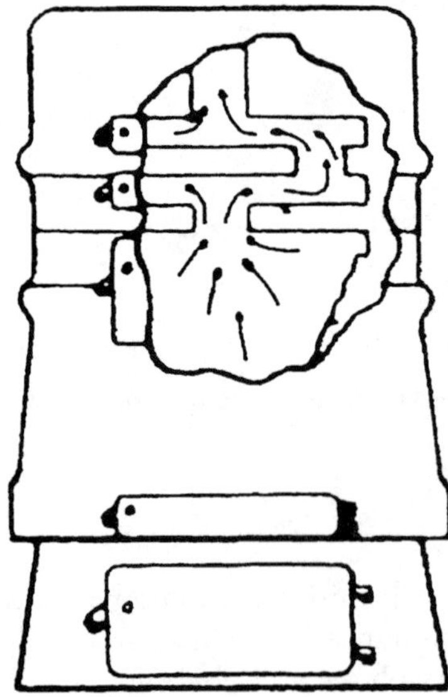

Figure 6.7 Horizontal-section cast iron boiler.

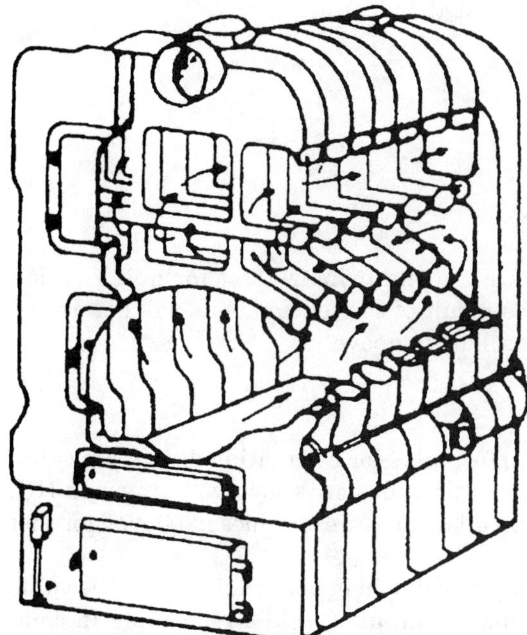

Figure 6.8 Vertical-section cast iron boiler.

The requirements of Part HC are applicable to steam heating boilers, hot water heating boilers, and hot water supply boilers that are fabricated primarily of cast iron. The requirements of Part HG will also be used in addition to these requirements.

6.3.1 Material requirement

Gray iron castings are used to produce cast iron parts for boiler construction. All materials used for cast iron boilers should meet the requirements of this Section.

Chemical composition. A manufacturer may use any procedure for melting. The cast material should have the following composition:

1. The manganese should be controlled to meet $Mn \geq (1.7 \times S) + 0.2$ where Mn = manganese percent, S = sulfur percent
2. The phosphorous content should not be more than 1.00%.

Tensile strength. Casting is listed by classes and according to the minimum tensile strength. The list below shows the classification of casting used for boiler construction:

Class (No.)	Tensile Strength (min. psi)
20	20,000
25	25,000
30	30,000
35	35,000
40	40,000

Tension test. The tension test is the primary test for qualification required by this Section. The result of the tension tests will determine various classes of castings as shown above.

6.3.2 Design requirements

Stress values are used in the formulas for calculating design pressure. Formulas for calculating design pressure are shown in paragraph HG. Table 6.1 shows the maximum allowable stress values in tension for cast iron.

The maximum allowable stress value in bending is 1.5 times the value permitted in tension. The maximum allowable stress value in compression is two times the value permitted in tension.

Heads. Heads with pressure on the concave side are designed according to the formula in paragraph HG. The maximum allowable stress values in tension can be found in Table 6.1. Cast iron pipes, flanges, and flanged fittings, Class 125 and Class 250, may be used for parts subject to pressure not more than ANSI ratings, and temperature not more than 450°F (232°C). For heads with pressure on the convex side, the thickness should not be less than the thickness required for heads with pressure on the concave side. Thickness should not be less than 0.01 times the inside diameter of the head skirt under any circumstances.

TABLE 6.1 Maximum Allowable Stress Values in Tension for Cast Iron, ksi (Multiply by 1000 to Obtain psi)

Class	Minimum Tensile Strength, ksi	Maximum Allowable Design Stress Values, in Tension, ksi
20	20.0	4.0
25	25.0	5.0
30	30.0	6.0
35	35.0	7.0
40	40.0	8.0

Openings and reinforcements. The design of openings and reinforcements in boilers and pressure parts should meet the requirements of Part HG. Cast iron flanges, nozzles, and opening reinforcements required in the design calculations should be cast integrally with the boiler or part.

Washout openings. All cast iron boilers are required to have washout openings for removal of sediments. Washout plug size is minimum NPS 1.5 (DN 40) for boilers with gross internal volume exceeding 5 cft (142 L). Washout plug size should be minimum 1 inch (25 mm) for boilers with gross internal volume less than 5 cft (142 L).

6.3.3 Assembly method

Cast iron boilers are assembled using internal connections such as nipples. External connections such as cast iron headers or threaded headers are also used to complete the assembly. The assembled boilers must be tested hydrostatically.

6.3.4 Hydrostatic test

A hydrostatic test is required to be performed on all boilers or boiler parts. The result of the test should be satisfactory in order to accept the boilers or boiler parts.

Steam boilers. Each individual section or boiler part is tested under hydrostatic pressure of minimum 60 psig (414 kPa). The test is performed at the shop where the section or part is fabricated. The complete boiler is required to be tested under hydrostatic pressure of minimum 45 psig (310 kPa). The hydrostatic test pressure should be controlled so that it can not exceed 10 psi (69 kPa) during the test.

Hot water boilers. Each individual section or part of hot water heating or hot water supply boilers with working pressure maximum 30 psi (207 kPa) is tested under hydrostatic pressure of minimum 60 psig (414 kPa). Each individual section or part of hot water heating or hot water supply boilers with working pressure more than 30 psi (207 kPa) is tested under hydrostatic pressure of 2.5 times the maximum allowable working pressure. The hydrostatic test is performed at the shop where the section or part is made. The complete boiler is required to be tested at hydrostatic pressure of minimum 1.5 times the maximum allowable working pressure.

6.3.5 Stamping requirements

Markings on cast iron boilers. Each boiler section designed, constructed of cast iron, and inspected in accordance with the rules of Part HC shall be cast with Code Symbol Stamp H as shown in Figure 10.3.

The following data shall be cast on each boiler section, at minimum 5/16 inch high (Figures 6.9 and 6.10):

1. The boiler or parts manufacturer's name or acceptable abbreviation, preceded by the words "Certified by"
2. Maximum allowable working pressure
3. Pattern number
4. Casting date
5. The shop assembler's name or acceptable abbreviation

A completed cast iron boiler may be marked as shown in Figure 6.11 or Figure 6.12 with the Code Symbol Stamp **H** and with the following data:

1. The shop assembler's name preceded by the words "Certified by"
2. Maximum allowable working pressure
3. Safety or safety relief valve capacity, lb/hr or MBH
4. Maximum water temperature

6.3.6 Manufacturer's data reports

When a cast iron boiler or parts of a cast iron boiler are completed, data reports and certificates are prepared by the manufacturer. A certified individual (CI) verifies all the reports and certificates.

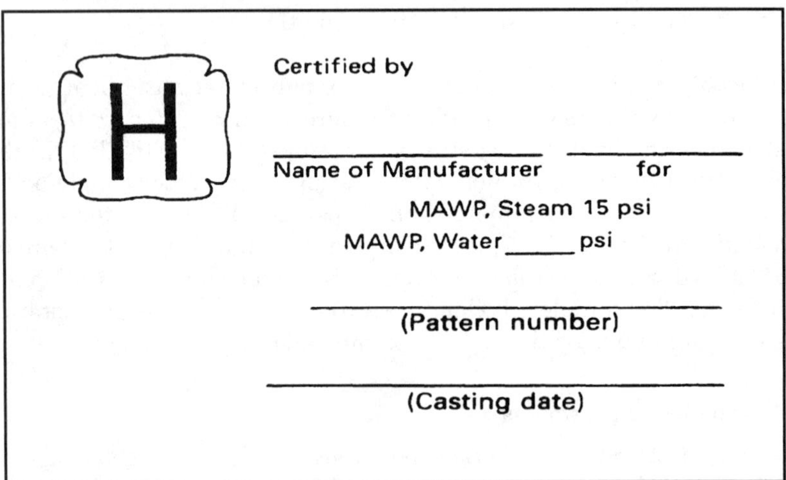

Figure 6.9 Steam and water boilers. *(Courtesy: ASME Section IV)*

FORM OF DATA CAST ON CAST IRON BOILER SECTIONS

H

Certified by

________________ __________

Name of Manufacturer for

MAWP, Water_____ psi

____

(Pattern number)

____

(Casting date)

Figure 6.10 Boilers suitable for water only. *(Courtesy: ASME Section IV)*

The CI is an employee of the manufacturer and qualified by the manufacturer for the job. The CI provides oversight of the activities that affect the proper utilization of the **H** symbol on cast iron boilers.

Master data report. Manufacturers of cast iron boilers, to which ASME Code Symbol Stamp H is applied, are required to prepare data reports.

FORM OF STAMPING ON COMPLETED CAST IRON BOILERS OR THEIR NAMEPLATES

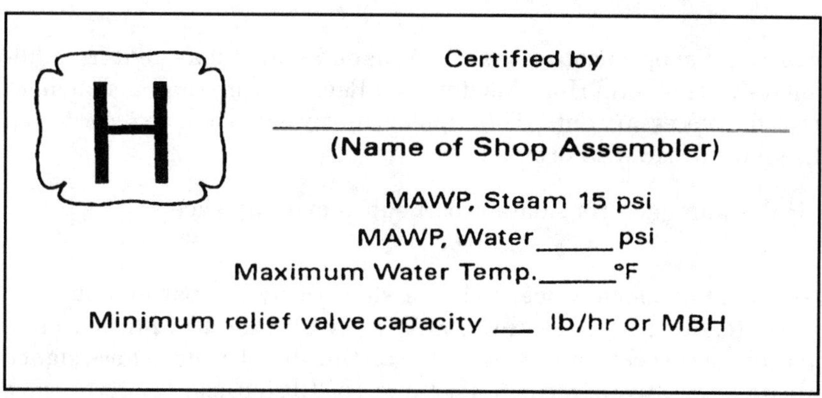

H

Certified by

____

(Name of Shop Assembler)

MAWP, Steam 15 psi

MAWP, Water_____ psi

Maximum Water Temp._____°F

Minimum relief valve capacity ___ lb/hr or MBH

Figure 6.11 Steam and water boilers. *(Courtesy: ASME Section IV)*

FORM OF STAMPING ON COMPLETED CAST IRON BOILERS OR THEIR NAMEPLATES

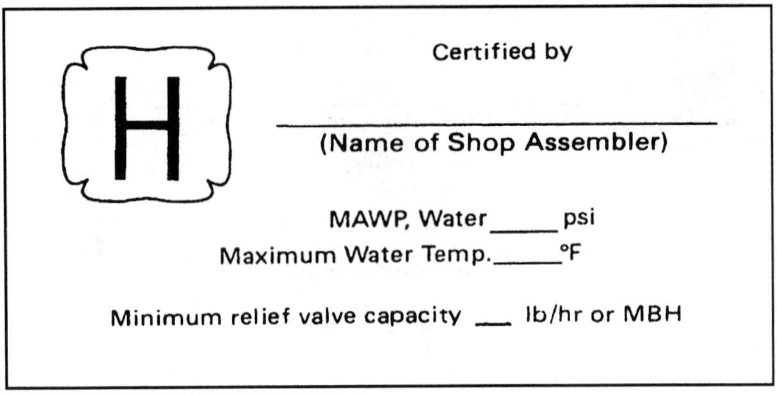

H

Certified by

____

(Name of Shop Assembler)

MAWP, Water_____ psi

Maximum Water Temp._____°F

Minimum relief valve capacity ___ lb/hr or MBH

Figure 6.12 Boilers suitable for water only. *(Courtesy: ASME Section IV)*

The following ASME Form shall be used for manufacturers' data reports:

- H-2—Manufacturers' Master Data Report Form for boilers constructed from cast iron (Appendix J).

The CI shall verify that cast iron sections marked with the H symbol have a current H-5. The manufacturers will distribute a copy of the data report to the owner, the insurance agency, and the jurisdiction. Manufacturers are required to keep a copy of the data report on file for a minimum of 5 years.

Data report supplementary sheet. A manufacturer may require additional sheets if Form H-5—Master Data Report Form—is not enough for recording necessary data. The supplementary data shall be recorded on the following ASME Form:

- H-6—Manufacturer's data report supplementary sheet

Certificates of conformance. The CI shall verify the pattern number, cast date, quantity of castings, MAWP, hydrostatic test pressure and quantity of cast sections listed on the certificates of conformance signed by a representative of the manufacturer. The CI shall sign the certificates of conformance if all data are correct. The following forms shall be used for certificates of conformance:

- HC-1—Manufacturer's Material Certificate of Conformance for cast iron boiler sections
- HC-2—Manufacturer's Material Certificate of Conformance for hydrostatic testing of cast iron boiler sections

6.4 Prevention of Thermal Shock

Thermal shock is a rapid increase in thermal stresses on a boiler which may cause a crack to form or to propagate. When a system is designed to modulate the water temperature based on outdoor temperature and provides a nighttime setback of zone temperature, a cast iron boiler may experience thermal shock.

When a cast iron boiler is installed for low return water temperatures, provisions should be made to protect the boiler from a thermal shock condition. A boiler may be protected from thermal shock by proper system piping design. This is accomplished by one of the following piping design methods:

- Primary/secondary piping system
- Three-way mixing valve

Primary/secondary piping. This is the preferred method to modulate the system water temperature, as cold water must never enter into a hot boiler. The primary loop runs between the boiler and secondary heating loop. The secondary heating loop supplies hot water to the heating zones.

A sensor is located in the secondary loop where a constant source of heat is desired. This sensor turns the boiler on and off in order to maintain the desired secondary loop temperature. When a zone calls for heat, the corresponding circulator starts, and the temperature at the sensor drops. The sensor turns on a boiler, which supplies hot water to the secondary loop via the primary loop.

Primary/secondary piping protects the boiler from thermal stresses, because the system is designed to maintain a high secondary loop temperature (140°F or higher) even during periods of night setback. This prevents a sudden rush of cool water into a hot boiler.

Three-way mixing valve. This is used for many installations for protecting a boiler from thermal shock. With minimal equipment investment, thermal shock condition can be reduced by using three-way valve installations.

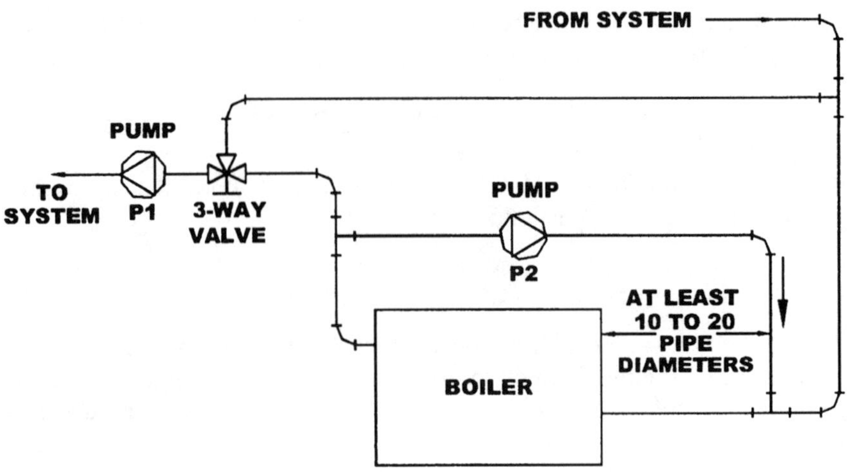

Figure 6.13 Bypass piping using three-way valve. *(Courtesy: Weil-McLain)*

The following steps are recommended for proper design of the bypass piping using three-way valve:

1. Piping selection
   - Use bypass piping arrangement as shown in Figure 6.13.
   - Follow basic boiler piping installations.
2. Three-way valve selection
   - Use valves with minimum timing of four minutes from one extreme position to the other.
   - Size to circuit design requirements. Generally, valve is one or two pipe sizes smaller than the boiler pipe size.
   - Avoid over-sizing to prevent poor control of mixed water.
   - Pressure drop across the valve should be high enough to provide accurate control without hunting.
3. Circulator selection
   - Size system circulator to circuit flow characteristics.
   - Size each bypass circulator.

6.5 Maintenance

Cast iron boilers can provide many years of service provided the equipment is maintained regularly. The best time to undertake thorough maintenance is during summer, when the boilers are idle. A complete check of the heating system should be done at the end of the heating season and problems listed so that necessary maintenance and repair can be completed before the next winter. Adequate provisions for expansion between the sections should be checked so that serious damage cannot result while the boiler is idle or when it is started up in the fall.

Technicians should undertake the following maintenance and repairs to ensure smooth operation and safety of cast iron boilers:

Cleaning. Before seasonal start-up, check the boiler for soot and scale in the flues; change oil filter and nozzle. Clean the burner and readjust burner input rate to maintain high operating efficiency. The venting system should be checked at the start of each heating season. Check the vent pipe from the boiler to the chimney for signs of deterioration by rust or sagging joints. Repair if necessary.

Heating surfaces of a cast iron boiler should be cleaned thoroughly. Inspect the firebox gas passages and breeching for soot accumulation and clean with a wire brush and vacuum cleaner. If the internal surfaces are dirty, the boiler should be cleaned. But first, allow the boiler to cool, drain, and then flush it out. Remove the blowdown valves and plugs and run high-pressure water stream through these openings so that any sludge or loose scale will be removed. Chemical cleaning is used to remove any hard scale from the internal surfaces.

Expansion between sections. During normal operation, cast iron boilers are subjected to a wide range of expansion and contraction due to the varying demands for heating and consequently an intermittent rate of firing. Check to make sure that there is ample provision for expansion between the sections. Check the tie-rods for incorrect installation, and look for any rust build-up between the sections. Either condition may result in cracking.

The tie-rods are necessary to hold the sections together. If the tie-rods of a push nipple-type boiler are installed incorrectly, one of the following procedures should correct the situation:

1. The solid nuts used on the tie-rods for drawing the sections together are backed off several times to allow ample room for expansions; or
2. The solid nuts are used in conjunction with compression washers; or
3. Split nuts are used instead of solid nuts, which in turn, may have been used in conjunction with compression washers.

Idle boilers. All idle cast iron boilers, whether used for steam or hot water heating, deteriorate less during idle periods if completely filled with water. Steam boilers should be filled at least to the top of the gage glass or preferably up to the steam outlet connection. Hot water heating boilers should be left entirely filled.

Servicing flooded boilers. Do not attempt to operate any boiler that has been submerged underwater either partially or fully. A flooded boiler should be checked thoroughly and conditions should be checked to determine whether the boiler should be replaced or repaired. It is advisable to replace a completely flooded boiler with a new boiler.

If a boiler has been subjected to partial flooding, thoroughly service the boiler using the following procedures:

- Replace all controls, gas valves, and electrical wiring on the boiler.
- Inspect all burner tubes, gas piping, manifolds, orifices, and flue ways for signs of rust and/or sediment from the flood waters.
- Replace all oil burners for oil-fired boilers.
- Replace all insulation that has been damaged by water.
- Where possible, inspect seal rings for damage from petroleum products.
- Thoroughly inspect all venting for signs of corrosion.

6.6 Inspection

Cast iron boilers operate under different conditions than steel boilers. An authorized inspector should look for water leaks at the connecting joints of sections. The sheet metal casing should be removed if there is evidence of leakage.

A common problem is cracking due to overheating or thermal shock. Overheating may occur due to low water conditions or poor circulation caused by sludge accumulation in the lower water passages of the boiler. Thermal shock may occur when cold water is added to raise the water level in an overheated boiler.

Cast iron sections use long rods, threaded on both ends, called draw bolts. When the boiler is under operating conditions, the heat causes the boiler to expand, which tightens the draw bolts. Thorough investigation should be carried out if loose draw bolts are found on cast iron boilers.

Like steel boilers, an authorized inspector is required to perform two inspections on cast iron boilers: internal inspection and external inspection.

Internal inspection. It is very difficult to perform internal inspections on cast iron boilers. Attempts should be taken to look inside by removing threaded plugs, valves, and so on. The authorized inspector should follow the jurisdictional requirements for internal inspections.

The inspector should observe the conditions of the float, float mechanism, and the float chamber when the low-water cutoff and water-feeding devices are disassembled. Threaded plugs in the piping connections for water gage glass, water columns, and low-water fuel cutoff should be inspected to ensure that there is no blockage.

External inspection. The external inspection is performed when the boiler is in operating conditions. The authorized inspector, upon entering in the boiler area should make a general assessment of the boiler, piping, controls, fuel system, and combustion air supply. The inspector should also observe if the boiler room is used for storage of any combustible liquids.

The following external inspection is carried out to ensure that the boiler functions safely and smoothly under the operating conditions:

- Look at the validity and other information of the current certificate issued by the local jurisdiction.
- Review and compare safety valve or safety relief valve nameplate data with the boiler nameplate information.
- Witness the test of the safety or safety relief valves.
- Witness the low-water fuel cutoff tests.
- Inspect the pressure and temperature controls.
- Check the accuracy of the pressure or altitude gages.
- Check the thermometer reading on hot water boilers.
- Check the water gage glass to observe clear indication of water level.
- Check for any types of leakages.
- Witness any pressure test, if required.
- Inspect the fuel burning system if required by the local jurisdiction.

6.7 Specifying a Cast Iron Boiler (Natural Gas-Fired Steam Boiler 100 HP, MAWP 15 psi)

Boiler capacity. Furnish and install one 100 HP wet base, cast iron sectional boiler with power burner that pressurizes the firebox and operates under forced and balanced draft. The boiler shall be designed for a maximum allowable working pressure of 50 psig water and 15 psig steam. The boiler shall have I=B=R Hydronics Institute net output at 100 percent firing rate—2609 MBH at natural gas at 1000 Btu/ft^3 with a minimum pressure of 10 psi. Electrical power available will be 115 V, 60 HZ, and single phase.

The unit must be completely factory assembled with controls and wiring and tested to ensure dependable performance.

Construction. The Boiler shall be designed and constructed in accordance with ASME Code Section IV—Rules for Construction of Heating Boilers. The main construction features are:

- Boiler sections assembled with short, individual rods.
- Cast with sealing grooves for high-temperature sealing rope to assure permanent gas-tight seal.

- Sealed watertight by elastomer sealing rings, not cast iron nipples. Each port opening is machined to completely capture sealing ring between sections.
- Must be hydro-wall designed to provide completely water-cooled combustion chamber.
- Provided with sufficient tappings to install required controls.
- Provided with cast-in air elimination to separate air from circulating water.
- Provided with expansion tank tapping to divert separated air to expansion tank.
- Constructed to provide balanced water flow through entire section assembly using single supply and return connections for water.
- Furnished with two observation ports (one in front and one in back) to allow visual inspection of flame.
- Provided with cast iron flue collar with a built-in adjustable damper capable of being locked into place after adjustment.
- Furnished with cast iron cleanout plates to cover cleanout openings on the front of the boiler.
- Shipped with insulated heavy gauge steel jacket with durable powdered paint enamel finish. Jacket designed to be installed after connecting supply and return piping.

Burner. The boiler burner package shall be UL listed. The burner shall be factory-mounted with fuel supply system and conform to burner manufacturer's installation instructions and ASME CSD-1 Code. The burner shall include:

- Forced draft combustion air blower.
- Control panel, National Electric Manufacturer's Association 1 (NEMA 1) with combustion control, motor starter, control switches, transformer and indicator lights.
- Panel options—power on/off on light:
  - Call for heat
  - Ignition on
  - Pilot failure
  - Low water
  - Flame failure
  - Silencing switch
  - Control fuse and holder
  - Post purge timer
  - Alarm bell

- Gas trim shall include main and pilot fuel valves, regulators, and high and low gas pressure switches.
- Burners and controls conform to the Factory Mutual (FM) standards.

Boiler trim. All electrical components to be of high quality and bear the UL label. Standard boiler trim shall include the following components:

- ASME certified safety valve set for 15 psi set pressure.
- Low-water cutoff water column, try cocks, gauge glass set with drain cock, feed water pump control, and blowdown valve.
- Secondary low-water cutoff with manual reset and blowdown valve.
- Steam pressure gauge.
- Low pressuretrol (operating) and high pressuretrol set at maximum pressure as a safety control.

Efficiency guarantee. The boiler must be guaranteed to operate at minimum fuel-to-steam efficiency of 83 percent from 30–100 percent of rating.

Shop tests. The completed fully assembled unit shall be factory test fired to check for proper operation of burner, controls, and safety devices.

Start-up services. A factory authorized representative of the manufacturer shall provide start-up service and training of the operator.

Warranty. The manufacturer shall warrant all furnished parts for 18 months from the date of boiler shipment. The pressure vessel parts shall carry a 5-year warranty against thermal shock damage.

Documents. The manufacturer shall provide the following documents:

- Three copies of perfect bound owner's manuals
- Three copies of the manufacturer's data reports
- Three copies of the certified prints

Chapter

7

Modular Boilers

A modular boiler is a group of small boilers, composed of either steel or cast iron, that collectively act as a large boiler with modulated output. These small boilers, known as *modules,* are manifolded together at the job site, without any intervening stop valves. Generally, modulation refers to the ability to adjust a boiler or water heater's input (firing rate) to meet the demand (output) of the system.

The modular boiler system is popular due to compact design and energy efficiency. Combined with a microprocess control unit, the system is suitable for any commercial or institutional buildings. The system is appropriate for either new construction or replacing a single large boiler with several small boilers. The boiler units can be placed in any arrangements making the best use of space. A modular boiler system is shown in Figure 7.1.

There are several modular systems in which the individual boilers are isolatable. In fact, the original U.S. Department of Energy tests that established that they had higher efficiency than conventional boilers, relied to some extent on the fact that each individual module could be isolated from the system and allowed to cool down to room temperature, thereby eliminating radiation losses from the jackets of nonoperating boilers. On the other hand, a boiler requiring repair can be isolated from the system, while the rest of the system remains online, providing heat.

Modular boiler systems have the following advantages:

- More efficient than traditional large boilers.
- Fire-up much quicker than conventional boilers.
- Boilers are activated sequentially, drawing water from the main loop—then the next water boiler unit and so on until the heating need is met.

Figure 7.1 Modular boiler system. *(Courtesy: Triad Boiler System)*

- Firing boilers remain isolated, so no heated water circulates through cold boilers.
- Unfired boilers provide additional backup.
- Outdoor temperatures and loop water temperatures are constantly monitored.

There is some confusion between modular boilers and multiple boilers. Modular boilers have no intervening stop valves, so some controls can be mounted on the manifold or header. Multiple boilers have stop valves, so every individual boiler should have all the controls required by the code.

7.1 Modular Systems

A modular system can be designed for steam heating boilers, hot water heating boilers, hot water supply boilers or a combination of more than one of these boilers.

7.1.1 Modular steam heating boilers

A steam heating assembly consisting of a grouping of individual steam boilers called modules intended to be installed as a unit with no intervening stop valves. Modules may be under one jacket or may be individually jacketed.

The individual modules shall be limited to a maximum input of 400,000 Btu/hr (gas), 3 gal/hr (11 L/hr) (oil), or 115 kW (electricity). An individual-packaged steam boiler is shown in Figure 7.2.

Each module of a modular steam heating boiler should be equipped with:

- A steam gage
- A water gage glass
- A pressure control that cuts off the fuel supply when the pressure reaches an operating limit (less than the maximum allowable pressure)
- Low water cutoff

The assembled modular steam boiler should also be equipped with a safety limit control that cuts off the fuel supply to prevent steam pressure from exceeding 15 psi (100 kPa) maximum allowable working pressure. The control should be constructed to prevent a pressure setting above 15 psi (100 kPa).

Figure 7.2 Modular steel package steam boiler. *(Courtesy: Triad Boiler Systems)*

7.1.2 Modular hot water boilers

A hot water boiler assembly consisting of a group of individual hot water boilers called modules is intended to be installed as a unit with no intervening stop valves. Modules may be under one jacket or may be individually jacketed.

The individual modules shall be limited to a maximum input of 4,000,000 Btu/hr (gas), 3 gal/hr (11 L/hr) (oil), or 115 kW (electricity). A modular hot water boiler is shown in Figure 7.3.

Each module of a modular hot water heating boiler should be equipped with:

- A pressure/altitude gage.
- A thermometer.
- Temperature control that cuts off the fuel supply when the temperature reaches an operating limit, which is less than the maximum allowable temperature.

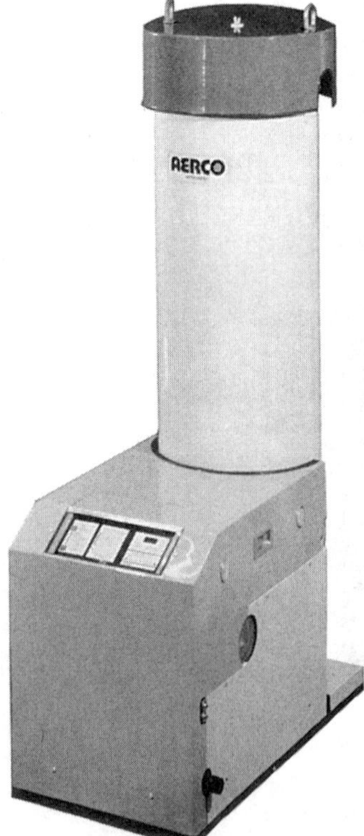

Figure 7.3 Modulating hot water boiler. *(Courtesy: Aerco International, Inc.)*

The assembled modular hot water heating boiler should also be equipped with:

- A safety limit control that cuts off the fuel supply to prevent the water temperature from exceeding the maximum allowable temperature at the boiler outlet. This control should be located within 3 feet (1.0 meter) of the fitting connecting the last module to the heating supply piping.
- Low water fuel cutoff.

7.1.3 Modular hot water heaters

In commercial facilities, generally one water heater is sized to meet the peak demand. But the demand for hot water varies with the time of day and day of the week, possibly by a factor of 10 or more. For a conventional water heater, the unit will cycle frequently most of the time. If modulating water heaters are used, they can reduce the cycling.

Instead of a single water heater, multiple modular units can be used to meet a range of hot water demand levels. Each modular unit has a much lower capacity than the single unit. Under low load condition, one modular unit can provide all of the required hot water. As the demand increases, additional modular units automatically come on line.

The modular water heaters can work as backup hot water heaters. In a single water heater application, failure of one unit can result in total loss of hot water. But in a modular installation, failure of one heater only reduces system capacity, but hot water is still available. Also, the modular system can reduce annual energy consumption by at least 10 percent.

7.1.4 Combination modular system

Generally, heating systems and domestic water systems are installed separately in a facility. However, a combination heating and domestic plant can be installed to share loads among common boiler modules. A three-module combination system is shown in Figure 7.4.

In the system, the only heating boilers should be specified with boiler management system (BMS) control option and the boilers for domestic water should be specified with combination control option. The boilers should be piped with a balanced flow piping arrangement. A two-position, two-way valve with an end switch should be installed in the supply piping to isolate the domestic water modules from the heating system.

7.2 Installation of Modular Boilers

A modular boiler is assembled in the field and consists of individual modules manifolded together with or without any intervening valves. It is

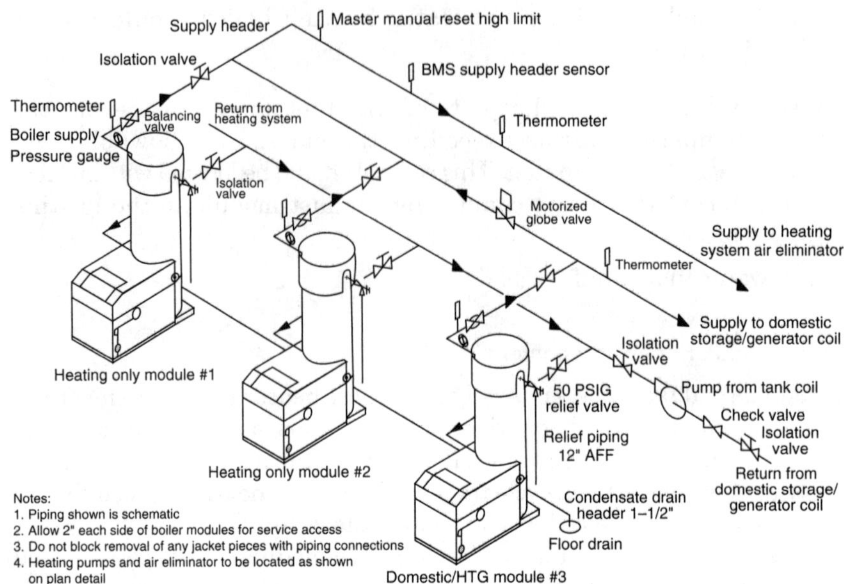

Figure 7.4 Combination modular system. *(Courtesy: Aerco International, Inc.)*

important that the manufacturer's installation guidelines be followed to ensure proper assembly, correct control location, and proper flow through each module.

Figures 7.5 and 7.6 show the two piping arrangements that are used for installation of modular boilers.

7.2.1 Individual modules

The individual modules should comply with all the requirements of Part HG (except as specified in HG-607 and HG-615) of ASME Code Section

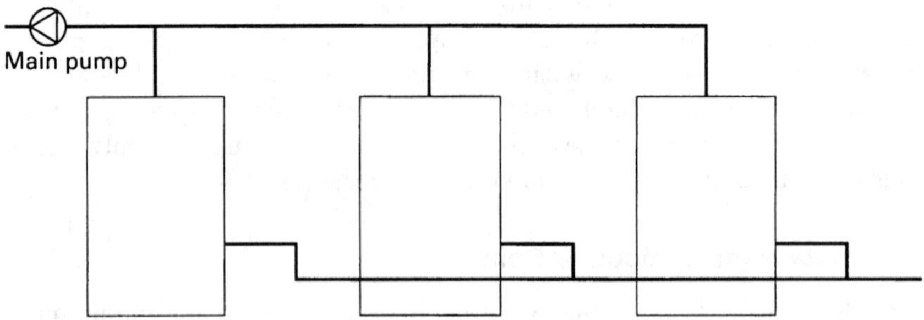

Figure 7.5 Modules connected with parallel piping. *(Courtesy: ASME Section VI)*

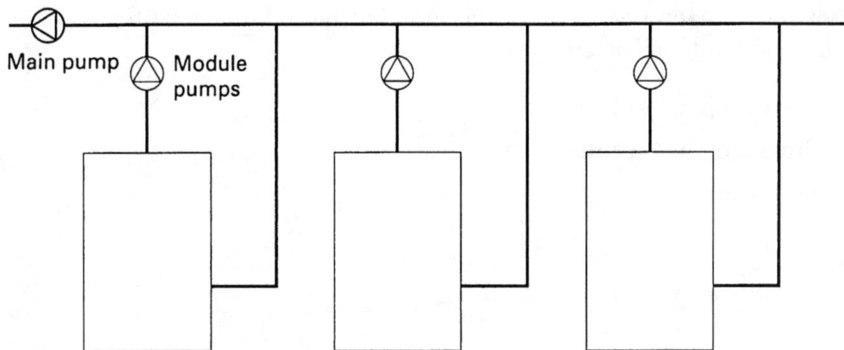

Figure 7.6 Modules connected with primary-secondary piping. *(Courtesy: ASME Section VI)*

IV and this paragraph. The individual modules should be limited to a maximum input of 400,000 Btu (gas), 3 gal/hr (11 L/hr) (oil), or 115 kW (electricity).

Steam heating boiler. Each module of a steam heating boiler should be equipped with:

- A safety valve
- A blow-off valve
- A drain valve

Hot water heating boiler. Each module of a modular hot water boiler should be equipped with:

- A safety relief valve
- A drain valve

7.2.2 Assembled modular boilers

The individual modules should be manifolded together at the field site or in the boiler room without any intervening valves. The header or manifold piping is field piping and is exempted from Article 2—materials of Part HG.

Steam heating boiler. The assembled modular steam heating boiler should also be equipped with:

- A feed water connection
- A return pipe connection

Hot water heating boiler. The assembled modular hot water heating boiler should also be equipped with:

- A makeup water connection
- Provision for thermal expansion
- Stop valves

Chapter 8

Boiler Design

The design of a heating boiler involves calculations, drawings, and specifications to satisfy the service conditions of the plants or facilities. During design of a boiler, two terms are frequently used: *design pressure*, used for calculating the minimum thickness requirement for the boiler parts, and *maximum allowable working pressure*, which is the maximum gage pressure or the pressure above atmospheric pressure that is permitted in the boiler and which is based on the lowest design pressure of any boiler part.

Heating boilers are designed in accordance with the rules of ASME Code Section IV—Rules for Construction of Heating Boilers. This section is divided into the following four parts and two subparts, together with the necessary figures and tables for each:

- Part HG—General Requirements for all Materials of Construction
- Part HF—Requirements for Boilers Constructed of Wrought Materials
  - Part HF, Subpart HW—Requirements for Boilers Fabricated by Welding
  - Part HF, Subpart HB—Requirements for Boilers Fabricated by Brazing
- Part HC—Requirements for Boilers Constructed of Cast Iron
- Part HLW—Requirements for Portable Water Heaters

The scope of Part HG covers material requirements, design, pressure-relieving devices, tests, inspection, stamping, instruments, fittings and controls, and installation requirements of low-pressure steam heating boilers, hot water supply boilers, hot water supply boilers and their appurtenances, but not potable water heaters. The rules of this Part are required to be used together with the specific requirements in Part HF—Boilers of Wrought Materials, and Part HC—Cast Iron Boilers, whichever is applicable. In this chapter, many references have been made to the ASME Code Section IV as designs cannot be discussed without referring to the actual code book.

This Part has provisions for mandatory requirements, specific prohibitions, and nonmandatory guidance for minimum construction requirements for the design, fabrication, installation, and inspection of steam heating, hot water heating, and hot water supply boilers. Enforcing or regulating authorities having jurisdiction at the location of an installation may establish applicability of these rules, in whole or in part. It becomes mandatory only if these rules are adopted by a jurisdiction.

The rules of ASME Code Section IV are restricted to steam boilers for operation at pressure not exceeding 15 psi (103 kPa), and hot water heating boilers and hot water supply boilers at pressures not exceeding 160 psi (1100 kPa) and/or temperatures not exceeding 250°F (121°C). When the service requirements exceed these limits, the rules of ASME Code Section I—Power Boilers will be applicable.

A heating boiler is designed for minimum pressure of 30 psi (kPa). Design pressure is calculated by formulas provided in the paragraphs HG-301 to HG-345.

The maximum allowable working pressure is stamped on a boiler as per HG-530 and must not exceed the design pressure of any parts. No boiler shall be operated at a pressure exceeding the maximum allowable working pressure except when the safety devices are discharging, at which maximum allowable working pressure shall not be exceeded by more than the amount provided in paragraphs HG-400.1 and HG-400.2.

8.1 Cylindrical Parts under Internal Pressure

The following formula is used to calculate the required thickness and the design pressure of any cylindrical parts under internal pressure (Figure 8.1):

$$t = \frac{PR}{SE - 0.6p}$$

$$p = \frac{SE_t}{R + 0.6t}$$

where P = design pressure, psi (not less than 30 psi or 207 kPa)

S = maximum allowable stress value from Appendices B and C, psi

t = required wall thickness, inch

R = inside radius of cylinder, inch

E = efficiency of longitudinal joint or ligament between tube holes, whichever is less

Figure 8.1 Firetube boiler shell. (*Courtesy: Industrial Boiler Co.*)

For seamless shells, $E = 1$, but for welded joints, efficiency in paragraph HW-702 should be used.

Example 8.1 What is the required thickness of a cylinder, welded by a full-penetration double butt joint, if the plates are of SA-285, Grade B; the inside diameter is 42 inches; and the working pressure is 150 lb?

Find the drum thickness, given the following:

$P = 150$ psi

$S = 10{,}000$ psi for SA-285, Grade B (Appendix B)

$R = 21$ inches (half the inside diameter of 42 inches)

$E = 0.85$ (paragraph HW-702)

$$t = \frac{PR}{SE - 0.6P}$$

$$t = \frac{150 \times 21}{10{,}000 \times 0.85 \; - 0.6 \times 150}$$

$$t = \frac{3{,}150}{8{,}410}$$

$$t = 0.37 \text{ inch}$$

8.1.1 Tubes

The formula shown in Example 8.1 is modified as shown in the following equation, to be used when determining the required thickness of tubes and pipes used as tubes, neither one of which is strength-welded to the tube sheet, header, or drum:

$$t = \frac{PR}{SE - 0.6P} + 0.04$$

where

Rolling and structural stability allowance = 0.04 inch (1 mm)

Rolling and structural stability allowance = 0 for tubes welded to tube sheets, headers, or drums

Seal-welded tubes would still require the addition of 0.4 inch

The minimum thickness of a tube or pipe used as a tube shall not be less than 0.061 inch (1.5 mm), although there is no minimum thickness requirement for nonferrous tubes installed by brazing.

8.1.2 Heads

Three types of heads used for construction of a boiler are ellipsoidal, torispherical, and hemispherical. The pressure may be on either the concave or convex side. The design pressure on the convex side shall be equal to 60 percent of that for heads of the same dimensions having pressure on the concave side.

The following symbols are used in the formulas to calculate required thickness of different types of heads under pressure on the concave side:

t = required wall thickness after forming, inch

P = design pressure, psi (not less than 30 psi or 207 kPa)

D = inside diameter of the head skirt; or inside length of the major axis of an ellipsoidal head or the inside diameter of a cone head at the point under consideration, measured perpendicular to the longitudinal axis, inch

S = maximum allowable stress value from Appendix B and Appendix C, psi

L = inside spherical or crown radius, inch

E = lowest efficiency of any joint in the head

For seamless head $E = 1$

For welded joints, use efficiency provided in HW-702.

Ellipsoidal heads. The following formulas are used for calculating the required thickness and the design pressure of a dished head of ellipsoidal form (Figure 8.2):

$$t = \frac{PD}{2SE - 0.2P}$$

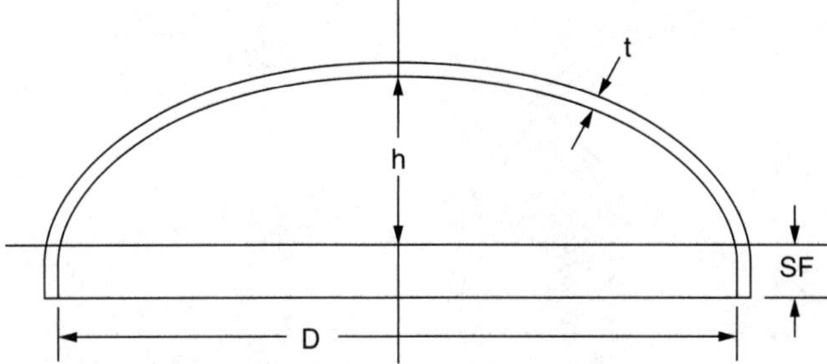

Figure 8.2 Ellipsoidal head.

or

$$P = \frac{2SEt}{D + 0.2t}$$

Note: Use of this formula is limited to 2:1 elliptical heads.

Torispherical heads. The following formulas shall be used for calculating the required thickness and the design pressure of a dished head of torispherical form (Figure 8.3):

$$t = \frac{0.885PL}{SE - 0.1P}$$

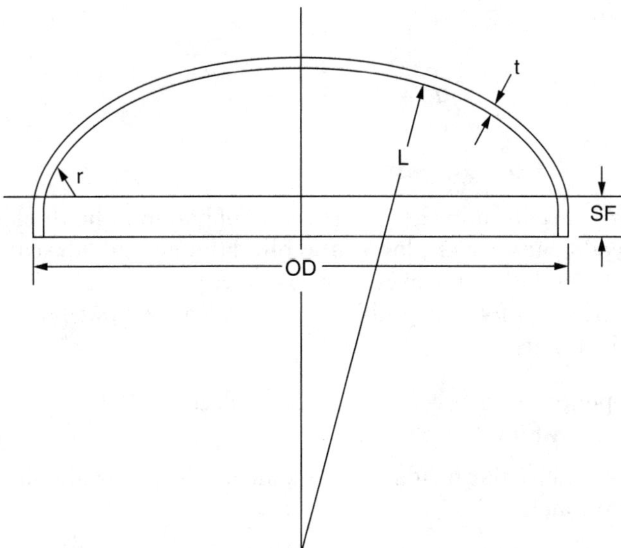

Figure 8.3 Torispherical head.

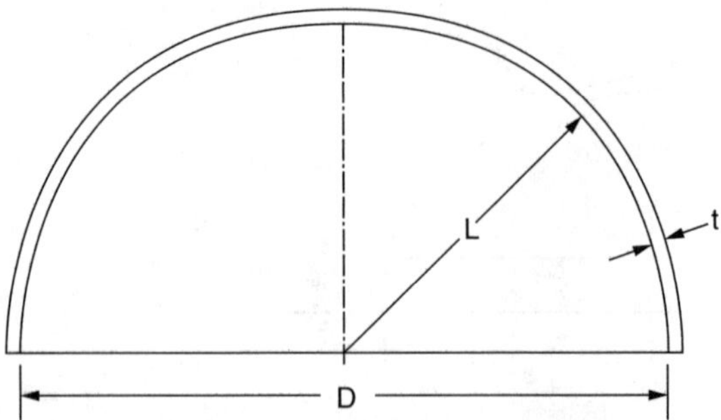

Figure 8.4 Hemispherical head.

or

$$P = \frac{SEt}{0.885L + 0.1t}$$

Hemispherical heads. The following formulas are used for calculating the required thickness and the design pressure of a dished head of hemispherical head (Figure 8.4) in which P does not exceed $0.665SE$:

$$t = \frac{PL}{2SE - 0.2P}$$

or

$$P = \frac{2SEt}{L + 0.2t}$$

8.1.3 Flat Heads

Flat heads and covers are used in the construction of boilers. The thickness of such unstayed heads, cover plates, and blind flanges shall be calculated as per the formulas specified in paragraph HG-307. The following symbols are used for acceptable types of unstayed flat heads and covers shown in Figure 8.5.

C = a factor depending on the method of attachment. Values of C may be found in Figure 8.5

D = long span of noncircular heads or covers measured perpendicular to short span, inch

d = diameter, or short span, inch

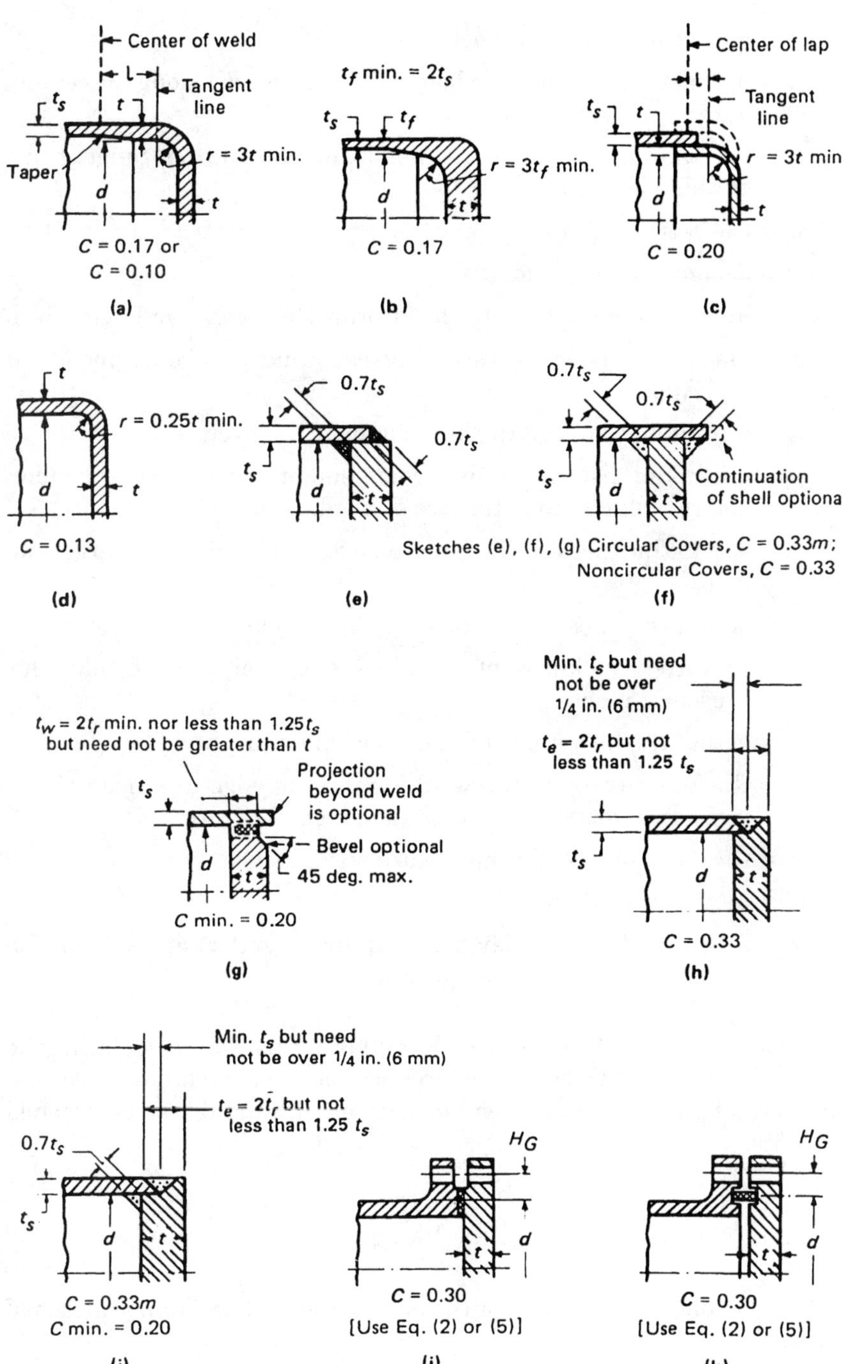

Figure 8.5 Examples of unstayed flat heads and covers. *(Courtesy: ASME Code Section IV)*

H_G = gasket moment arm, inch

L = perimeter of noncircular bolted head measured along the centers of the bolt holes, inch

l = length of flange or flanged heads, measured from tangent line of knuckle, inch

m = the ratio t_r/t_s

P = design pressure, psi

r = inside corner radius on the head formed by flanging or forging, inch

S = maximum allowable working stress value, psi (from Appendix B and Appendix C)

t = minimum required thickness of head or cover, inch

t_e = minimum distance from beveled end of drum, pipe, or header, before welding, to outer face of head, inch

t_f = actual thickness of the flange on a forged head, at the large end, inch

t_h = actual thickness of flat head or cover, inch

t_r = required thickness of seamless shell, pipe, or header, for pressure, inch

t_s = actual thickness of shell, pipe, or header, inch

t_w = thickness through the weld joining the edge of a head to the inside of a drum, pipe, or header, inch

t_1 = throat dimension of the closure weld, inch

W = total bolt load, lb

Z = a factor of noncircular heads and covers that depends on the ratio of short span to long span

Circular flat plates. The following formulas are used for calculating the minimum required thickness of circular, flat, unstayed heads, covers, and blind flanges: except when head, cover, or blind flange is attached by bolts.

$$t = d\sqrt{\frac{CP}{S}}$$

Use the following formula for the head, cover, or blind flange attached by bolts:

$$t = d\sqrt{CP/S + 1.9H_G/Sd^3}$$

The thickness shall be determined for both operating conditions and gasket seating, and the value greater of these two shall be applicable.

Noncircular flat heads. The following formula shall be used to calculate the required thickness for noncircular, flat, unstayed heads, covers, and blind flanges:

$$t = d\sqrt{ZCP/S}$$

where

$$Z = 3.4 - \frac{2.4d}{D} \qquad (Z \text{ should not exceed } 2.5)$$

or

$$t = d\sqrt{ZCP/S + 6WH_G SLd^2}$$

When the noncircular heads, covers, or blind flanges are attached by bolts causing a bolt edge moment. The thickness shall be determined for both operating conditions and gasket seating, the value greater of these two shall be applicable.

8.2 Cylindrical Parts under External Pressure

Cylindrical pressure parts—furnaces, tubes, and so on—are subjected to external pressure. Section II, Part D, Appendix 3 (Basis for Establishing External Pressure Charts) is used as the basis of these rules.

Furnaces. Plain-type furnaces, tubes, ring reinforced–type furnaces, corrugated furnaces, combination furnaces, and semicircular furnaces are the examples of cylindrical parts used for construction. Formulas and procedures for calculation of wall thicknesses for these cylindrical parts can be found in paragraph HG-312. The following symbols are used in the formulas:

- A = factor determined from Figure G in Subpart 3 of ASME Code Section II, Part D (Appendix G)
- B = factor determined from the applicable material in Subpart 3 of ASME Code Section II, Part D for maximum design, metal temperature, psi (Appendix G)
- D_o = outside diameter of furnace, inch
- L = design length of plain furnace taken as the distance from center to center of weld attachment, inch

P = design pressure, psi

t = minimum required wall thickness of furnaces, inch

Plain furnaces. The following procedures are followed to determine the required minimum wall thickness of the furnace:

1. Assume a value for t, then determine the ratios L/D_o and D_o/t.
2. Enter Appendix G at the value of L/D_o. If L/D_o is less than 0.05, enter the chart at a value of $L/D_o = 0.05$. If L/D_o is more than 50, enter the chart at a value of $L/D_o = 50$.
3. Move horizontally to the line for the value of D_o/t as determined above. From this point of intersection, move vertically downward to determine the value A.
4. Use value A and move vertically to an intersection with the material/temperature line for the design temperature. From this point, move horizontally and read the value B.
5. Calculate the value of the maximum allowable external working pressure P_a using the following formula:

$$P_a = \frac{B}{D_o/t}$$

6. Compare P_a with P. If P_a is less than P, a different value of t or different value of L, or some combination of both must be selected so that P_a is equal to or greater than P.

Tubes. The above procedure is used to calculate the wall thickness of ferrous tubes under external pressure. Additional thickness allowance of 0.04 inch (1 mm) is added as an additional allowance for rolling and structural stability. This additional thickness is not required for tubes welded to tube sheets, headers, or drums.

Ring reinforced–type furnace. The ring reinforced furnace (Figure 8.6) is a plain furnace with a circular stiffening ring welded to it. The stiffening ring is rectangular in cross-section and its thickness should not be less than 5/18 inch (8 mm) or more than 13/16 inch (21 mm) and in no case 1 1/4 times thicker than the furnace wall.

The ratio of height of the ring to its thickness (H_r/T_r) should not be less than $3n$ or more than 8. The thickness of the furnace wall or tube wall and the design of the stiffening rings are determined by the procedures as applied to a plain furnace.

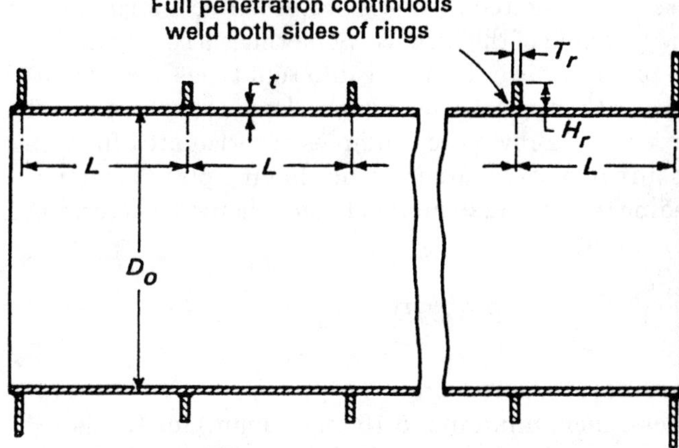

Figure 8.6 Ring reinforced furnace. (*Courtesy: ASME Code Section IV*)

The moment of inertia of the stiffening ring shall be calculated by the following formula:

$$I_s = \frac{D_o^2 L\left(t + \frac{A_s}{L}\right)A}{14}$$

where I_s = required moment of inertia about its neutral axis parallel to the axis of the furnace, in.^{4}

A_s = cross-sectional area of the stiffening ring, in.^{2}

The procedure for determining the moment of inertia is as follows:

1. Assume that D_o, L, and t are known. Select a rectangular-shaped ring and determine its area A_s.
2. Calculate moment of inertia I by the above formula. Use the following formula to calculate B:

$$B = \frac{PD_o}{t + \frac{A_s}{L}}$$

3. Enter the right-hand side of the chart in Appendix G for the material at the value B calculated above.
4. Follow horizontally to the material line and move down vertically to read A.
5. With values of all the symbols known, calculate I_s. If I_s is greater than I, a new section of the ring must be selected and new I_s calculated. This section of the ring is satisfactory if I_s is smaller than I.

Corrugated furnaces. Corrugated furnaces are used in construction for easy expansion and cleaning of the furnace. Sometimes a corrugated furnace is welded to a plain furnace. There are different types of corrugated furnaces depending on the type of construction. Leeds suspension bulb, Morison, Fox, Purves, and Brown are examples of corrugated furnaces.

The design pressure of a corrugated furnace having plain portions at the ends not exceeding 9 inches (229 mm) in length should be determined by the following formula:

$$P = C_t/D$$

where P = design pressure, psi

t = thickness, inch, minimum 5/16 inch (8 mm) for Leeds, Morison, Fox, and Brown, and minimum 7/16 inch (11 mm) for Purves and other furnaces corrugated by sections not over 18 inches (457mm) long

D = mean diameter, inch

Take inside diameter plus 2 inches as mean diameter for the Morison furnace.

C = 17,300 for Leeds furnaces, 15,600 for Morison furnaces, 14,000 for Fox, Purves, and Brown furnaces.

8.2 Example A corrugated furnace (Brown type, 42 inches mean diameter) is found in a boiler. The plain ends of the furnace are 8 1/2 inches long, the desired working pressure is 125 psi, and the corrugations are 8 3/4 inches from center to center and 1-5/8 inches deep, with sections 17-1/2 inches long. Given the following, what is the required thickness?

$P = 125$ psi

$D = 42$ inches

$C = 14{,}000$

$$P = \frac{Ct}{D}$$

$$t = \frac{PD}{C}$$

$$t = \frac{125 \times 42}{14{,}000}$$

$t = 0.375$ or 3/8 inch

Combination-type furnaces. Combination-type furnaces are widely used. This furnace is designed in such a fashion that each type of furnace is self-supporting and does not require support from other furnaces at the connecting point. The formulas in paragraph HG-312.1 and HG-312.3 are used for plain sections and corrugated sections shall be designed as per the formula in paragraph HG-312.6. Full-penetration welding must be used to connect a plain self-supporting section to a corrugated self-supporting section.

Semicircular furnaces or crown sheets. The thickness of the semicircular furnace or crown sheet shall be minimum 5/16 inch (8 mm). The allowable working pressure of a semicircular furnace or crown sheet shall not exceed 70 percent of P_a as calculated by the formula and using chart.

8.3 Openings in Boilers

Shape of openings. Opening shapes shall be circular, elliptical or obround for cylindrical, spherical, or conical portions of boilers or in formed heads. An obround opening means an opening formed by two parallel sides and semicircular ends.

Size of opening. With adequate reinforcement, openings in cylindrical and spherical shells are not limited to any size. The rules are intended to apply to openings of the following dimensions:

- One-half of the boiler diameter but not exceeding 20 inches (508 mm) for boilers 60 inches (1520 mm) in diameter and less.
- One-third of boiler diameter but not exceeding 40 inches (1000 mm) for boilers over 60 inches (1520 mm) in diameter.

Design of finished openings. Finished openings shall be designed as per paragraph HG-320.3. The following symbols are used in Figure 8.7, the chart showing limits of sizes of openings with inherent compensation in cylindrical shells:

P = design pressure, psi

d = maximum allowable diameter of opening, inch

D = outer diameter of the shell, inch

t = nominal thickness of the shell, inch

S = maximum allowable stress value, psi from Appendix B and Appendix C

$K = PD/2St$

Calculate K from the formula. For any K value <50 percent, the opening will be inherently compensated by the excess thickness in the shell (i.e., $A_1 > A_r$)

Calculate Dt. Locate the intersecting point of K and Dt. Go horizontally left from this intersecting point and determine maximum diameter of the opening.

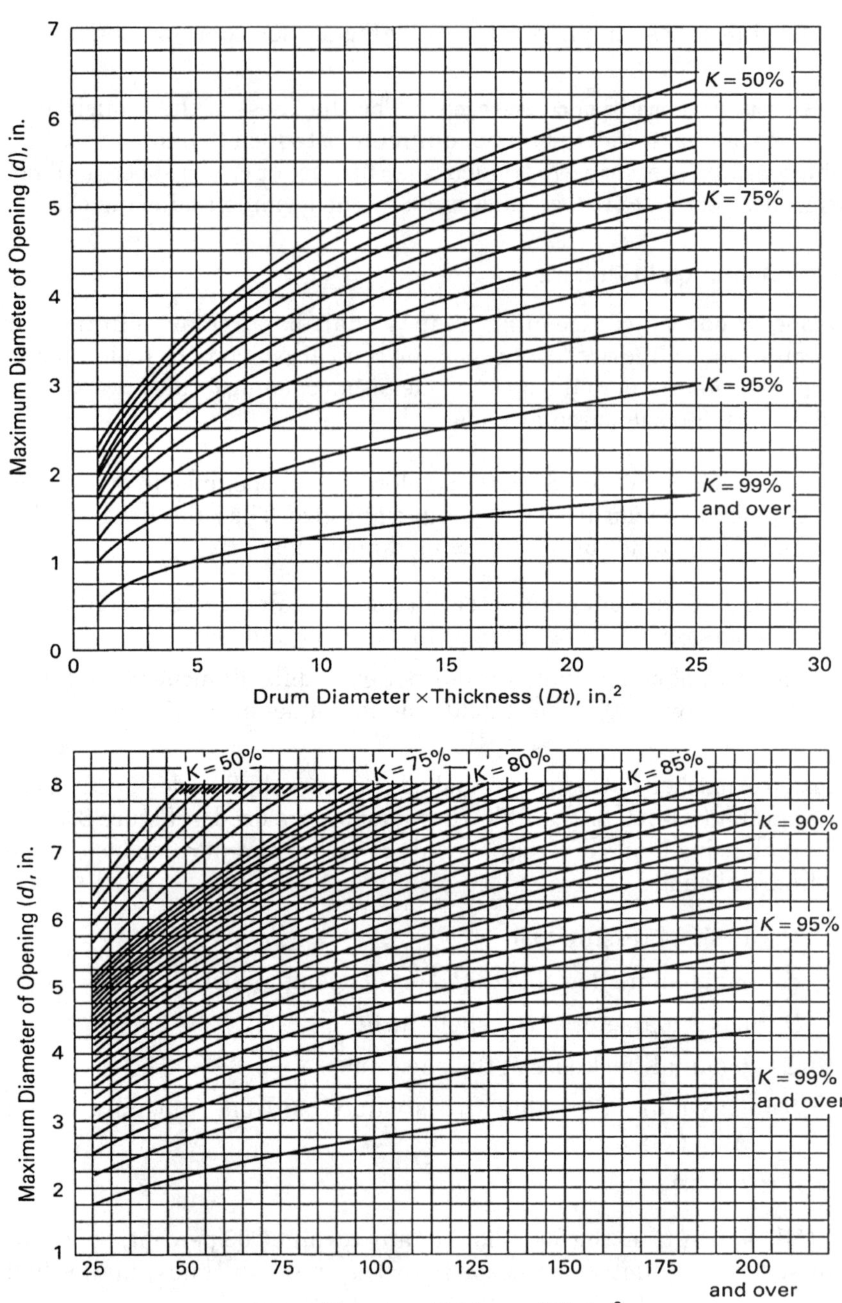

Figure 8.7 Sizes of openings with inherent compensation in cylindrical shell. (*Courtesy: ASME Code Section IV*)

8.4 Reinforcement Requirements for Openings

When a hole for a nozzle is cut in a shell, the vessel is weakened. Extra metal is provided to compensate for this weakness and strengthen the vessel. A certain amount of metal in the shell and in the nozzle is required to support the internal pressure. Any metal in excess of that required to support the internal pressure in the shell and nozzle is available as compensation for that nozzle opening. The common rules for reinforcement are:

- Add enough reinforcement to compensate for weakening yet preserve the general dilation or strain pattern.
- Place material adjacent to the opening for optimum distribution.

Reinforcement is required for all openings, except openings in a definite pattern and small openings. Reinforcement should be provided in such amount and distribution that the requirements are satisfied for all planes through the center of the opening and normal to the boiler surface. The total cross-sectional area of reinforcement A required should be determined by the following formula by using the symbols as shown in Figure 8.8.

$$A = dt_r F + 2t_n t_r F (1 - f_{r1})$$

where d = the diameter in the given plane of the finished opening, inch
F = a correction factor, a value 1.00 may be used
t_r = the required thickness of a seamless shell or head, inch
A_1 = area in excess of thickness in the boiler shell available for reinforcement, in.^{2}
A_2 = area in excess of thickness in the nozzle wall available for reinforcement, in.^{2}
E_1 = 1.0 when an opening is in the solid plate or when the opening passes through a circumferential joint in a shell or cone, or the joint efficiency obtained when any part of the opening passes through any other welded joint
D_p = outside diameter of reinforcing element, inch
R_n = inside radius of the nozzle under consideration, inch
S = maximum allowable stress value, psi from Appendix B and Appendix C
S_n = allowable stress in nozzle, psi
S_v = allowable stress in vessel, psi
S_p = allowable stress in reinforcing element (plate)
f_r = strength reduction factor, not more than 1.0
$f_{r1} = S_n/S_v$ for nozzle inserted through the vessel wall
f_{r1} = 1.0 for nozzle wall abutting the vessel wall
$f_{r2} = S_n/S_v$

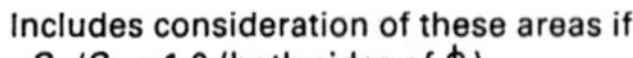

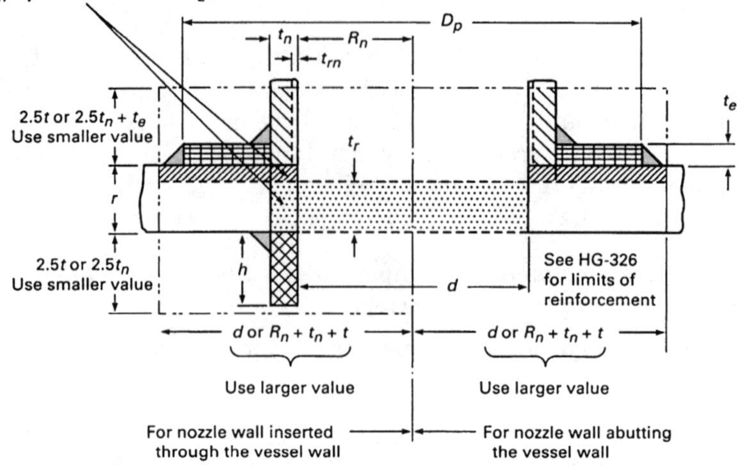

Without Reinforcing Element

A	=	$dt_r F + 2t_n t_r F(1 - f_{r1})$	Area required
A_1	=	$d(E_1 t - Ft_r) - 2t_n (E_1 t - Ft_r)(1 - f_{r1})$ $2(t + t_n)(E_1 t - Ft_r) - 2t_n (E_1 t - Ft_r)(1 - f_{r1})$	Area available in shell; use larger value
A_2	=	$5(t_n - t_{rn}) f_{r2} t$ $5(t_n - t_{rn}) f_{r2} t_n$	Area available in nozzle projecting outward; use smaller value
A_3	=	$2t_n f_{r2} h$	Area available in inward nozzle
A_{41}	=	outward nozzle weld = $(\text{leg})^2 f_{r2}$	Area available in outward weld
A_{43}	=	inward nozzle weld = $(\text{leg})^2 f_{r2}$	Area available in inward weld

If $A_1 + A_2 + A_3 + A_{41} + A_{43} > A$ — Opening is adequately reinforced

If $A_1 + A_2 + A_3 + A_{41} + A_{43} < A$ — Opening is not adequately reinforced so reinforcing elements must be added and/or thickness must be increased

With Reinforcing Element Added

A	=	same as A above	Area required
A_1	=	same as A_1 above	Area available
A_2	=	$5(t_n - t_{rn}) f_{r2} t$ $2(t_n - t_{rn})(2.5t_n + t_e) f_{r2}$	Area available in nozzle projecting outward; use smaller area
A_3	=	same as A_3 above	Area available in inward nozzle
A_{41}	=	outward nozzle weld = $(\text{leg})^2 f_{r3}$	Area available in outward weld
A_{42}	=	outer element weld = $(\text{leg})^2 f_{r4}$	Area available in outer weld
A_{43}	=	inward nozzle weld = $(\text{leg})^2 f_{r2}$	Area available in inward weld
A_5	=	$(D_p - d - 2t_n) t_e f_{r4}$ [Note (1)]	Area available in element

If $A_1 + A_2 + A_3 + A_{41} + A_{42} + A_{43} + A_5 > A$ — Opening is adequately reinforced

NOTE:
(1) This formula is applicable for a rectangular cross-sectional element that falls within the limits of reinforcement.

Figure 8.8 Formulas for reinforced openings. (*Courtesy: ASME Code Section IV*)

f_{r3} = (lesser of S_n or S_p)/S_v
f_{r4} = S_p/S_v
h = distance nozzle projects beyond the inner surface of the vessel wall, inch
t_e = thickness of attached reinforcing pad, inch
t = nominal thickness of the boiler shell, inch
t_r = required thickness of a seamless shell or head, inch
t_n = nominal thickness of nozzle wall, inch
t_{rn} = required thickness of a seamless nozzle wall, inch
d = diameter in the plane under consideration of the finished opening, inch

8.5 Inspection and Access Openings

Manhole, handhole, inspection hole, or washout plug openings are provided with all boilers for the purpose of inspection and removal of accumulated deposits. An electric boiler with more than internal gross volume of 5 cu ft (exclusive of casing and insulation) shall have an opening for inspection. An electric boiler equipped with immersion-type resistance elements without a manhole shall have an inspection opening not less than 3 inches pipe size (DN 75) located in the lower portion of the shell or head. An electric boiler for steam service shall be provided with an inspection opening or manhole at or near the normal waterline. Access doors shall be provided for the furnaces of the internally fired boilers.

Manholes. A manhole is required in the front head below the tubes of a horizontal-return tubular boiler 60 inches (1520 mm) or over in diameter. A manhole shall be provided in the upper part of the shell or in the head of a firetube boiler more than 60 inches (1520 mm) in diameter, except in a vertical firetube boiler.

An elliptical manhole opening size shall be minimum 12 in. × 16 in. (305 mm × 406 mm) in size. A circular manhole size shall be minimum 15 inches (381 mm) in diameter. The width of bearing surface for a gasket on a manhole opening shall not be less than 1/16 inch (18 mm).

Handholes and washout plugs. Each boiler shall be provided with handholes, inspection openings, or washout openings. Location of these openings is given in paragraph HG-330.4. Locomotive or firebox-type boilers must have one handhole or washout plug in the lower part of the waterleg and at least one opening near the line of crown sheet. In addition to that the boilers for steam service shall have at least one inspection opening above the top row of tubes.

The minimum size of the inspection opening shall be NPS 3 (DN 75), washout plug 1.5 inches, pipe size (DN 40), and handhole opening 2.75 in. × 3.5 in. (70 mm × 89 mm).

Access doors. A fire door or access door shall be provided for the furnace of an internally fired boiler with furnace dimension 28 inches or more. Size of the door shall be a minimum of 11 in. × 15 in. (280 mm × 381 mm) or 10 in. × 16 in. (254 mm × 406 mm) or 15 inches (381 mm) in diameter. The access door for use in a boiler setting shall have equivalent size of 12 in. × 16 in. (305 mm × 406 mm) or equivalent area.

8.6 Stayed Surfaces

Flat plates with stays or staybolts of uniform cross-section are used for boiler construction. Figures HG-340.1 and HG-340.2 in ASME Section IV show acceptable methods of staying. The thickness and design pressure for the stayed plates are calculated by the following formulas:

$$t = p\sqrt{P/SC}$$

$$P = t^2\, SC/p^2$$

where t = required thickness of plate, inch
- P = design pressure, psi
- S = maximum allowable stress value from Appendix B and Appendix C
- p = maximum pitch measured between straight lines passing through the centers of the stays in the different rows, inch
- C = a factor depending on type of stays, given in paragraph HG-340.1. While connecting two plates by staying and one of them requires staying, the value of C shall be governed by the thickness of required staying.
- C = 2.7 for stays welded to plates
- C = 3.1 for stays screwed through plates
- r = radius of firebox corner, inch

Two flat stayed surfaces may intersect at an angle, in which case the pitch shall be calculated by the following formula:

$$p = \frac{90t}{\beta}\sqrt{\frac{CS}{P}}$$

The maximum pitch shall not exceed 8.5 inches (216 mm) except for welded in stays in which case it shall not exceed 15 times the diameter of the stay.

8.7 Staybolts

Threaded staybolts. The ends of the treaded staybolts should extend a minimum of two threads beyond the plate and after that they will be riveted over or fitted with extended nuts. The outside ends of solid staybolts

TABLE 8.1 Copper Plates and Copper Staybolts

Copper plate thickness, in.	Min. staybolt diameter, in.
Not exceeding 1/8	1/2
Over 1/8, but not over	5/8
Over 3/16	3/4

maximum 8 inches (203 mm) in length should be drilled with telltale holes a minimum of 3/16 inch (4.8 mm) in diameter to a depth extending at least 0.5 inch (13 mm) beyond the inside of the plate. Solid staybolts over 8 inches (203 mm) long should not be drilled. Hollow staybolts may be used instead of solid staybolts with drilled ends. Telltale holes are not required for staybolts attached by welding. Annealing should be done after upsetting on the ends of the threaded stays upset for threading. The ends of the staybolts fitted with nuts should not be exposed to radiant heat.

Welded-in staybolts. Installation requirements of welded-in staybolts are given in HW-7.10.

Area of stays. The required area of a stay is determined based on the minimum cross-section at the root of the thread. Corrosion allowance is not considered. The required area is calculated by dividing the load on the stay by the allowable stress value for the material.

Load carried by stays. The area supported by a stay is the area based on full pitch dimensions minus the area occupied by the stay. The load carried by a stay is calculated by multiplying the area supported by the design pressure.

Stays fabricated by welding. Normally, stays of one-piece construction are used. Stays made of parts may be attached by welding, and joint efficiency of 60 percent will be used for calculating strength. Ferrous stays welded in by fusion welding will have minimum cross-sectional areas of 0.44 in.^{2} (284 mm^2).

Nonferrous stays. There are requirements for minimum diameters for nonferrous materials like copper and copper-nickel stays. The minimum diameters of nonferrous stays is shown in Table 8.1 and 8.2.

TABLE 8.2 Copper-Nickel Plate and Copper-Nickel Staybolts

Copper-nickel plate thickness, in.	Min. staybolt diameter, in.
Not exceeding 1/8	3/8
Over 1/8, but not over 3/16	7/16
Over 3/16	1/2

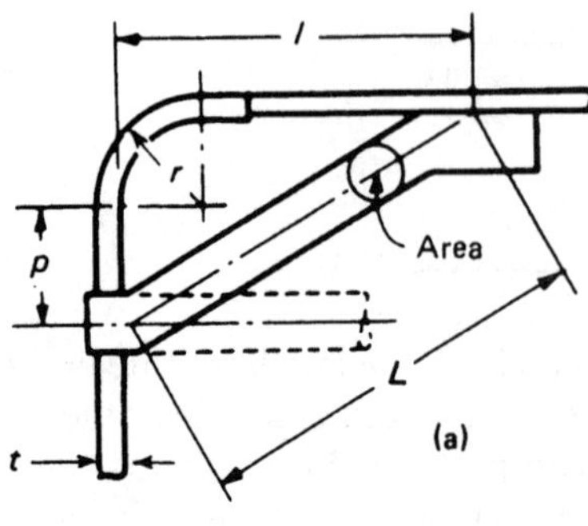

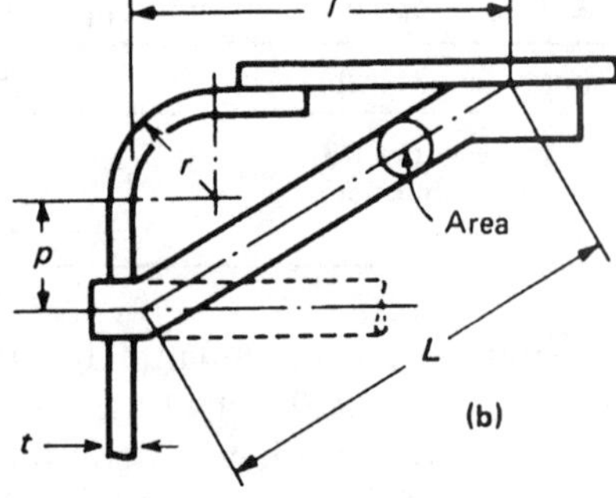

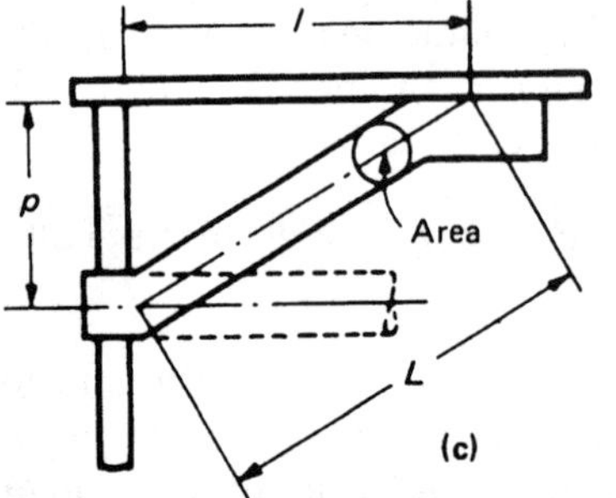

GENERAL NOTES:
1. Determine area of diagonal stays per HG-343.1
2. Determine weld details and weld size for stay-to-tube-sheet and stay-to-shell per HW-710
3. Determine diagonal stay pitch per HG-340.1
4. Max. $r = 8\,t$
   Min. $r = 3\,t$
   t = nominal thickness of tube sheet

Figure 8.9 Installation of diagonal stays. *(Courtesy: ASME Code Section IV)*

Diagonal stays. Diagonal stays are used to support the tube sheet and shell diagonally. Details for the installation of diagonal stays are shown in Figure 8.9. The required area of a diagonal stay is calculated by the following formula:

$$A = \frac{aL}{l}$$

where A = sectional area of diagonal stay, in.^{2}
a = sectional area of direct stay, in.^{2}
L = length of diagonal stay, inch
l = length of line, drawn perpendicular to boiler head or surface supported, to center of palm of diagonal stay, inch

8.8 Staying of Heads

The portions of the heads that require staying are stayed as flat heads. For unflanged heads in boilers less than 30 psi (207 kPa) pressure, staying is not required if the greatest distance measured along a radial line from the inner surface of the shell to a point does not exceed 1.25p. For unflanged heads in boilers for over 30 psi (207 kPa) or flanged heads of any pressure, staying is not required if the greatest distance measured as above does not exceed $1.5p$. The maximum distance between the inner surface of the shell and the centers for unflanged heads should not be more than the allowable pitch as calculated by HG-3.40, the value of C given for thickness of plate and type of stay. For a flanged head welded to the shell, the greatest distance between the inner surface of the supporting flange and the lines parallel to the surface of the shell should be p as calculated by HG-3.40, plus the inside radius of the supporting flange, the value of C given for thickness of plate and the type of stay. The greatest distance between the edges of the tube holes and the center of the first row of stays will be calculated by HG-3.40, the value of C given for the thickness of plate and type of stay.

Horizontal fire tube boilers with manhole openings on the heads are shown in Figure HG-345.1(a) and Figure HG-345.1(b). The area to be stayed may be reduced by 100 in.^{2} (645 cm^2) provided an unflanged manhole ring meeting the requirements of HG-3.21 is provided in a flat stayed head under the following conditions:

1. The distance between the manhole opening and the inner surface of the supporting flange does not exceed one-half the maximum allowable pitch for an unflanged manhole or one-half the maximum allowable pitch plus the inside radius of the supporting flange for a flanged-in manhole in a flanged head.
2. The distance between the centers of the first row of stays, or the edges of the tube holes, and the manhole opening does not exceed one-half the maximum allowable pitch as determined by HG-340.

8.9 Tube Sheets

Firetubes in a firetube boiler may be used as stays. The required thickness, maximum pitch, and the design pressure for tube sheets with firetubes used as stays may be calculated by using the following formulas:

$$t = \sqrt{\left(\frac{P}{CS}\right)\left(p^2 - \frac{\pi D^2}{4}\right)}$$

$$p = \sqrt{\left(\frac{CSt^2}{P}\right) + \left(\frac{\pi D^2}{4}\right)}$$

$$P = \frac{CSt^2}{p^2 - \left(\frac{\pi D^2}{4}\right)}$$

where

t = required thickness of plate, inch
p = maximum pitch measured between centers of tubes in different rows, inch
C = 2.7 for firetubes welded to plates not over 7/16 inch (11 mm) in thickness
C = 2.8 for firetubes welded to plates over 7/16 inch (11 mm) in thickness
S = maximum allowable stress values given in Appendix B and Appendix C
P = design pressure, psi
D = outside diameter of the tubes, inch

The pitch of firetubes used as stays will not exceed 15 times the diameter of the tubes. Firetubes welded to tube sheets and used as stays will meet the requirements of HW-7.13.

8.10 Ligaments

A ligament—sometimes referred to as webbing—is the type of metal between the holes in a tube sheet. The three types of ligaments are:

1. Longitudinal, which are located between the front and lengthwise holes along the drum.
2. Circumferential, which are located between the holes and which encircle the drum.
3. Diagonal, which constitute a special case because they are located between the holes and are offset at an angle to each other.

Rules of ligaments are applicable to groups of openings in cylindrical pressure parts, which form a definite pattern. The rules also apply to openings spaced not exceeding two diameters, center to center. The following symbols are used in the formulas for calculation of efficiency of ligaments:

p = longitudinal pitch of adjacent openings, inch

p' = diagonal pitch of adjacent openings, inch

p'' = transverse pitch of adjacent openings, inch

p_1 = pitch between corresponding openings in a series of symmetrical groups of openings, inch

d = diameter of openings, inch

n = number of openings in length p_1

E = efficiency of ligament

Openings parallel to shell axis. These openings may have equal pitch in every row or unequal pitch in symmetrical groups. The ligament efficiency is determined by the following formulas:

1. For equal pitch of openings in every row (Figure 8.10):

$$E = \frac{p - d}{p}$$

2. For unequal pitch in symmetrical groups of openings:

$$E = \frac{p_1 - nd}{p_1}$$

If openings are not in symmetrical groups, then the efficiency will be determined as follows for the group of openings which give lowest efficiency:

- Efficiency as calculated by above formula (2), using p_1 equal to the inside diameter of the shell or 60 inches (1520 mm), whichever is less.
- 1.25 times the efficiency calculated by above formula (2), using p_1 equal to the inside radius of the shell or 30 inches (760 mm), whichever is less.

Opening transverse to shell axis. The ligament efficiency of openings spaced at right angles to the axis is equal to two times the efficiency of similarly spaced holes parallel to the shell axis as calculated by the above formulas.

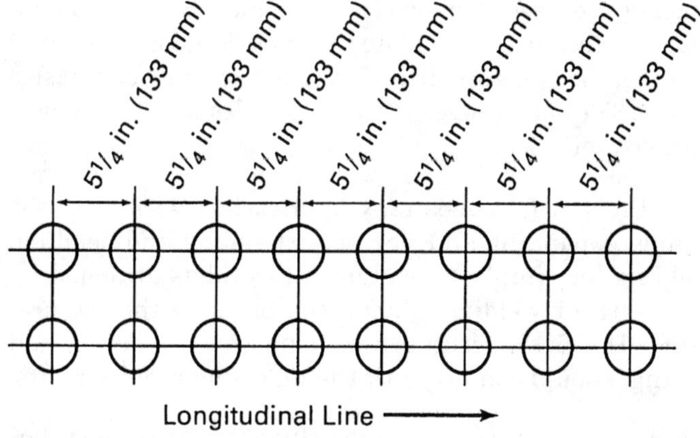

Figure 8.10 Tube spacing with pitch of holes equal in every row. *(Courtesy: ASME Code Section IV)*

Holes along a diagonal. The ligament efficiency of openings that are equally spaced diagonally is calculated by the following formula:

$$E = \frac{p' - d}{p'F}$$

where F is the factor shown in Figure HG-321 for the angle which the diagonal makes a plane through the longitudinal axis.

8.3 Example The shell of a vessel is drilled for tube holes of 3-1/32 inches diameter spaced longitudinally in groups of two tubes pitched 4-7/8 inches with 6 inches of spacing between the groups. Circumferentially, the tube holes are symmetrical. What is the efficiency of the longitudinal tube ligaments?

$P_1 = 10.875$ inches

$n = 2$

$d = 3.03125$ inches

$$E = \frac{P_1 - nd}{P_1}$$

$$E = \frac{10.875 - (2)(3.03125)}{10.875}$$

$$E = \frac{4.813}{10.875}$$

$$E = 0.4425$$

The efficiency of the longitudinal tube ligaments is 44.25%.

8.11 Tube Holes and Tube Attachments

Tube holes. Tube holes are drilled to full size from the solid plate. They also may be punched 0.5 inch (13 mm) smaller in diameter than full size when the plate thickness exceeds 3/8 inch (10 mm) and 1/8 inch (3.2 mm) smaller in diameter than full size when plate thickness is 3/8 inch (10 mm) or less. After punching, the holes are drilled, reamed, or finished full size with rotating cutters. The sharp edges of the tube holes will be removed with a file or suitable tool.

Tube attachments. Ends of firetubes may be attached with the tube sheets by expanding, expanding and flaring, expanding and beading, expanding or welding, or simply by welding. Firetubes attached by expanding and welding, or welding, should comply with the requirements of paragraph HW-7.13. Firetubes attached by expanding and flaring, or expanding, should comply with the following requirements:

1. If firetube ends are in contact with primary furnace gases, the tubes should extend not less than tube thickness or 1/8 inch (3.2 mm)

whichever is greater. Tubes should not extend more than 1/4 inch (6 mm) or the tube thickness, whichever is greater.

2. If firetube ends are not in contact with primary gases, the tubes should extend not less than the tube thickness or 1/8 inch (3.2 mm), whichever is greater. Tubes should not extend more than 3/8 inch (10 mm) or the tube thickness, whichever is greater.

Watertubes may be attached with the drums by expanding, expanding and flaring, expanding and beading, expanding and welding, or welding. For attaching water- tubes by other than expanding and beading, the tubes should extend not less than 1/4 inch (6 mm) nor more than 1/2 inch (13 mm). Water tubes in hot water boilers may be installed into headers by using O-ring seals instead of expanding, welding, or brazing under the conditions specified in the Code.

8.12 External Piping

Threaded connections. Threaded pipe connections should be tapped into material having minimum thickness as shown in Table 8.3. When a curved surface is tapped, the minimum thickness should be sufficient to permit at least four full threads.

Flanged connections. Flanged connections to external piping should meet the requirements of ANSI B16.5—*Steel Pipe Flanges and Flanged Fittings*. Steel flanges which do not meet the requirements of ANSI B16.5, should be designed in accordance with Appendix II of ASME Code Section VIII, Division 1.

TABLE 8.3 Minimum Thickness for Material for Threaded Connections to Boilers

Size of pipe connections, in.	Min. thickness of material required, in.
Under 3/4	1/4
3/4 to 1, incl.	5/16
1-1/4 to 2-1/2, incl.	7/16
3 to 3-1/2, incl.	5/8
4 to 5, incl.	7/8
6 to 8, incl.	1
9 to 12, incl.	1-1/4

Chapter

9

Pressure Relief Valves

A pressure relief valve is a safety device used on heating boilers to protect life and property when all other safety measures fail. The ASME *Boiler and Pressure Vessel Code Section IV* requires that all heating boilers must be protected by pressure relief valves. The Code further states that:

- At least one pressure relief valve should be set at or below the maximum allowable working pressure (MAWP) of the boiler.
- Relieving pressure should not exceed MAWP by more than 3 percent for fired steam boilers.

The primary purpose of a pressure relief valve is to prevent pressure in the boiler system from increasing beyond the safe design limits. The secondary purpose is to minimize damage to other system components as a result of operation of the pressure relief valve itself.

The following are advantages of pressure relief valves:

- Most reliable—if properly sized and operated
- Versatile—can be used for many services

9.1 Types of Pressure Relief Valves

A pressure relief valve is a spring-loaded device, which is designed to relieve excess pressure and to reclose and prevent further flow of fluid after normal conditions have restored (Figure 9.1). It may be used for either compressible or incompressible fluids, depending on design, adjustment, or application.

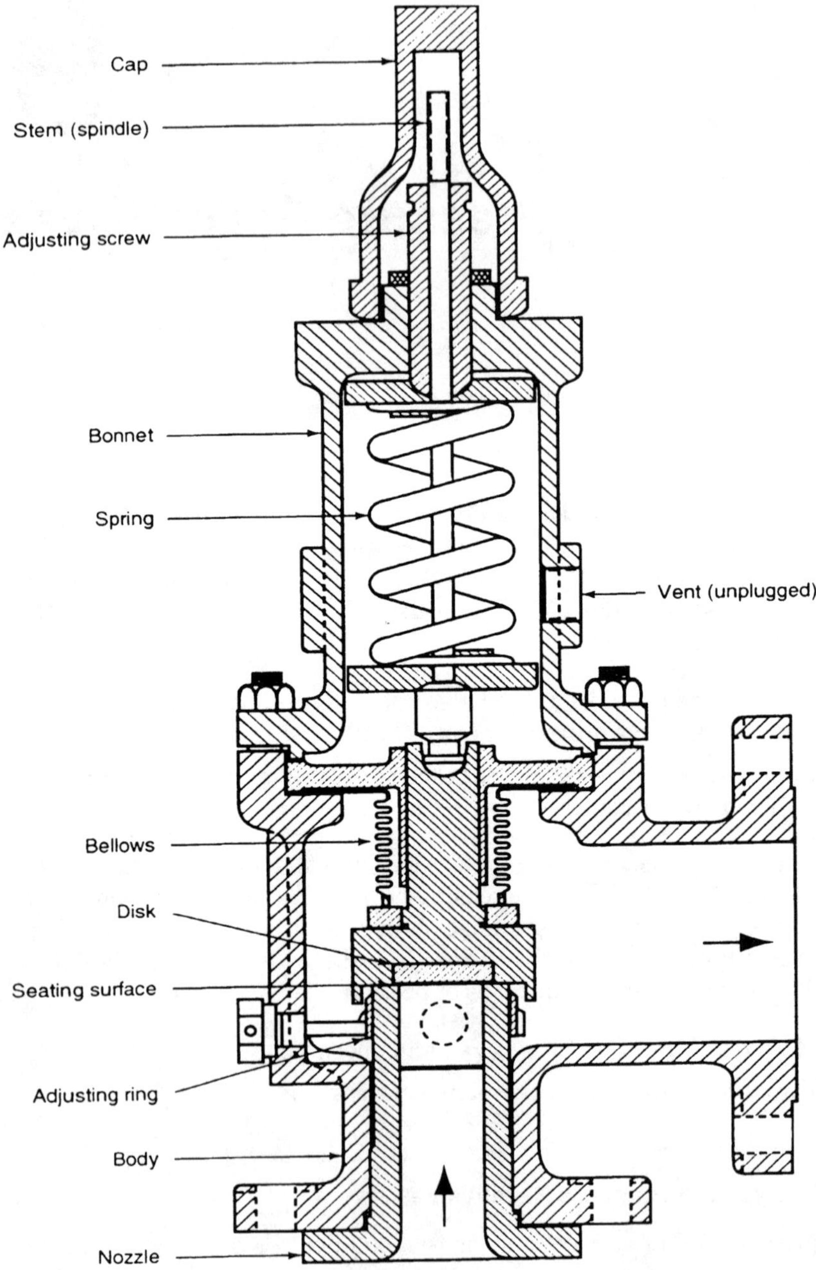

Figure 9.1 Pressure relief valve. *(Courtesy: API RP 520)*

There are many types of pressure relief valves, based on design and construction. They are generally classified as safety valves, safety relief valves, and relief valves.

9.1.1 Safety valves

A safety valve is a pressure relief valve actuated by inlet static pressure and characterized by rapid opening or pop action (Figure 9.2). Safety valves are used primarily with compressible gases and in particular for steam and air.

Safety valves are classified according to lift and bore of the valves. Types of safety valves are low-lift, full-lift, and full-bore safety valves.

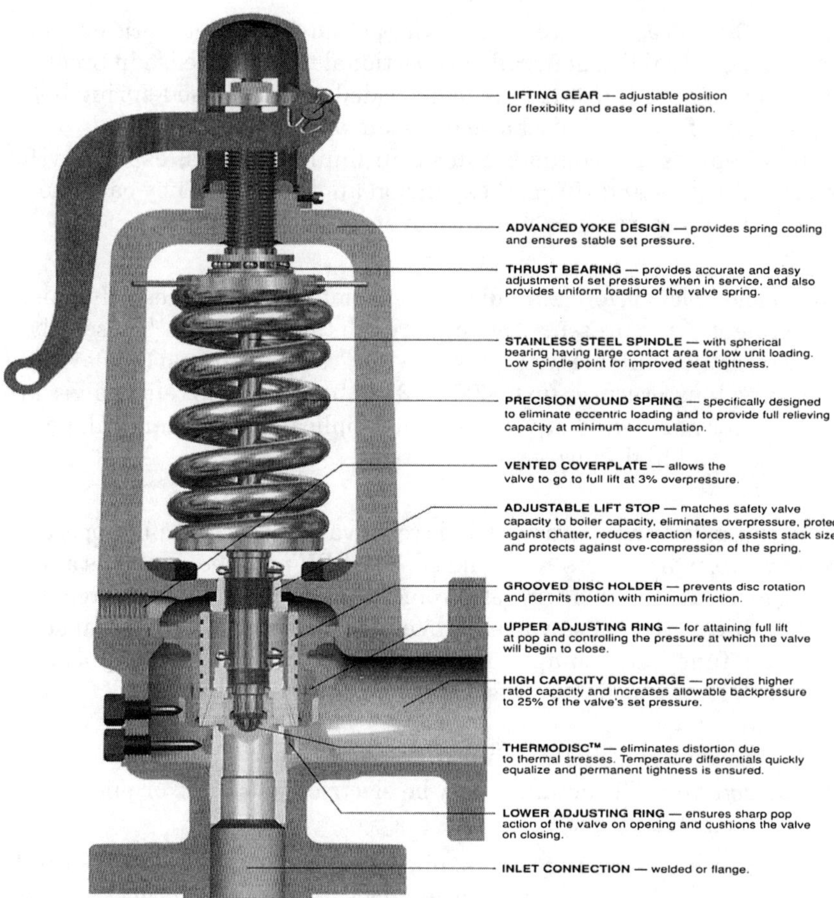

Figure 9.2 Safety valve for steam boiler. *(Courtesy: Dresser Flow Control)*

Low-lift safety valve. A low-lift safety valve is that in which the disk lifts automatically such that the actual discharge area is determined by the position of the disk.

Full-lift safety valve. A full-lift safety valve is that in which the disk lifts automatically such that the actual discharge area is not determined by the position of the disk.

Full-bore safety valve. A full-bore safety valve is that which has no protrusion in the bore and wherein the valve disk lifts to an extent sufficient for the minimum area, at any section at or below the seat, to become the controlling orifice.

9.1.2 Relief valves

A relief valve is a pressure relief device actuated by inlet static pressure having a gradual lift, generally proportional to the increase in pressure over opening pressure. It may be provided with enclosed spring housing suitable for closed discharged system application.

Relief valves are commonly used in liquid systems, especially for lower capacities and thermal expansion applications. They can also be used in pump systems as pressure-overspill devices.

Adjustable relief valve. An adjustable relief valve features convenient adjustment of the pressure setting through the outlet port. These valves are generally available with pressure ranges up to 508 psi (35 bar), and operating temperatures up to 600°F (315°C). Adjustable relief valves are suitable for nonvented or vented inline applications in chemical, petrochemical, and high-purity gas industries.

Electronic relief valve. An electronic relief valve (ERV) is a pilot-operated relief valve, which offers zero leakage. The ERV package combines a zero-leakage isolation valve with electronic controls to monitor and regulate system pressure. These valves provide protection either in a capacity-relieving function or simply in an overpressure protection application.

The electronic relief valve system consists of:

1. *The valve.* Generally a metal seated ball valve is used.
2. *The actuator.* The actuator may be electric, hydraulic, or pneumatic, and operated by gears.
3. *The control system.* The ERV is supplied with or without remote controls and display. Numerous pressure ranges from 0 to 5000 psi (34.5 Mpa) are available. Accuracy of 0.25 percent is achieved for 1000 to

3000 psi and 0.1 percent for 5000 psi units. Standard units operate from 115 V ac or 125 dc and control ac, dc, or pneumatic actuators.

9.1.3 Safety relief valves

A safety relief valve is a pressure relief valve characterized by rapid opening, or "pop action," or by opening in proportion to the increase in pressure over the opening pressure, depending on the application, and which may be used either for liquid or compressible fluid.

In general, the safety relief valve performs as a safety valve when used in a compressible gas system. This valve opens in proportion to the overpressure when used in liquid systems like a relief valve.

Safety relief valves are classified as conventional, pilot-operated, balanced-bellows, power-actuated, and temperature-actuated.

Conventional pressure relief valves. The conventional pressure relief valve is characterized by a rapid-opening pop action or by opening in a manner generally proportional to the increase in pressure over the opening pressure (Figure 9.3).

The basic elements of a conventional pressure relief valve consist of:

- An inlet nozzle connected to the vessel or system to be protected
- A movable disc, which controls flow through the nozzle
- A spring, which controls the position of the disc

Under normal operating conditions, the pressure at the inlet is below the set pressure and the disc is seated on the nozzle, prevent to flow through the nozzle.

Conventional pressure relief valves are for applications where excessive variable or built-up back pressure is not present in the system. The operational characteristics are directly affected by changes of the back pressure on the valve.

Pilot-operated pressure relief valves. A pilot-operated pressure relief valve is that in which the major relieving device is combined with and controlled by a self-actuated auxiliary pressure relief valve.

The primary difference between a pilot-operated pressure relief valve and a spring-loaded pressure relief valve is that a pilot-operated valve uses process pressure instead of a spring to keep the valve closed. A pilot is used to sense process pressure and to pressurize or vent the dome pressure chamber, which controls the valve opening or closing.

A pilot-operated pressure relief valve consists of: the main valve, a floating unbalanced piston assembly, and an external pilot. The pilot controls the pressure on the top-side of the main valve unbalanced moving

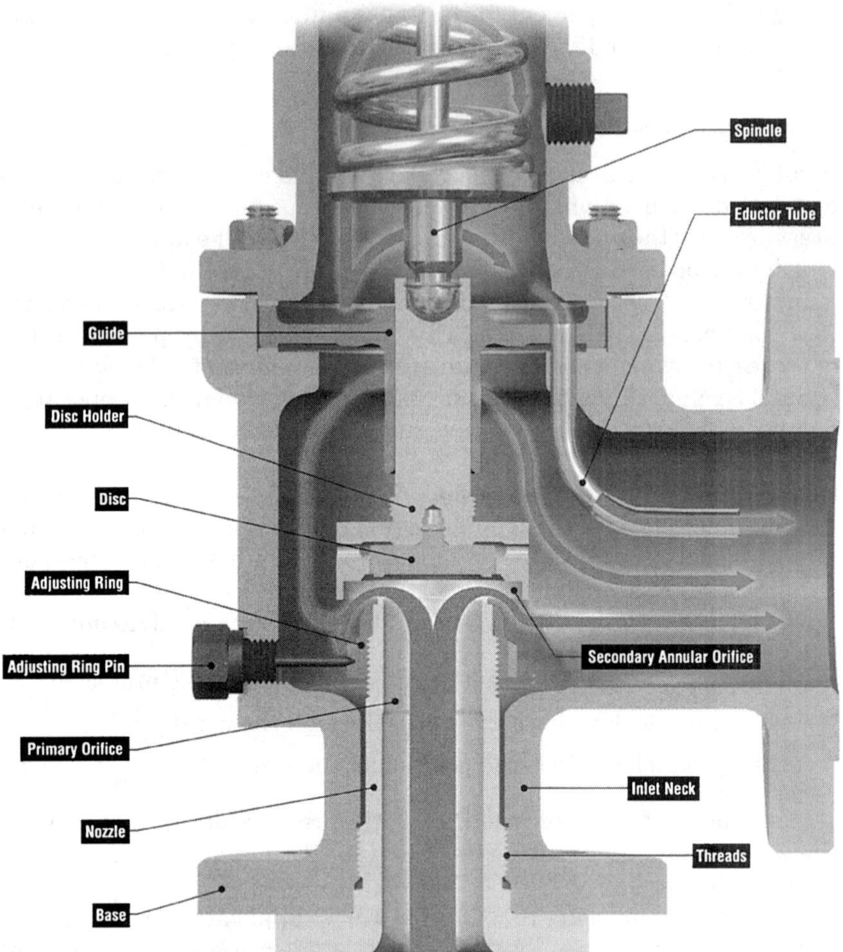

Figure 9.3 Conventional pressure relief valve. *(Courtesy: Dresser Flow Control)*

chamber. A resilient seat is normally attached to the lower end of this member.

- At pressures below the set pressure, the pressure on opposite sides of the moving members is equal.
- When the set pressure is reached, the pilot opens, depressurizes the cavity on the top-side and the unbalanced member moves upward, causing the main valve to relieve.
- When the process pressure decreases to a predetermined pressure, the pilot closes, the cavity above the piston is depressurized, and the main valve closes.

Balanced-bellows pressure relief valves. A balanced-bellows pressure relief valve is a spring-loaded safety valve, which incorporates a bellows or other means of balancing the valve disc to minimize the effects of back pressure on the performance characteristics of the valve. The term balanced means the set pressure of the valve is not affected by back pressure. Balanced pressure relief valves should be selected where the built-up back pressure is too high for a conventional relief valve.

Back pressure which occurs in the downstream system while the valve is closed, is called *superimposed back pressure*. This back pressure is the result of the valve outlet being connected to a pressurized system or caused by other pressure relief valves venting to a common header. Compensation for superimposed back pressure is provided by reducing the spring force. The force of the spring and the back pressure acting on the disc should be equal to the force of the inlet pressure acting to open the disc.

When superimposed back pressure is variable, a balanced pressure relief valve is recommended. The bellows are designed with an effective pressure area equal to the seat area of the disc. The bonnet is vented to ensure that the pressure area of the bellows will always be exposed to atmospheric pressure and to provide a telltale sign if the bellows begin to leak. Variations in back pressure will have no effect on set pressure. However, back pressure may affect the flow.

Back pressure, which occurs after the valve is open and flowing, is called *dynamic* or *built-up* back pressure. This type of back pressure is caused by the fluid flowing from the pressure relief valve located in the downstream piping system. Built-up back pressure does not affect the valve opening pressure, but will have an effect on valve lift and flow. On application of 10 percent overpressure, balanced-bellows designs are recommended when built-up back pressure is expected to exceed 10 percent of the cold differential test pressure (CDTP).

The bellows offsets the effects of variable back pressure, and seals the process fluid from escaping into the atmosphere and isolates the spring, bonnet, and guiding surfaces from contacting the process fluid.

Power-actuated pressure relief valves. A power-actuated pressure relief valve is that in which the main relieving device is combined with and controlled by a device requiring an external source of energy.

The power-actuated pressure relief valve is one whose function to open or close is fully controlled by a source of power such as electricity, air, steam, or hydraulic. The valve may discharge to the atmosphere or to a container at lower pressure. The discharge capacity may be affected by downstream conditions, and such effects should be taken into account.

If power-actuated pressure relieving valves are positioned in response to other control signals, the control impulse to prevent overpressure should be responsive only to pressure and should override any other control function.

Power-actuated valves are mostly used for forced-flow steam generators with no fixed steam and waterline. These valves are also used in nuclear power plants.

Temperature-actuated pressure relief valves. A temperature-actuated pressure relief valve is that which may be actuated by external or internal temperature or by pressure on the inlet side. This is also called a T&P safety relief valve.

The thermal sensing elements for this valve should be so designed and constructed that they will not fail in any manner which could obstruct flow passages or reduce capacities of the valve when elements are subjected to saturated steam temperature corresponding to capacity test pressure. T&P safety relief valves incorporating these elements should comply with a nationally recognized standard such as ANSI Z21.22, *Relief Valves for Hot Water Supply Systems*.

9.2 ASME Code Requirements

The required rules for pressure relieving devices have been prescribed in Article 4 of Section IV. These rules are applicable for steam boilers, hot water boilers (hot water heating and hot water supply), tanks, and heat exchangers.

9.2.1 Safety valve requirements for steam boilers

The safety valve should relieve all the steam generated by a steam-heating boiler. Each boiler shall have at least one or more officially rated safety valves that are identified with ASME Code Symbol **V.** The valves should be the spring-loaded, pop-type (Figure 9.4), adjusted and sealed to discharge all steam at a pressure not exceeding 15 psi (103 kPa). The size of the safety valve shall be a minimum of NPS 1/2 (DN 15) and a maximum of NPS 4 1/2 (DN 115).

The minimum capacity required by a safety valve can be determined by either of the following methods:

1. Determine maximum Btu output at the boiler nozzle and divide that output by 1000. This is applicable for a boiler heated by any type of fuel.
2. Determine minimum pounds (kg) of steam generated per hour per ft^2 (m^2) of boiler heating surface, as shown in Table 9.1.

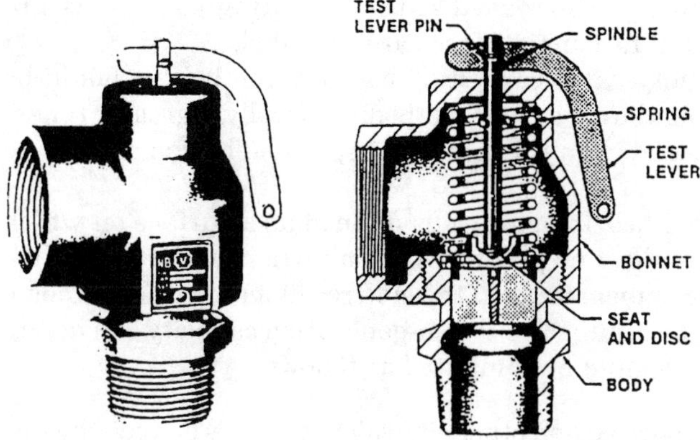

Figure 9.4 Spring-loaded pop safety valve for low-pressure steam boiler.

General notes

1. When a boiler is fired only by a gas having a heat value not more than 200 Btu/ft^3 (7400 KJ/m^3), the minimum safety valve or safety relief valve capacity may be based on the values as assigned for hand-fired boilers.
2. The minimum safety valve or safety relief valve capacity for electric boilers shall be 3.5 lb/hr/kW (1.6 kg/hr/kW) input.
3. The manufacturer may determine the minimum lb/hr/ft^2 for extended heating surfaces.

The safety valve capacity of each steam boiler should be such that the pressure cannot rise more than 5 psi (35 kPa) above the MAWP with the fuel burning equipment operated at maximum capacity. The safety

TABLE 9.1 Minimum Pounds of Steam per Hour (kg/hr) per Square Foot (m^2) of Heating Surface

Boiler Heating surface	Firetube boilers	Watertube boilers
Hand fired	5 (24)	6 (29)
Stoker fired	7 (34)	8 (39)
Oil, gas, or pulverized Fuel fired	8 (39)	10 (59)
Waterwall heating surface:		
Hand fired	8 (39)	8 (39)
Stoker fired	10 (49)	12 (59)
Oil, gas, or pulverized Fuel fired	14 (68)	16 (78)

valve capacity should be increased if the operating conditions are changed, or additional heating surfaces are installed.

The minimum safety valve capacity for a cast-iron boiler should be determined by the maximum output method. Generally, a greater relieving capacity is provided than the minimum specified by the rules.

Heating surface. The heating surface is defined as a surface on which there is water on one side and products of combustion on the other side. The heating surface is measured on the side receiving heat. This measurement is used to determine the steam-generation capacity of a boiler. The heating surface should be computed as follows:

- The boiler heating surface and other equivalent surfaces outside the furnace should be measured circumferentially, plus any extended surfaces.
- The waterwall heating surface and other equivalent surfaces within the furnace are measured as the projected tube area (diameter × length) plus any extended surfaces on the furnace side. Heating surfaces of the tubes, fireboxes, shells, tube sheets, and the projected area of headers are considered for this purpose.
- The manufacturer may determine the maximum designed generating capacity based on the total surface when the extended surfaces or fins are used. This generating capacity should be included in the total minimum relief-valve capacity marked on the stamping or nameplate.

9.2.2 Safety relief valve requirements for hot water boilers

There should be at least one safety relief valve (Figures 9.5 and 9.6), of the automatic-reseating type, for each hot water heating or supply boiler. The valve shall be identified with ASME Code Symbol **HV** and shall be set at or below the maximum allowable working pressure. The size of the safety relief valve shall not be less than NPS $^3/_4$ (DN 20) or more than NPS $4^1/_2$ (DN 115). A safety relief valve of size NPS $^1/_2$ (DN 15) may be used for a boiler with heat input not more than 15,000 Btu/hr (4.4 kW).

If the water temperature in hot water-heating or supply boilers is limited to 210°F (99°C), one or more T&P safety relief valves may be used in lieu of safety relief valves. Such T&P safety relief valves shall be ASME-rated with the **HV** symbol, of automatic-reseating type, and set at or below the maximum allowable working pressure.

When more than one safety relief valve is used for hot water boilers, the additional valves should also be ASME-rated. These valves should have a set pressure within a range not exceeding 6 psi (40 kPa) above the MAWP of the boiler up to 60 psi (400 kPa), and 5 percent more for boilers having a MAWP more than 60 psi (400 kPa).

Figure 9.5 Safety relief valve on a hot water boiler.

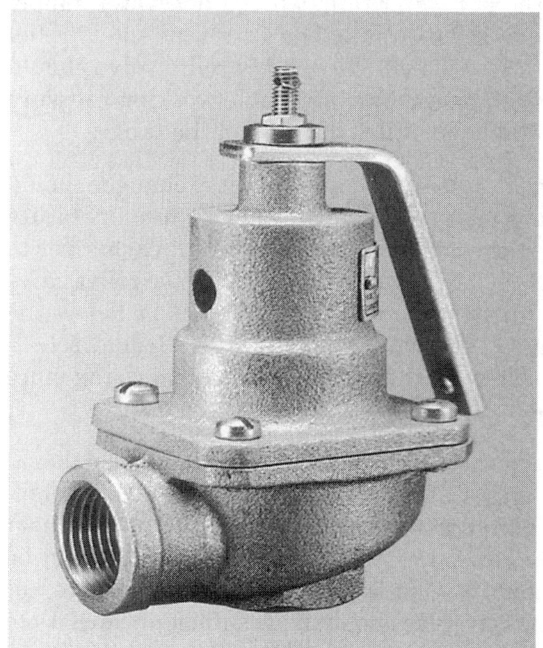

Figure 9.6 Bronze safety relief valve for hot water boiler.

The relieving capacity in pounds per hour (kg/hr) of the pressure relieving devices in a boiler should be greater than that determined by dividing the maximum output by 1000 in Btu at the boiler nozzle. Alternatively, the relieving capacity may be determined on the basis of pounds (kg) of steam generated per hour per ft^2 (m^2) of boiler heating surfaces as given in Table 9.1. The minimum relieving capacity for a cast-iron boiler shall be determined by the maximum output method.

When a single safety relief valve is used in a boiler, the relieving capacity should be such that pressure cannot rise more than 10 percent above the MAWP with the fuel-burning equipment operating at the maximum capacity. When more than one safety relief valve is used, the overpressure should be limited to 10 percent above the set pressure of the highest set valve.

9.2.3 Safety and safety relief valves for tanks and heat exchangers

When safety and safety relief valves are required for tanks and heat exchangers, the following three conditions may be considered: steam to hot water supply, high-temperature water to water-heat exchanger, and high-temperature water to steam-heat exchanger.

Steam to hot water supply. The pressure of steam should not exceed the safe working pressure of the hot water tank when a hot water supply is heated directly by steam in a coil or pipe. The size of the safety relief valve should be at least NPS 1 (DN 25). The safety relief valve should be set to relieve at or below the maximum allowable working pressure of the tank. The valve should be installed directly on the tank.

High-temperature water to water-heat exchanger. A heat exchanger should be equipped with one or more safety relief valves when high-temperature water is circulated through the coils or pipes of the heat exchanger to heat water for space heating or hot water supply. A safety relief valve should be ASME-rated with the symbol HV, and set at or below the MAWP of the heat exchanger. The valve should have sufficient relieving capacity to prevent the heat exchanger pressure from rising more than 10 percent above the MAWP of the vessel.

High-temperature water to steam heat exchanger. A heat exchanger should be equipped with one or more safety valves (Figure 9.7) when high-temperature water is circulated through the coils or tubes of the heat exchanger to generate low-pressure steam. A safety valve should be ASME-rated with the symbol V, and set to relieve at a pressure not exceeding 15 psi (100 kPa). The valve should have sufficient capacity to prevent the heat exchanger from rising more than 5 psi (35 kPa) above the MAWP of the vessel.

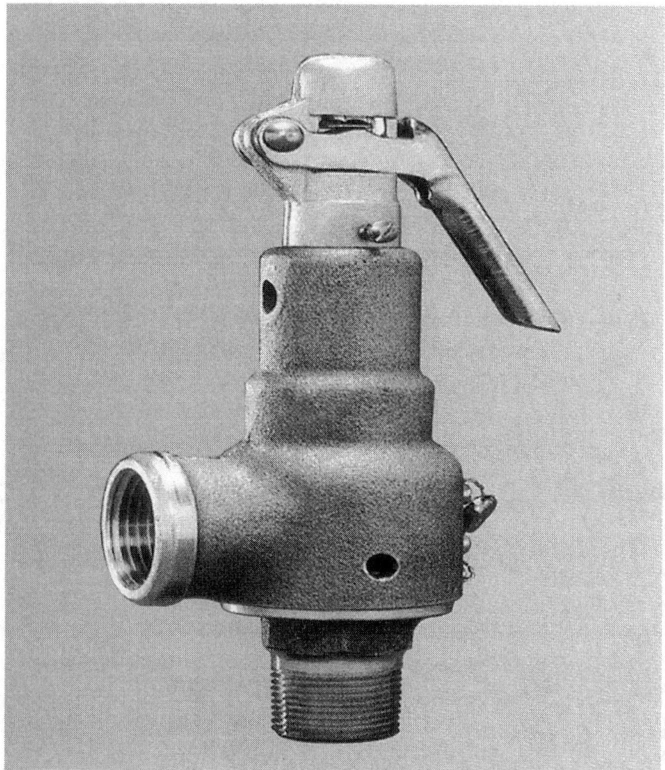

Figure 9.7 Bronze safety valve for steam heat exchanger.

9.2.4 T&P safety relief valves for hot water heaters

A water heater is designed in accordance with the Part HLW of Section IV. The requirements of safety relief valves have been specified in Article 8 of Part HLW. Each water heater shall have at least one T&P safety relief valve (Figures 9.8 and 9.9) or at least one safety relief valve. The valve shall be marked with the ASME Code Symbol **HV**. The minimum size of the valve not should be less than NPS $^{3}/_{4}$ (DN 20).

The pressure setting of a T&P pressure relief valve should be less than or equal to the MAWP of the water heater. If any components in the hot water system (such as expansion tanks, storage tanks, and piping) have lower working pressure than the water heater, the valve should be set at the pressure of the component with the lowest maximum allowable working pressure. If more than one valve is used, the additional valve may be set within a range not exceeding 10 percent over the set pressure of the first valve.

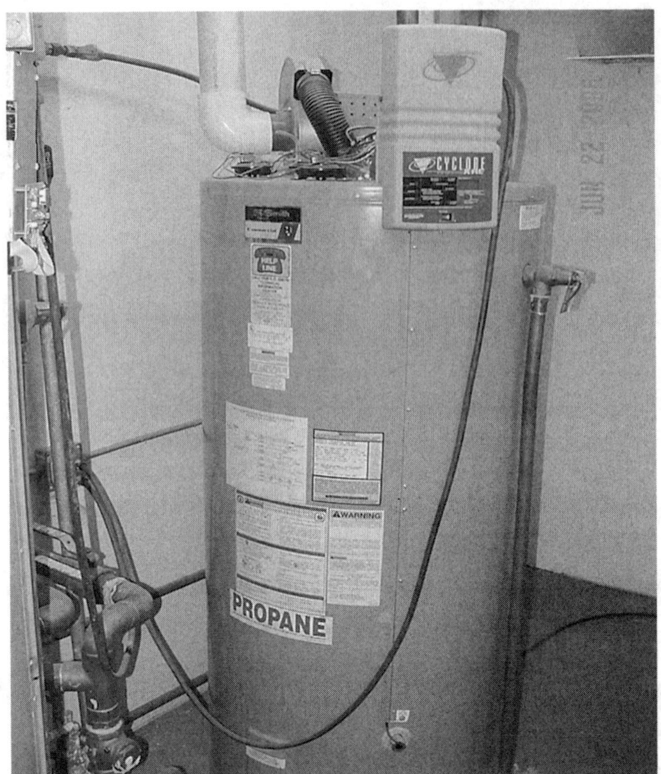

Figure 9.8 T&P installed on a hot water heater.

The relieving capacity in Btu/hr of the T&P safety relief valve should not be less than the maximum allowable input of the water heater. The relieving capacity for electric heater should be 3500 Btu/hr (1.0 kW) per kW of input. The T&P safety relief valve capacity for each water heater should be such that the pressure cannot rise more than 10 percent above the MAWP with the fuel-burning equipment operated at maximum capacity.

The T&P safety relief valves should be installed either by the installer or the manufacturer before a water heater is placed in operation.

9.2.5 Minimum requirements for safety and safety relief valves

Mechanical requirements. Safety valves shall be designed to have control blowdown of 2 to 4 psi (13.8 to 28 kPa). Safety valves shall be spring-loaded and the full-lift spring compression shall not exceed 80 percent of the solid deflection. The set pressure tolerance shall not exceed 2 psi

Figure 9.9 T&P safety relief valve for a hot water heater. *(Courtesy: Conbraco Industries, Inc.)*

(13.8 kPa) for the safety valve and 3 psi (20.6 kPa) for the safety relief valves, with the boiler pressure not exceeding 60 psig. The tolerance is five percent for pressures above 60 psig.

Material selection. Materials for the construction of valve bodies, bonnets, or pressure parts should conform to ASME Code Section II. The manufacturer can use materials other than the listed in ASME Code Section II, if specifications requiring equivalent chemical and physical properties are maintained. Materials used for seats and discs should withstand heat and provide resistance to steam cutting.

Manufacture and inspection. An ASME designee is required to inspect manufacturing, inspection, and test operations—including capacity. The manufacturing, production, test facilities, and quality control procedures are required to be acceptable by an ASME designee.

A manufacturer may obtain permission from ASME to produce pressure relief valves with ASME Code Symbol **HV**. The permission is issued for 5 years. In order to obtain permission, the manufacturer shall successfully demonstrate the following tests:

1. Two relief valves will be selected by an ASME designee from sample production.
2. An ASME authorized laboratory will conduct operational and capacity tests in the presence of an ASME designee.
3. In case of failure of capacity or performance test, the tests shall be repeated at the rate of two replacement valves for each valve failed.
4. If any of the replacement valves fails, the manufacturer's Code Symbol **HV** on the particular type of valve will be revoked.

Manufacturer's testing. A manufacturer shall have a well-established program for testing safety valves. Safety valves are tested on steam or air. Safety relief valves are tested on water, steam, or air. Every safety valve is tested to demonstrate its popping point, blowdown, and tightness. The tightness tests are very important for safety relief valves. Tightness tests are conducted at maximum operating pressure, not exceeding the reseating pressure.

Design requirements. A manufacturer is required to submit valves for capacity certification or testing by a test laboratory. At that time, the ASME designee has the authority to review the design of the valves to conform with Code requirements.

9.2.6 Discharge capacities of safety and safety relief valves

Valve markings. A manufacturer and/or an assembler shall posses a valid certification of authorization from ASME to apply the Code Symbol to each safety valve. Each safety valve is required to be marked with the data as per paragraph HG-402.1 and marking shall include the following:

1. The name or an acceptable abbreviation of the manufacturer
2. Manufacturer's design and type number
3. NPS size ________ inch (the nominal pipe size of the valve inlet)
4. Set pressure ____________ psi
5. Capacity ________ lb/hr, or capacity __________ Btu/hr
6. Year built
7. ASME Code Symbol (Figure 9.10)

9.2.7 Calculation of capacity to be stamped on valves

Capacity tests of the valves shall be conducted in the presence of, and certified by, an observer authorized by ASME. One of the following test methods is required to be used:

Figure 9.10 ASME Symbol for pressure relief valve. *(Courtesy: ASME Section IV)*

1. *Coefficient method.* This method based on coefficient calculation is used for safety valves. Three valves, three sizes of each (total nine valves), are tested. A coefficient of discharge is determined by the following formula:

   K_D = actual steam flow

   Theoretical steam flow K = average $K_D \times 0.90$ where K = coefficient of discharge for the design

   K_D = average coefficient of nine tests

   The stamped capacity will be determined by the following formulas: For 45 deg. seat,

   $$W = 51.5\pi DLP \times 0.707\ K$$

   For flat seat,

   $$W = 51.5\pi DLPK$$

   For nozzle,

where $W = 51.5\ APK$
W = weight of steam/hr, pound
D = seat diameter, inch
L = lift, inch
P = (1.10 × set pressure + 14.7 psia for hot water application
= (5.0 psi = 15 psi set + 14.7 psia for steam boilers)
A = nozzle-throat area, in.^{2}

2. *Slope method.* This method based on slope calculation is used for pressure relief valves. Four valves of each combination of pipe and orifice are used for the test. The slope is calculated by the following formula:

$$slope = w/p = \frac{measure\ capacity}{absolute\ flow\ pressure,\ psia}$$

The slope shall be calculated for each test point and average slope shall be determined.

$$\text{Minimum slope} = 0.95 \times \text{average slope}$$
$$\text{Maximum slope} = 1.05 \times \text{average slope}$$

Testing values should be between the minimum and maximum slope value range. The authorized observer may require additional valves tested if values are not within the range. The relieving capacity to be stamped is determined by the following:

$$\text{Stamped capacity} \leq \text{rated slope} \times (11.10 \times \text{set pressure} + 14.7 \text{ psia for hot water applications})$$

$$\text{Rated slope} = 0.90 \times \text{average slope}$$

3. *Three-valve method.* When one or more sizes of a design is set at a pressure, a manufacturer may submit three valves of each size, of each design, set at a pressure for testing. In that case, the stamped capacity should not exceed 90 percent of the average capacity of the three valves tested.

Capacity testing. Testing shall be conducted conforming to the requirements of ASME PTC 25-1994. The tests are required to be witnessed by an authorized observer of ASME. Data reports on the tests shall be signed by the manufacturer and the authorized observer. The testing facilities are subject to review by ASME within each five years. Safety valves are tested at 5 psi (35 kPa), over the set pressure. Safety relief valves are tested at 110 percent over the set pressure. The testing medium should be dry saturated steam at minimum 98 percent quality, and maximum 20°F (11°C). The capacity of a safety relief valve in terms of Btu can be calculated by multiplying the capacity in pounds per hour, W, by 1000.

Test record data sheet. A data sheet on the test record is prepared and signed by the ASME-authorized observer. A manufacturer will use that data sheet for construction and stamping of valves of the corresponding design and construction. New tests shall be conducted when design, affecting the flow path, lift, or performance characteristics, is changed.

9.3 Specifying Presure Relief Valves

9.3.1 Specifying a safety valve

Following are some typical specifications for a safety valve for a low-pressure steam-heating boiler:

Number of valve	1
Valve size	2 in. × 2 in.
Set pressure	15 psig
Required capacity	3150 lb/hr
Service	Steam
ASME Code	Section IV
Body	One piece, all bronze construction
Spring	Rust-proof steel spring
Disc	Nozzle-type, chrome-plated brass disc
O-ring	PTFE-coated O-ring for positive seal

9.3.2 Specifying a safety relief valve

The safety relief valve protects against excessive water pressure (from thermal expansion) and steam generated when Btu input controls fail. Following are some typical specifications for a safety relief valve for a hot water boiler:

Number of valve	1
Valve size	0.75 in. × 0.75 in.
Set pressure	150 psig
Maximum temperature	250°F
Required capacity	2,597,000 Btu/hr
Service	Hot water
ASME Code	Section IV
Body	Brass body with polished chrome finish
Spring	Stainless steel

Chapter

10

Boiler Construction

A heating boiler is constructed according to the drawings, designs, calculations, and specifications. The construction may be performed partly or completely in the manufacturer's boiler shop or in the field. A heating boiler is fabricated either by welding or brazing method in accordance with the rules of *ASME Boiler and Pressure Vessel Code Section IX*—Welding and Brazing Qualifications. Whatever may be the method, the construction procedures are required to be in compliance with the applicable rules of *ASME Code Section IV*—Rules for Construction of Heating Boilers.

The word *construction* is sometimes used instead of the word *fabrication*. Fabrication means assembly of the components of the boiler by using suitable methods. On the other hand, construction is an all-inclusive terms comprising materials, design, fabrication, examination, inspection, testing, stamping, and certification of a boiler.

This chapter describes construction procedures, except design in accordance with the *ASME Boiler and Pressure Vessel Code Section IV*. The chapter covers the following requirements:

- Materials
- Fabrication
- Inspection and testing
- Stamping and data reports

10.1 Material Requirements

Material requirements for construction of boilers are found in *ASME Code Section II*—Part A, B, C, and D. Most of the materials are covered in this section of the Code. Materials not covered by specifications in Section II are not allowed to use for construction.

Material subject to stress due to pressure shall be selected from the materials specifications of *ASME Code Section II*—Materials Part A, B, C, and D. If not found in *ASME Code Section II*, that materials shall not be used. However, such materials can be used after obtaining authorization from the Boiler and Pressure Vessel Committee of the American Society of Mechanical Engineers.

Material specifications in *ASME Code Section II* are not limited by method of production if product complies with the requirements of the specification. Materials exceeding the thickness limit of *ASME Code Section II* may be used if they comply with the other requirements of the specification and with thickness requirements of this code. Materials not identified by mill test reports may be used for nonpressure parts provided the failure of such parts will not affect the pressure parts to which they are attached.

Wrought materials and cast iron are also used for construction of boilers. The rules for construction of boilers by wrought materials are given in Part HF and the rules for cast iron boilers are given in Part HC.

10.1.1 Stress values

The maximum stress value for material is used for calculations of boiler design. The maximum allowable stress values for ferrous and nonferrous materials are given in Appendices B and C.

10.1.2 Minimum thickness

The minimum thickness is calculated on all pressure parts to make sure that materials can withstand the pressure for which they are designed.

Ferrous plates. The minimum thickness of any ferrous plate used under pressure shall be 1/4 inch (6 mm). If any pipe is used instead of plate, its thickness also shall be 1/4 inch (6 mm). The minimum allowable thickness of ferrous shell plates, tube sheets, or heads for various shell diameters is given in Table HF-301.1 of Section IV.

Nonferrous plates. The minimum thickness of any nonferrous plate under pressure shall be 1/8 inch (3.2 mm) for copper-admiralty, and 3/32 inch (2.4 mm) for copper-nickel. The minimum allowable thickness of nonferrous shell plates, tube sheets, or heads for various shell diameters are shown in Table HF-301.2 of Section IV.

10.2 Fabrication

Forming plates. The longitudinal joints of boiler shells are formed from plates. The ends of the plates are formed to the proper curvature by pressure, not by sledging.

Base metal preparation. The base metals have to be prepared for joints by machining, thermal cutting, chipping, grinding, or a combination of these methods prior to welding. The mechanical and metallurgical properties of the base metal should not be destroyed if a thermal-cutting method is used. Cast surfaces should be machined, chipped, or ground properly before welding. Surfaces should be free from all scale, rust, oil, grease, and other foreign materials before welding.

Assembly. Parts to be welded are properly fitted, aligned, and retained in the right position during the welding operation. Bars, jacks, clamps, or tack welds may be used to hold the edges of parts, and alignment tolerance should be maintained. The edges of the plates for butt joint should not be offset from each other more than the following amount shown in Table 10.1, where t is the plate thickness:

Tack welds may be used to secure the alignment. Tack welds should be removed completely or prepared properly by grinding or other suitable means so that they may be incorporated into the final weld. Tack welds should be made by a fillet weld or butt weld procedure qualified in accordance with *ASME Code Section IX*. Tack welds, if left in place, should be made by welders qualified in accordance with *ASME Code Section IX* and be inspected visually for defects.

Distortion. The cylinder of a drum or shell subjected to internal pressure should be circular at any section within a limit of one percent of the mean diameter. Suitable methods such as reheating, rerolling, or reforming may be used to make distorted cylinder circular. Cylindrical parts subjected to external pressure should be circular with maximum deviation of not more than 0.25 inch (6 mm) from a true circle.

10.2.1 Specific welding requirements

Longitudinal and circumferential joints. Finished butt-welded joints should have complete penetration and fusion. A longitudinal seam welding by

TABLE 10.1 Direction of Joints in Cylindrical Vessels

Plate thickness, in	Longitudinal	Circumferential
Up to 1/2, in.	1/4t	1/4t
Over 1/2 to 3/4, in.	1/8 in.	1/4t
Over 3/4 in.	1/8 in.	3/16 in.

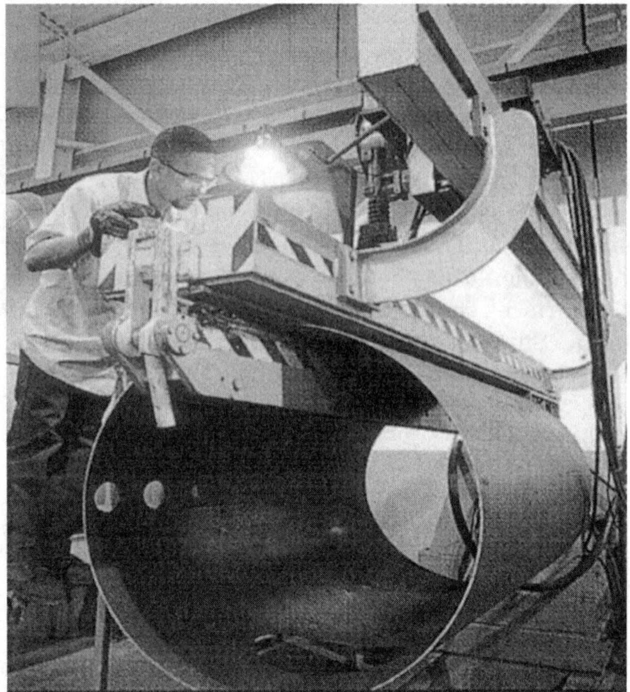

Figure 10.1 Submerged arc longitudinal seam welder. *(Courtesy: Lochinvar)*

a submerged arc welding machine to form a full-penetration weld on both the inner and outer surface of the shell is shown in Figure 10.1.

The joint surface may be left as welded if the weld is free of ripples, grooves, overlaps, abrupt ridges, or valleys. The welding process should not reduce the thickness more than 1/32 inch (0.8 mm) or 10 percent of the nominal thickness of the adjoining surface, whichever is less. For double-welded butt joints, the root will be prepared by suitable methods such as chipping or grinding before applying weld metal on the second side to be welded.

Fillet welds. Adequate penetration into the base metal at the root of the weld will be obtained while making a fillet weld. The thickness of the base metal will not be reduced due to the welding process.

Stud welding. There are two types of stud-welding processes: arc-stud welding and resistance-stud welding. If stud welding is used for attaching load-carrying studs, a production stud-welding test of the procedure and welding operator is performed on five studs, welded and tested in accordance with the requirements of *ASME Code Section IX*. Procedure and performance qualification tests are also made in accordance with *ASME Code Section IX*. If stud welding is used for joining nonpressure-bearing attachments to pressure parts by an automatic welding process in accordance with a welding procedure specification complying with *ASME Code Section IX*, procedure and performance qualification testing is not required. If stud welding is used to attach non–load-carrying studs, a production weld test is specified by the manufacturer and performed.

Welding by non–certificate holders. The manufacturer is the holder of the certificate of authorization. The manufacturer may use welders, including brazers, and welding and brazing operators not employed by him provided:

1. The manufacturer takes responsibility of all code construction.
2. All welders are qualified by the manufacturer in accordance with *ASME Code Section IX*.
3. All welding is performed in accordance with the manufacturer's *ASME Code Section IX*—Qualified Welding Procedure Specifications.
4. The manufacturer's quality control system includes supervision of welders, termination of welders, and assigning identification symbols to welders.
5. The manufacturer is responsible for code symbol stamping and completing data report forms.

Welding defects repairs. Welding defects such as cracks, pinholes, incomplete fusion, and any other defects shown during leakage tests should be repaired. The defects have to be removed by mechanical means or by thermal grooving processes. Once the defects have been removed, the joint will be rewelded and reexamined.

Post-hydro test. Sometimes nonpressure parts are required to be welded to the pressure parts after the final hydrostatic test. The following are the requirements if nonpressure parts, other than insulation attachment pins, are welded to the pressure parts after hydrostatic tests:

1. The welding is limited to P-No. 1 material.
2. The attachments will be done by stud welding or fillet welds and the throat is not to exceed the lesser of 1.5 times the thickness of the pressure part or 0.25 inch (6 mm).
3. The completed weld is inspected by the authorized inspector.
4. The Manufacturer's Data Report Form will be signed after completion of welding.

Inspection by the authorized inspector is not required if insulation attachment pins are stud-welded to pressure parts after hydrostatic testing. A production stud-weld test also is not required. Insulation attachment pins should have a diameter not more than 50 percent of the plate thickness or 3/16 inch (4.8 mm), whichever is less. The insulation attachment pins will be installed prior to applying the code symbol stamp and signing the Manufacturer's Data Report.

10.3 Inspection and Testing

A boiler is inspected and tested in the shop during construction in accordance with paragraph HG-515. A manufacturer is responsible for design, construction, and quality control in accordance with the requirements of *ASME Code Section IV*. An authorized inspector is responsible for inspection and testing of a boiler for compliance of the code. A boiler can be marked with ASME Code Symbol **H** only if all the requirements of *ASME Code Section IV* are complied with.

Manufacturer's responsibility. A manufacturer is required to manufacture a boiler in accordance with all the requirements of the Code. The manufacturer will follow the quality control system for the shop and cooperate with an authorized inspector at the various stages of construction to make an acceptable product under the Code. The manufacturer shall provide the following documents/information to the authorized inspector:

1. Certificate of authorization from the ASME for manufacturing the type of boiler
2. Design calculations and drawings for the boiler or part
3. Identifications of the materials used for construction
4. Manufacturer's Data Reports
5. Allow access for the authorized inspector to various areas of the shop
6. Keep the authorized inspector informed of the progress work.
7. Material identification, dimension check, and material test reports
8. Permission from the authorized inspector for correction of nonconformities
9. Qualification of the welding and/or brazing procedures
10. Qualification of welders, welding operators, or brazers used in production
11. Records of examination of parts prior to joining by welding or brazing
12. Records of examination of the parts for marking, surface defects, and dimensions

13. Records of hydrostatic tests, if applicable
14. Application of the required stamping and/or nameplate to the boiler
15. Preparing the Manufacturer's Data Reports, and
16. Retention of Manufacturer's Data Reports

In cases of multiple, duplicate boiler fabrication, the manufacturer will specify detailed procedures in the quality control manual. These procedures will be accepted by the inspection agency, jurisdiction, and ASME designee. The authorized inspector employed by the inspection agency will carry out the inspection procedures.

Authorized inspector's responsibility. An authorized inspector is employed by an authorized inspection agency accredited by ASME. The authorized inspection agency may be either a jurisdiction, or an insurance company, authorized to write boiler and pressure vessel insurance. The National Board of Boiler and Pressure Vessel Inspectors commissions these authorized inspectors. An authorized inspector is charged with the following responsibilities:

1. Check validity of the Manufacturer's Certificate of Authorization
2. Make sure manufacturer follows the quality control system accepted by ASME
3. Review design calculations, drawings, specifications, procedures, records, and test results
4. Check material to ensure it complies with the requirements of the Code
5. Check all the welding and brazing procedures to make sure they are in compliance with Code
6. Check qualification of welders, welding operators, brazers, and brazer operators
7. Check for proper joint factors for brazed joints
8. Make visual inspection for transfer-of-material identification
9. Witness the proof tests
10. Inspect each boiler and water heater during and after construction
11. Verify the stamping and/or nameplate information and attachment
12. Sign Manufacturer's Data Reports

The authorized inspector will make all necessary inspections to enable him or her to certify that the boilers have been designed and constructed in accordance with *ASME Code Section IV* and the code symbol stamp can be applied.

10.4 Stamping of Boilers

Each heating boiler designed and constructed in accordance with the rules of *ASME Code Section IV*—Boilers and Pressure Vessels is required to be stamped with the appropriate code symbol stamp. Stamping is done after the hydrostatic test, and in the presence of an authorized inspector.

10.4.1 Code symbol stamp

ASME has three code symbols such as **H**, **HLW**, and/or **HV** for the design, construction, and assembly of boilers and their parts in accordance with *ASME Code Section IV*. ASME will Issue a Certificate of Authorization and grant use of the code symbol stamps to organizations qualified as per paragraph HG-540. The Certificates of Authorization and the code symbol stamps are the property of ASME.

Application. Any organization interested in obtaining a certificate of authorization for one or more of the ASME Code Symbols will apply to the Society on prescribed forms. A separate application has to be submitted for each plant where code items will be built. An organization is required to sign an agreement with an authorized inspection agency for code inspection at each location. If approved, and the administrative fee paid, a Certificate of Authorization (Figure 10.2) and code symbol stamp will be issued to the qualified organization.

The certificate of authorization for the symbols **H**, **HLW**, and/or **HV** remains valid for 3 years; Certificate of Authorization for H (cast iron) remains valid for 1 year. An organization may apply to renew six months prior to the date of expiration of any such certificate.

Inspection agreement. A manufacturer (except for cast-iron heating boilers) is required to have an inspection agreement signed with an authorized inspection agency for providing inspection services. The agreement specifies the mutual responsibilities of the manufacturer and the authorized inspectors. The manufacturer will notify the Society if there is any cancellation or change to the inspection agreement. A manufacturer of pressure relief valves does not require to have an inspection agreement.

Quality control system. A quality control system is a written document to establish that all code requirements for production will be met. The quality control system describes authority and responsibility, organization, drawings, design calculations, and specification control, material control, examination and inspection program, correction of nonconformities, welding, calibration of measurement and test equipment, sample forms, and authorized inspector. An example of the mandatory quality control system is shown in Appendix F of *ASME Code Section IV.*

CERTIFICATE OF AUTHORIZATION

This certificate accredits the named company as authorized to use the indicated symbol of the American Society of Mechanical Engineers (ASME) for the scope of activity shown below in accordance with the applicable rules of the ASME Boiler and Pressure Vessel Code. The use of the Code symbol and the authority granted by this Certificate of Authorization are subject to the provisions of the agreement set forth in the application. Any construction stamped with this symbol shall have been built strictly in accordance with the provisions of the ASME Boiler and Pressure Vessel Code.

COMPANY:

RAYPAK, INC.
2151 EASTMAN AVE.
OXNARD, CALIFORNIA 93030

SCOPE:

HEATING BOILERS EXCEPT CAST IRON AT THE ABOVE LOCATION ONLY (THIS AUTHORIZATION DOES NOT COVER WELDING OR BRAZING)

The American Society of Mechanical Engineers

AUTHORIZED: **DECEMBER 30, 2003**
EXPIRES: **DECEMBER 16, 2006**
CERTIFICATE NUMBER: **3,502**

Chairman of The Boiler
And Pressure Vessel Committee

Director, Accreditation and Certification

Figure 10.2 ASME Certificate of Authorization. *(Courtesy: Raypak, Inc.)*

10.4.2 Stamping of boilers

Boilers except cast-iron boilers. Each boiler designed, constructed, and inspected in accordance to the rules in Part HG shall be stamped with Code Symbol Stamp **H** as shown in Figure 10.3.

The following data as shown in Figure 10.4 or 10.5 shall be stamped with letters a minimum of 5/16 inch high on the boiler proper, or on a name plate, minimum 3/64 inch thick, permanently fastened to the boiler:

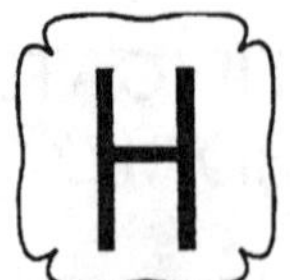

Figure 10.3 ASME Code Symbol Stamp H for heating boilers. *(Courtesy: ASME Section IV)*

FORM OF STAMPING ON COMPLETED BOILERS OR THEIR NAMEPLATES

(Not Applicable for Boilers Constructed Primarily of Cast Iron)

Certified by

H

____

(Name of Manufacturer)

[1]MAWP, Steam 15 psi

[1]MAWP, Water ____ psi

[1]Maximum Water Temp. ____ °F

[2]Heating surface ____ sq ft, boiler

[3]Heating surface ____ sq ft, water wall

[3]Heating surface ____ sq ft, extended

[3,4]Ext HS generating cap ____ lb/hr/sq ft

Minimum relief valve capacity ____ lb/hr or MBH

Manufacturer's serial no. ____________

[5]Year built ____________

GENERAL NOTE: Acceptable abbreviations to any of the stamp wording may be used.

NOTES:

(1) For steam only boilers, MAWP Water and Maximum Water Temperature markings are optional.

(2) Kilowatt power input for electric boilers.

(3) List each type of surface separately. May be omitted if type heating surface is not present.

(4) Generating capacity for extended heating surface [see HG-403(d)].

(5) May be omitted when year built is prefix to serial number (see HG-530.1).

Figure 10.4 Stamping on steam and water boilers. *(Courtesy: ASME Code Section IV)*

FORM OF STAMPING ON COMPLETED BOILERS OR THEIR NAMEPLATES
(Not Applicable for Boilers Constructed Primarily of Cast Iron)

H

Certified by

____
(Name of Manufacturer)
[1]MAWP, Water ____ psi
[1]Maximum Water Temp. ____ °F

[2]Heating surface ____ sq ft, boiler
[3]Heating surface ____ sq ft, water wall
[3]Heating surface ____ sq ft, extended
[3,4]Ext HS generating cap ____ lb/hr/sq ft
Minimum relief valve capacity ____ lb/hr or MBH
Manufacturer's serial no. ______________
[5]Year built ____________________

GENERAL NOTE: Acceptable abbreviations to any of the stamp wording may be used.

NOTES:

(1) For steam only boilers, MAWP Water and Maximum Water Temperature markings are optional.

(2) Kilowatt power input for electric boilers.

(3) List each type of surface separately. May be omitted if type heating surface is not present.

(4) Generating capacity for extended heating surface [see HG-403(d)].

(5) May be omitted when year built is prefix to serial number (see HG-530.1).

Figure 10.5 Boilers suitable for water only. *(Courtesy: ASME Code Section IV)*

1. The boiler manufacturer's name, preceded by the words "Certified by"
2. Maximum allowable working pressure
3. Safety or safety-relief valve capacity, lb/hr or MBH (i.e., thousands Btu/hr)
4. Heating surface, ft^2 (or kilowatt power input for electric boilers)
5. Manufacturer's serial number

6. Year built
7. Maximum water temperature

Stamping should be located in a conspicuous place so that it is easily visible. Stamp locations for different types of boilers can be found in paragraph HG-530.

Stamping of parts and accessories. Markings are required for the parts and accessories for which partial data reports have been furnished. The following markings are required on such parts and accessories:

1. The official code symbol shown in Figure 10.3 above the word "Part"
2. The part manufacturer's name
3. The part manufacturer's serial number

Field-assembled boilers. More than one party such as boiler unit manufacturer, authorized assembler and welder, and authorized inspector is involved in completing the unit in the field. The boiler unit manufacturer will complete a Manufacturer's Data Report Form H-2 or H-3. If a boiler is assembled by anyone other than the boiler unit manufacturer, the assembler will complete the data report as the "assembler" or "assembly organization" instead of "manufacturer." Welding on an assembly or subassemblies, if required, will be made by a heating boiler stamp holder. An authorized inspector will make inspections as required to certify that the boiler has been constructed in accordance with the Code. An assembler's **H** stamp will be applied in the field. A unit manufacturer's **H** stamp will be applied in the field only if the Certificate of Authorization indicates the stamp was applied in the field. The authorized inspector will sign the field inspection certificate portion of the data report.

10.5 Manufacturer's Data Reports

When a boiler or parts of a boiler is completed (Figure 10.6), data reports are prepared by the manufacturer and signed by the authorized inspector. There are three types of data reports:

Master data reports. A manufacturer of heating boilers to which are ASME Code Symbol Stamp **H** is applied is required to prepare data reports. A data report may include serial numbers of more than one identical boiler if completed, inspected, and stamped in a continuous 8-hour period. The following ASME forms are used for manufacturer's data reports:

Figure 10.6 Packaged firetube hot water supply boiler. *(Courtesy: York-Shipley Global)*

- H-2 Manufacturer's Data Report for All Types of Boilers Except Watertube and those Made of Cast Iron (Appendix H)
- H-3 Manufacturer's Data Report for Watertube Boilers (Appendix I)

The manufacturer will distribute a copy of the data report to the owner, the insurance agency, and the jurisdiction. The manufacturer is required to keep a copy of the data report on file for minimum 5 years or the boiler may be registered and the original data report may be filed with the National Board of Boiler and Pressure Vessel Inspectors.

Partial data report. A manufacturer may need to buy parts from another manufacturer to complete the boiler. In this case, the parts manufacturer will submit a partial data report in duplicate to the manufacturer of the finished boiler. A partial data report is completed on ASME form:

- H-4 Manufacturer's Partial Data Report

Supplementary data report. A supplementary sheet is used to record additional data when the data report form is insufficient. Additional data is recorded on the following form and attached to the Manufacturer's Data Report.

- H-6 Manufacturer's Data Report Supplementary Sheet

Chapter

11

Combustion Systems

Combustion of fuels is required to generate heat for heating boilers. A fuel system should be in place for supplying fuel in proper condition for the combustion process. The principal fuels used are gas, oil, coal, and electricity as a source of heat. The fuel system includes necessary devices for storing, heating, filtering, and supplying fuel to the boiler in proper condition for safe and efficient burning.

Combustion is a process of chemical combination of oxygen with fuel that results in the production of heat. Combustion occurs when elements in gas, oil, or coal combine with oxygen to produce heat. The process of combustion is designed to achieve complete combustion—the burning of all the fuel with a minimum amount of excess air. The highest combustion efficiency is obtained only from complete combustion. The combustion equipment includes a pilot, burner, and controls. This equipment should also reduce air pollution by controlling the NO_x formation during the combustion process.

The *ASME Boiler and Pressure Vessel Code Section VI*—Recommended Rules for the Care and Operation of Heating Boilers provides guidelines for fuels, fuel burning equipment, and controls. In addition, requirements for fuel trains, controls, and safety devices for the fuel burning system can be found in the ASME Code CSD-1.

11.1 Fuels

Fuels commonly used for combustion in boilers are gas, oil, coal, and wood products. These fuels act as a source of heat in boilers either to generate steam or hot water for space heating and hot water use. In addition, electricity and solar energy are used as fuels for the boilers. Whatever the type of fuel, heat energy from fuel is converted into thermal energy in the boilers.

Fuel selection, which was previously based on economics, in today's world is more likely to be governed by air pollution authorities. It is the state and local actions, along with the Federal Environmental Protection Agency (EPA), that directly impact boiler plants. The *Federal Clean Air Act, 1963* was amended in 1970, 1977, and 1990. The 1990 amendments consist of 11 titles, provisions of which have the potential to affect every source of air pollution.

11.1.1 Gas

Gas used for boilers may be in the form of *natural, manufactured, mixed,* or *liquefied petroleum gas. Natural gas* consists of methane (75 to 99 percent), ethane (5 to 23 percent) and a small percentage of other gases (0.008 to 11 percent). This type of gas is widely used because it burns clean and causes less pollution; it is popular due to its ease of use, low maintenance requirements and reasonable cost. *Manufactured gas* is a by-product of industrial processes and its heat content varies, based on its manufacturing. Blast furnace gas, coke-oven gas, and refinery gases are several examples of manufactured gas. *Mixed gas* is a combination of different types of gases and is used for specific purpose.

Heating value, which is expressed in British Thermal Units (Btu's), indicates how much heat a gas can produce. The heating values of various gas fuels in *Btu per cubic feet* are given in Table 11.1.

11.1.2 Liquefied petroleum gas (LPG)

Liquefied petroleum gas is obtained from hydrocarbon industries and is stored in a liquid state in tanks under high pressure. When this pressure is reduced, the liquid is changed to gas at the pressure for the burner. LPG should be stored and handled according to the regulations of local jurisdiction. Two LPGs, propane and butane, have the properties given in Table 11.2.

Because propane and butane are considered to be expensive for a primary fuel, their typical uses have been either as a secondary fuel, or primary in the areas where other fuels are not available. It is necessary to ensure that propane vaporizers are kept in good working condition

TABLE 11.1 Heating Values of Gas Fuels

	Low	High
Natural gas	950	1150
Manufactured gas	350	600
Mixed gas	600	800

TABLE 11.2 Properties of Propane and Butane

	Propane	Butane
Boiling temp	–40°F	32°F
Heat value	2500 to 3260 Btu/cu ft	2500 to 3260 Btu/cu ft
Specific gravity	1.5 to 2.01	1.5 to 2.01

so that liquid propane does not enter a burner, which is typically designed to burn gaseous propane. Also, consult the burner manufacturer prior to changing from one gaseous fuel to another. Special precautions are necessary when using LPG as a fuel, due to the fact that it is heavier than air. It is particularly important to avoid installing propane-system piping or propane-fired equipment below grade.

11.1.3 Fuel oils

Fuel oils are classified by their *viscosity.* The viscosity of an oil is its internal resistance to flow and is related to the number of seconds required for a specific amount of oil, at a given temperature, to pass through an orifice. This property is important when considering pumping requirements. The unit of viscosity is seconds saybolt universal (SSU) for light oils and seconds saybolts furol (SSF) for heavy oils.

Besides viscosity, other important properties of fuel oils are specific gravity, flash point, and pour point. Specific gravity is the relationship of the density of the oil to the density of water at standard temperature.

Specific gravity of fuel oils can be determined by the following formula:

$$\text{Specific gravity} = \frac{\text{lb/gal oil at 60°F}}{\text{8.34 lb/gal water}}$$

or

$$\text{Specific gravity } 60°/60°\text{F} = \frac{141.5}{131.5 + °\text{API}}$$

where API means American Petroleum Institute. When a hydrometer reads specific gravity in degrees, the American Petroleum Institute degree (°API) are used to measure specific gravity. If °API is known, the following formula is used to determine approximate heating value of fuel oils:

$$\text{Heating value, Btu/lb} = 17{,}780 + (54 \times °\text{API})$$

Properties of fuel oils determine their grade, classification, and suitability for specific uses. Table 11.3 shows the properties of fuel oils.

TABLE 11.3 Properties of Fuel Oils

	Grade No.1	Grade No.2	Grade No.4	Grade No. 5	Grade No.6
Color	Light	Amber	Black	Black	Black
°API 60°F	40	32	21	17	12
Specific gravity, (60/60°)	0.8215	0.8654	0.9279	0.9529	0.9861
Viscosity, SSU	31	35	77	232	...
Viscosity, SSF	...	...	...	...	170
Pour point	<0	<0	10	30	65
Temp. for atomizing, °F Atmospheric	Atmospheric	25 (min.)	130	200	
lb/gal, 60/60°F	6.870	7.206	7.727	7.935	8.212
Btu/lb	19,850	19,500	19,100	18,950	187,500

Since fuel oils are prepared for combustion in low-pressure boiler burners by atomization, the oil delivered to the burner should be preheated to the proper temperature for atomization.

11.1.4 Coal

Coal is widely used in the electric utility industry for steam generation. The conditions under which coal is formed determine the properties of coal. Based on its rank, which is the degree of hardness, coal is mainly classified into three types: *anthracite, bituminous,* and *lignite.*

Anthracite coal is commonly known as hard coal and is shinny black in color. Semi-anthracite is not as hard as anthracite and is gray in color. *Bituminous* coal is known as soft coal and is widely used in the utility industry. These coals ignite easily and burn with a high flame. Sub-bituminous coal is also a soft coal, but it has a lower carbon content than bituminous coal. *Lignite coal,* which has a low heating value and a high moisture content, is rarely used as a boiler fuel.

A coal analysis is necessary to determine its suitability as a fuel for a particular application. Either an aproximate or an ultimate analysis is required to identify the characteristics of a particular type of coal. Aproximate analysis is used to determine the moisture, volatile matter, ash, fixed carbon, and sulfur content in a coal specimen. An ultimate analysis is used to determine elements such as nitrogen, oxygen, carbon, ash, sulfur, and hydrogen in a coal specimen. The amount of each element present in the coal specimen determines the heating value or calorific value, which is expressed in Btu/lb. In fact, a bomb calorimeter measures the heating value of coal.

Coal analyses are performed in three different ways, i.e., as received or as fired, moisture-free or dry coal, and moisture- and ash-free or combustible. The analysis, which includes actual moisture, is referred to *as received.* This moisture is removed in moisture-free analysis. When the moisture and ash are not included, the analysis is called *combustible analysis.*

11.1.5 Electricity

Electricity is used as a source of heat for heating boilers. The electricity is applied for generating heat for either electrode or immersed direct-resistance, element–type boilers. In an electrode-type boiler, heat is generated by the passage of an electric current using water as the conductor, whereas, in immersion-resistance element–type boiler, heat is generated by the passage of an electric current through a resistance heating element immersed in water.

11.2 Fuel-Burning Equipment

Fuel burning equipment is used to burn the fuel efficiently and release the heat energy in fuel. The main fuel-burning equipment is the burner, which passes fuel and air to the furnace for safe and efficient combustion, after a prepurge period designed to ensure that any residual combustibles are removed from the furnace, breeching, and stack prior to ignition.

A *pilot* is used to ignite the flame prior to the burner ignition. The basic ignition sequence is that first a pilot flame is lit, it is detected by a flame sensor, and then the main flame is ignited. Pilots are classified into the types based on the time a pilot operates with respect to the main burner. The types used are *continuous, intermittent,* and *interrupted.* A continuous pilot stays lit all the time, an intermittent pilot is lit prior to the main burner ignition and stays lit while the burner is on, and an interrupted pilot continues to be lit for a sufficient length of time so that the main flame burner ignites and then goes off. Most of the pilots use natural gas or propane, although oil may be used in some pilot flame ignition.

The kind of fuel used determines the fuel-burning equipment. For instance, the main types of fuel-burning equipment are *gas burning equipment, oil burning equipment,* and *coal burning equipment.* Moreover, operating controls, limit controls, safety controls, and programming controls are required for controlling functions of the fuel-burning equipment.

11.2.1 Gas-burning equipment

Gas burners are used to supply the proper mixture of gas and air to the furnace for complete combustion. A typical gas fuel train for input greater than 2,500,000 Btu/hr and less than or equal to 5,000,000 Btu/hr, and according to the requirements of ASME Code CSD-1 is

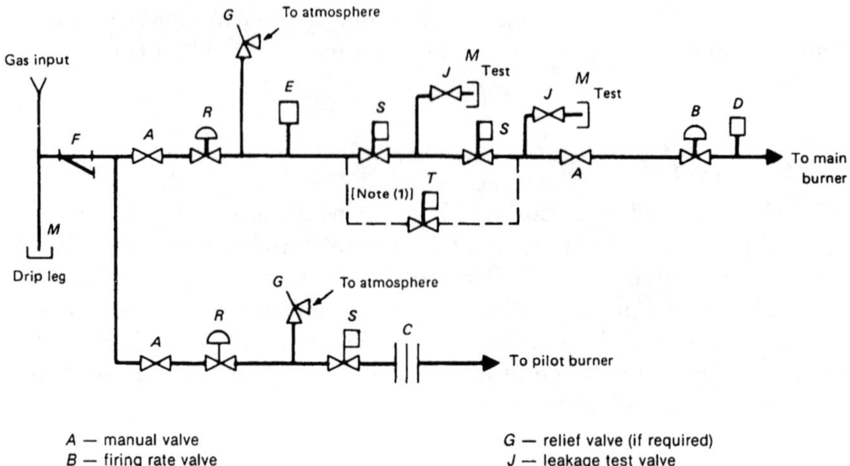

A — manual valve
B — firing rate valve
C — orifice
D — high gas pressure switch
E — low gas pressure switch
F — gas filter or strainer (if required)
G — relief valve (if required)
J — leakage test valve
M — pipe cap
R — pressure regulator
S — safety shutoff valve
T — safety shutoff valve with proof of closure

GENERAL NOTE: Since boiler design may vary, American National Standard for Gas-Fired Low-Pressure Steam and Hot-Water Boilers, ANSI Z21.13, does not contain a typical fuel train; however, through laboratory testing procedures, ANSI Z21.13 does determine that safe lighting of the boiler will be accomplished; this Standard illustrates a typical fuel train for boilers. The specific fuel train diagram for boilers complying with ANSI Z21.13 is supplied in the boiler manufacturer's instructions.

NOTE:
(1) Alternate arrangement — *T* may be used in place of two *S* type valves.

Figure 11.1 Typical fuel gas train. (*Courtesy: ASME Code CSD-1*)

shown in Figure 11.1. The type of gas burner may be selected on the basis of gas pressure available. There are two types of gas burners: atmospheric and power burners.

Atmospheric gas burners. Atmospheric burners with a venturi tube use the energy in the gas to aspirate primary air. This premixed blend is delivered to the flame zone. Once the flame is established, secondary air is drawn from the surrounding space to complete the combustion. These types of burners operate at higher excess air than other types of burners. Atmospheric burners depend upon natural draft for combustion air (Figure 11.2).

Power gas burners. In power burners, air is forced to mix with the gas by using a fan to maintain suitable draft for complete combustion. Normally, either an induced draft fan or a forced draft fan is used to maintain proper draft condition in the furnace. Based on the type of draft, power burners are classified as induced draft and forced draft.

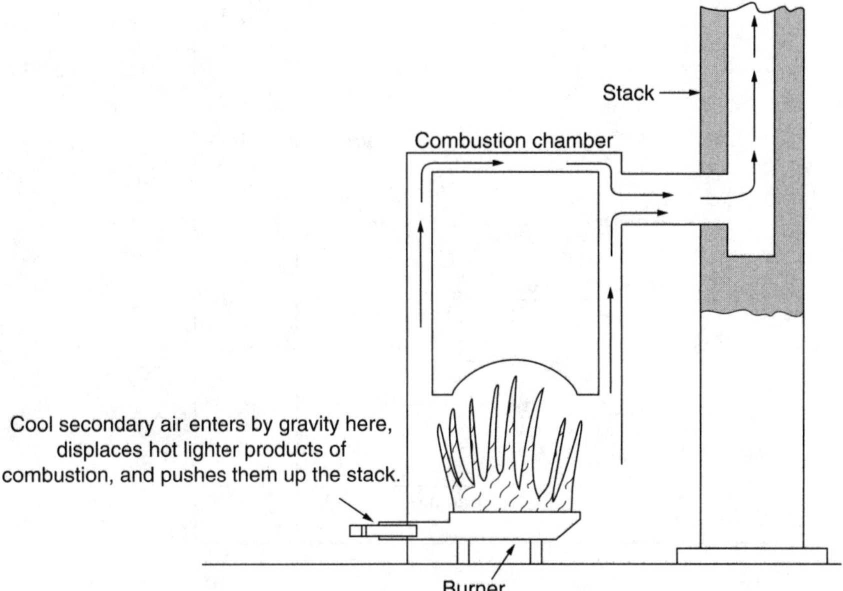

Figure 11.2 Atmospheric gas burner.

1. *Induced draft burners.* These types of burners operate with a furnace pressure slightly less than atmospheric pressure so that surrounding air can flow into the boiler. Once combustion takes place, the gases of combustion flow into the chimney and then rise to the atmosphere. An induced draft fan is used to maintain proper draft condition in the furnace. An induced draft burner is shown in Figure 11.3.
2. *Forced draft burners.* These types of burners are designed to operate with a furnace pressure higher than the atmospheric pressure. Fans or blowers are used to produce the flow of air for combustion.

Combination gas-fuel burners. Combination fuel burners are used to burn more than one fuel in a burner. Generally, the burners are designed for burning gas (natural gas or liquefied petroleum gas) in combination with any grades of oils, with provisions to switch over from one fuel to another. A combination burner is basically the same as connecting a gas burner with a fuel oil burner capable of switching over from one fuel to another if necessary. A combination gas-and-oil burner is shown in Figures 11.4.

11.2.2 Oil-burning equipment

Fuel oil burners are used to mix fuel oil with air mechanically so that a fine spray form is delivered to the furnace for efficient combustion. As a

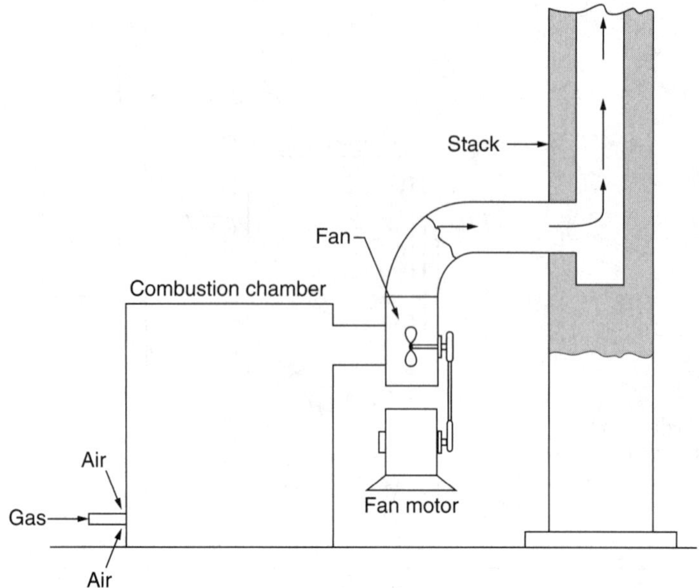

Figure 11.3 Induced draft burner.

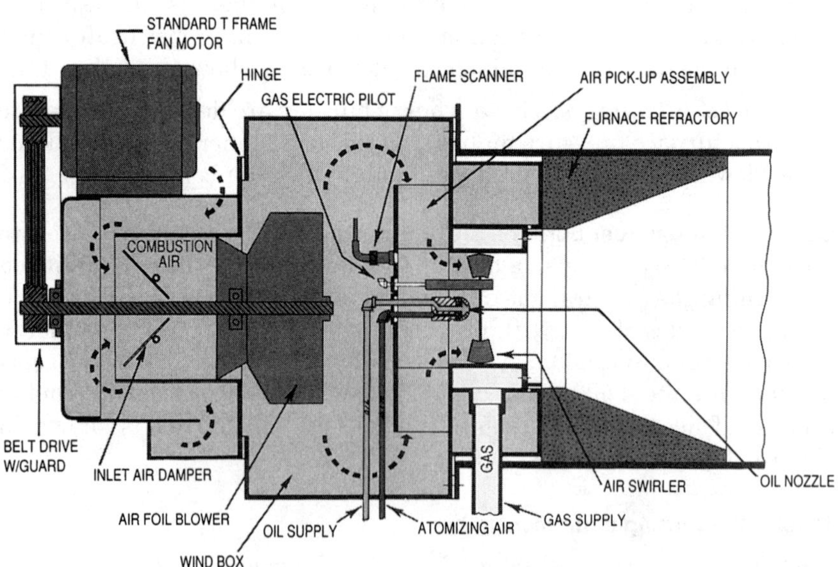

Figure 11.4 Combination gas and oil burner. *(Courtesy: Johnston Boiler Company)*

Figure 11.5 Fuel oil burner. *(Courtesy: Gordon-Piatt Energy Group, Inc)*

result of forcing fuel oil through a nozzle by using air or steam, atomization of fuel oil is accomplished. A fuel oil burner is shown in Figure 11.5.

Based on the type of atomization of the fuel, burners are classified as pressure-atomizing burners and air-atomizing burners:

Pressure-atomizing burners. In these types of burners, fuel oil is pumped into the burner and atomization of fuel is obtained without using air or steam. Pressure-atomizing burners are generally classified into two types: high-pressure mechanical atomization burners and low-pressure mechanical atomization burners.

1. *High-pressure mechanical atomization burner.* This type of burner is characterized by an air tube and a pressurized oil supply so that a spray of atomized oil is mixed with the air stream emerging from the air tube. The oil is supplied to the burner at or above 100 psig. A typical arrangement is shown in Figure 11.6. Electric ignition is obtained by a high-voltage transformer located above the nozzle. A gas pilot is used for gas ignition on a larger burner.
2. *Low-pressure mechanical atomizing burner.* This type of burner uses low-pressure atomizing for supplying a mixture of oil and air to the burner nozzle. The air pressure before mixing is in the range of 1 to 15 psig, and varies between 2 to 7 psig for an air-oil mixture.

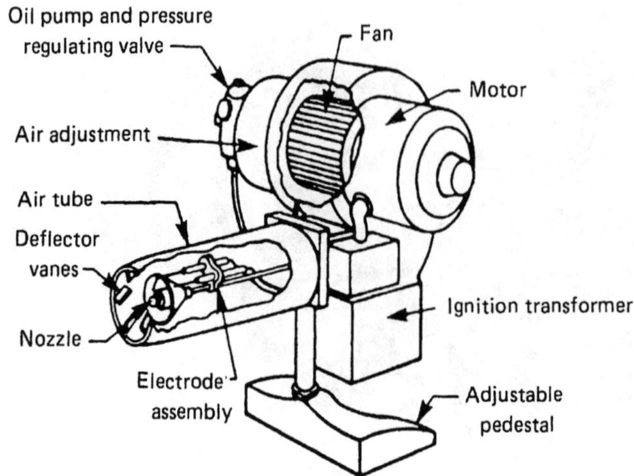

Figure 11.6 High-pressure atomizing burner. *(Courtesy: ASME Section VI)*

Air-atomizing burners. Air-atomizing burners are the same as the steam atomizing burners except that compressed air, instead of steam, is used for atomization of fuel oil. Atomizing air at the correct pressure is delivered by an air compressor. These types of burners may be used for burning all grades of fuel oils.

11.2.3 Coal-burning equipment

Coal is introduced into the furnace either by manual- or hand-firing, or by mechanical devices. However, the hand-firing method is obsolete now and has been almost completely replaced by mechanical devices like stokers. A stoker is a device consisting of a mechanically operated fuel-feeding mechanism and a grate. The stoker feeds solid fuel into a furnace and distributes it over the grate. It admits air to the fuel for the purpose of combustion, and provides a means for removal or discharge of refuse. The most common types of stokers are underfeed stokers, spreader stokers, and chain-grate stokers.

Underfeed stokers. In these types of stokers, coal is fed in from the bottom of the burning fuel bed. Coal is loaded into a hopper and delivered into the retort. There a distributing bar with pusher blocks distributes the coal back and up. The coal is delivered by a piston-type ram, and the lateral movement of the bar slowly moves the ash toward the dump grates. Single retort stokers are used in small boilers, but

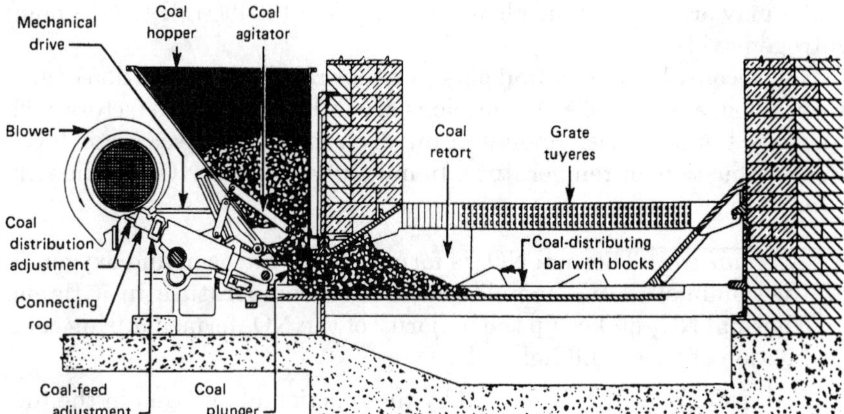

Figure 11.7 Underfeed single retort stoker.

multiple retort stokers are used in large boilers to obtain high rates of combustion. A typical underfeed stoker is shown in Figure 11.7.

Spreader stokers. Spreader stokers are generally designed for burning small size coal. In this type of stoker, some of the coal is burned in suspension and the remainder falls into the grate, where final combustion takes place. The stoker installation consists of a variable feeding device, a mechanism for distributing the coal into the furnace, and grates with suitable openings for admitting air. The hopper delivers coal to the conveyor feeder, then to the distributor, which distributes coal to be burned in suspension or on the grate. Air is supplied both from under the grate and over the fire for complete combustion.

Chain-grate stokers. These types of stokers use a moving grate to feed coal to the furnace. An endless chain travels over the two sprockets, one at the front and the other at the rear of the furnace. Coal is fed from the hopper, deposited on the front of the grate, travels past an ignition point, and begins combustion. Sufficient air is introduced through windboxes under the grates to complete combustion before the end of the grate is reached.

11.3 NO_x Control

The nitrogen pollutants generated by boilers are nitric oxide (NO) and nitrogen dioxide (NO_2), collectively referred to as NO_x. Ninety-five percent of NO_x produced during combustion is NO. Once it is emitted into the atmosphere, NO reacts to form NO_2, It is NO_2, which reacts with other pollutants to form ozone. A typical natural gas–fired industrial boiler may produce uncontrolled emissions of 0.133 lb/MMBtu (112 ppm) nitrogen oxides.

NO_x is considered toxic and plays a major role in the formation of acid rain, smog, and ozone. NO_x emissions are influenced by the factors such as flame temperature, amount of nitrogen in the fuel, excess air level, and combustion air temperature. In industrial boilers, NO_x is primarily formed in two ways:

1. *Thermal NO_x.* Thermal NO_x is formed when nitrogen and oxygen in the combustion air combine at a high temperature in a flame. Thermal NO_x makes up the majority of the NO_x formed during combustion of gases and light oils.
2. *Fuel NO_x.* Fuel NO_x is formed by the reaction of nitrogen in the fuel with oxygen in the combustion air. It is not a serious problem for gaseous fuels. In oils containing nitrogen, fuel NO_x can go up to 50 percent of the total NO_x emissions. Fuel NO_x in commercial and industrial boilers is economically reduced by switching to cleaner fuels.

11.3.1 NO_x control technologies

By controlling NO_x levels, along with other primary pollutants, the levels of acid rain and ozone can be reduced. NO_x controls can be classified into two types: combustion control techniques, and post-combustion methods.

Combustion control techniques. Combustion control techniques reduce the amount of NO_x emission by controlling the NO_x formation during the combustion process. These techniques are more economical than post-combustion techniques and are used on industrial boilers. The techniques are:

1. *Fuel selection.* A coal with low fuel nitrogen (less than 1.5%) may eliminate the need for any NO_x reduction techniques. Natural gas has no nitrogen in the fuel. Fuel oil typically has lower nitrogen than coal. A fuel with low nitrogen content should be selected for a boiler with a high temperature furnace.
2. *Low excess air combustion.* Boilers are fired with excess air for complete combustion. But high excess air levels (>45%) may increase NO_x formation, because excess nitrogen and oxygen in the air entering the flame will combine to form NO_x. Limiting the excess air entering a flame may reduce NO_x reduction by 5 to 10 percent when firing natural gas.
3. *Burner selection.* Selecting a burner relative to the furnace side limits the oxygen availability to form NO_x while simultaneously creating a

larger flame. Modern low-NO_x forced-draft combustion systems are available for power boilers (Figure 11.8).

This is a factory packaged and tested low NO_x combustion system, which incorporates both flue gas recirculation (FGR) and secondary gas combustion (SGC) to reduce NO_x emission levels. The low NO_x adapter can be applied directly to the burner. It is possible to reduce NO_x in excess of 60 percent on gas and 40 percent on oil. Similarly, burners may be modified to achieve low NO_x levels.

4. *Water or steam injection.* Water or steam may be introduced into the flame to reduce the flame temperature, thereby reducing thermal NO_x formation. Water or steam injection can reduce NO_x up to 80 percent when firing natural gas. Many times, water or steam injection is used in conjunction with other NO_x reduction techniques.
5. *Flue gas recirculation.* Flue gas reduction (FGR) is the most effective method of NO_x reduction from industrial boilers. FGR is the method of recirculating a portion of relatively cool exhaust gases back to into the combustion zone to lower the flame temperature. It is currently the most effective and popular technology for firetube and watertube boilers. FGR method can reduce NO_x 80 percent for natural gas

Figure 11.8 NO_x reduction burner for gaseous fuels. *(Courtesy: Gordon-Piatt Energy Group, Inc.)*

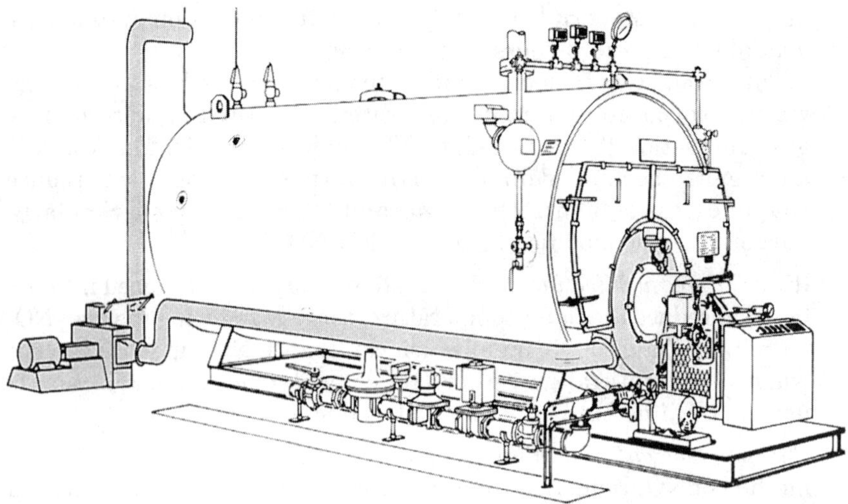

Figure 11.9 Flue gas circulation system for NO_x reduction. *(Courtesy: Gordon–Piatt Energy Group, Inc.)*

and 20 to 25 percent for fuel oils. A flue gas circulation system for NO_x reduction is shown in Figure 11.9.

Post-combustion methods. Post-combustion methods address NO_x emissions after formation of NO_x during the combustion process. Post-combustion methods are more expensive than combustion control techniques and are generally used on boilers with inputs more than 100 MMBtu/hr. The following post-combustion methods are used:

1. *Selective noncatalytic reduction.* This method involves the injection of a NO_x reducing agent like ammonia or urea, in the boiler exhaust gases at a temperature about 1400 to 1600°F. The ammonia or urea breaks down the NO_x in the exhaust gases into water and atmospheric nitrogen. This method can reduce NO_x up to 50 percent if applied properly.
2. *Selective catalyst reduction.* This method involves the injection of ammonia in the boiler exhaust gases in the presence of a catalyst. The catalyst allows the ammonia to reduce NO_x levels at lower exhaust temperatures than selective noncatalytic reduction. Selective catalytic reduction can be used at exhaust gas temperatures between 500 to 1200°F. This method can result in NO_x reduction up to 90 percent if applied properly.

Chapter

12

Boiler Appurtenances

Appurtenances are defined as devices that are attached directly to a boiler so that the boiler can be operated safely and efficiently. These devices are also known as fittings (trim) and mountings. Examples of appurtenances are safety valves, pressure gages, gage glass, blowdown valves, stop valves on the main steam line, and the like.

The appurtenances warrant careful attention and necessary maintenance to keep them in good working condition. *ASME Boiler and Pressure Vessel Code Section VI*—Recommended Rules for the Care and Operation of Heating Boilers provides useful information about these devices.

12.1 Safety Valves and Safety Relief Valves

The safety or safety relief valve(s) is mounted to prevent the pressure in the boiler from exceeding its maximum allowable working pressure (MAWP). This is the final mechanical device protecting the boiler against overpressure and is considered the most important valve on the boiler. This is so important that Chapter 9 has been devoted to pressure relief valves.

12.2 Low-Water Fuel Cutoffs and Water Feeders

A low-water fuel cutoff is the primary low water control, and is provided to shut off the burner in case of low water level in a boiler. It is located slightly below the normal operating water level (NOWL).

An auxiliary low-water cutoff is wired in series with the primary low-water cutoff and is located in such a manner that it will de-energize the limit circuit at a level slightly below the primary low-water cutoff. The purpose of an auxiliary cutoff is to act as a backup safety control in case the primary low-water cutoff fails to shut off the burner. There are generally two types of low water controls, i.e., float-operated or probe- operated controls set to de-energize the burner limit circuit and shut down the burner when a low-water condition in a boiler is detected.

12.2.1 Float-type low-water fuel cutoffs

A float-type cutoff actuates a switch or several switches that are designed to perform certain function(s). If the water level drops below the safe operating level, the first function is to de-energize the burner control circuit and this will shut down the burner. Another function will sound the low-water alarm when the burner is shut down. Also, a pump on-off control may start and stop the boiler feed pump to maintain a safe water level. In addition, a high-water alarm control function may be added to sound an alarm when water level exceeds a predetermined level. A typical float-type low-water cutoff is shown in Fig. 12.1.

A float-type low-water cutoff in combination with a water feeder is used on steam boilers. The cutoff when constructed as a separate unit, may be installed either on hot water boilers or used as a second cutoff on steam boilers.

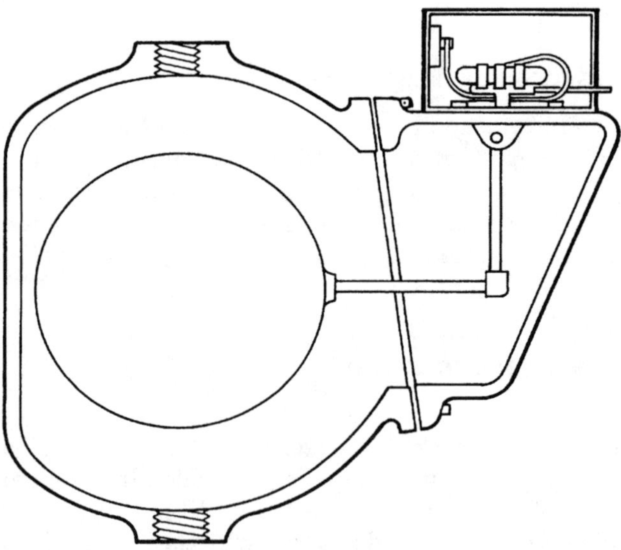

Figure 12.1 Float-type low-water cutoff. *(Courtesy: ASME Code Section VI)*

12.2.2 Electric probe-type low-water fuel cutoffs

The probe-type control performs the same functions as the float-type control, but the probe depends on the conductivity of water to complete the circuit. Metal probes are suspended in a chamber or directly in the boiler shell and a low current utilizes the conductivity of water to complete a circuit and energize a control relay. When water levels drop below the end of the probe (electrode), the circuit is de-energized and the control relay is de-energized. In cases of high water, the probe circuit is energized and the control relay is energized. A typical electric probe-type cutoff is shown in Figure 12.2.

12.3 Steam Traps

Steam traps are automatic devices that are used to remove condensate and air from the steam lines without loss of steam. If condensate is not removed from the steam lines, it could result in water hammer, which leads to rupture. The main steam headers, ends of the main steam branch lines, bases of steam risers, radiators, heat exchangers, and steam jackets require steam traps to remove condensate. A steam strainer should be installed at the trap inlet to prevent scale or other solid particles from entering the trap.

The various types of steam traps generally used are thermostatic steam traps with a bellows element; thermostatic steam traps with a bimetallic element; inverted-bucket steam traps-float and thermostatic steams trap; and thermodynamic disc steam traps.

12.3.1 Thermostatic steam trap with a bellows element

A thermostatic trap uses temperature differential to distinguish between condensate and live steam. A bellows trap (Figure 12.3) includes a valve element that expands and contracts in response to temperature changes. A volatile chemical such as alcohol or water is used inside the element. Evaporation provides the force to change the position of the valve. During start-up, the trap is open due to the relative cold condition. This operating condition permits air to escape and provides maximum condensate removal when the load is the highest.

12.3.2 Thermostatic steam trap with a bimetallic element

This steam trap depends on the bending of a composite strip of two dissimilar metals to open and close a valve. Air and condensate pass through the valve until the temperature of the bimetallic element

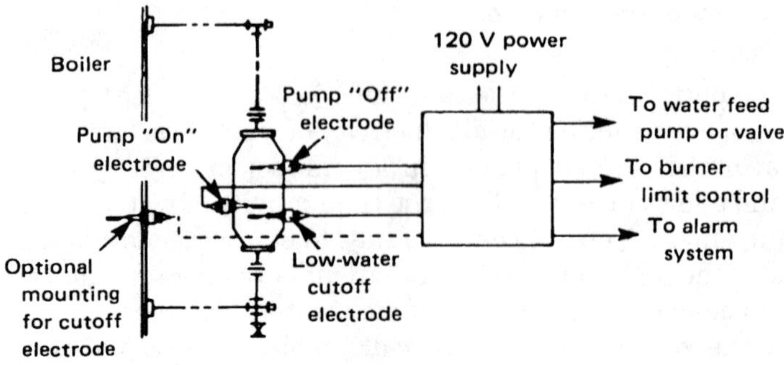

Automatic Boiler Water Feed and Low-Water Cutoff – For Steam

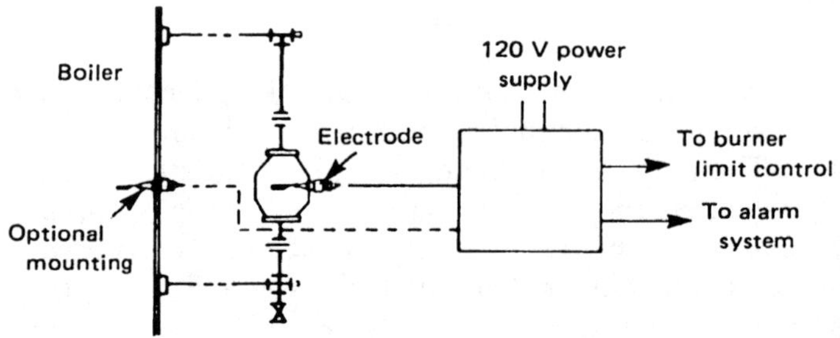

Low-Water Cutoff – For Steam

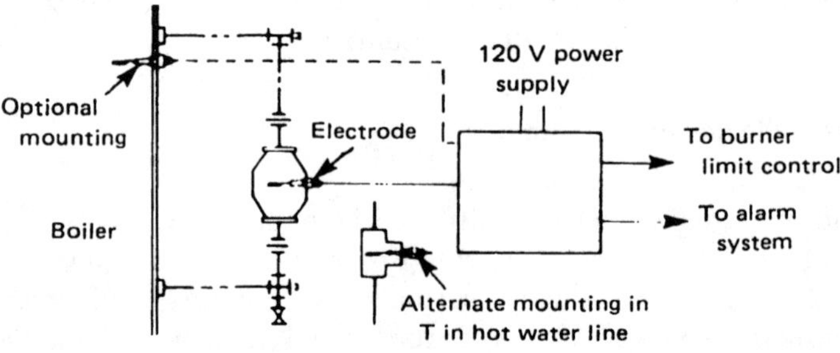

Low-Water Cutoff – For Hot Water

Figure 12.2 Electric probe-type low-water control. *(Courtesy: ASME Code Section VI)*

reaches the steam temperature. The steam heats the bimetallic element and causes the valve to close and the trap to remain to closed until the temperature of the condensate cools to permit the bimetallic element to return to its original shape and open the valve. A thermostatic steam trap with a bimetallic element is shown in Figure 12.4.

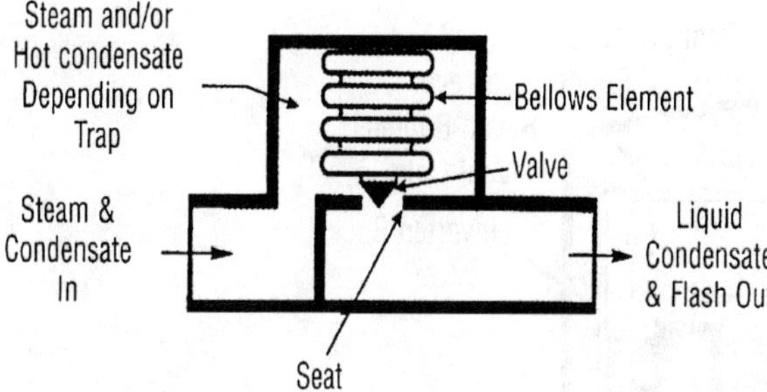

Figure 12.3 Thermostatic steam trap with a bellows element. *(Courtesy: Improving Steam System Performance)*

12.3.3 Inverted-bucket steam traps

In this type, as steam enters the trap and is captured inside the bucket, it causes the bucket to move upward. The upward movement closes the valve, which prevents steam from escaping. When condensate is collected and cools the steam, the bucket moves downward. This movement causes the valve to open and allow the condensate to escape. An inverted-bucket steam trap is shown in Figure 12.5.

12.3.4 Float and thermostatic (F&T) traps

The float trap depends on the movement of a spherical ball connected to a lever to open and close the outlet opening of the trap body. The float trap also includes a thermostatic element that allows air to escape at the start-up and during operation. The thermostatic elements used in this trap are the same as those used in thermostatic trap. A float and thermostatic steam trap is shown in Figure 12.6.

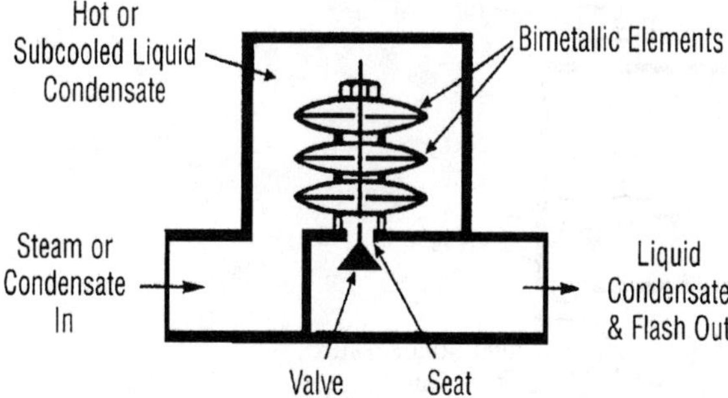

Figure 12.4 Thermostatic steam trap with a bimetallic element. *(Courtesy: Improving Steam System Performance)*

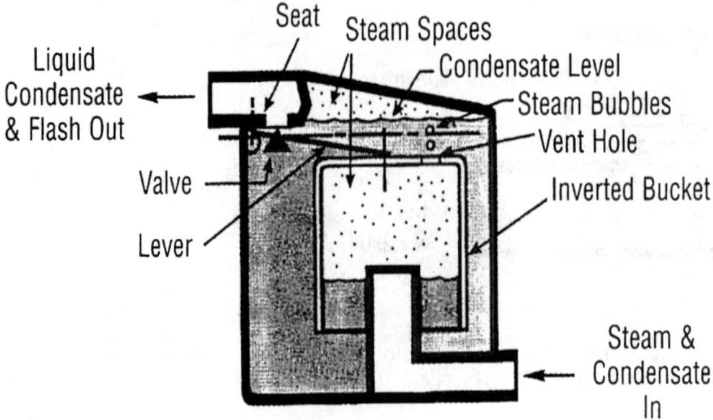

Figure 12.5 Inverted-bucket steam trap. *(Courtesy: Improving Steam System Performance)*

12.3.5 Thermodynamic disc steam traps

This trap uses the position of a float disc to control steam and condensate flow. When condensate flows through the trap, the disc is raised and causes the trap to open. As steam and air pass through the trap, the disc moves downward. The force for moving the disc downward is generated by the pressure difference between the low-velocity steam above the disc and high-velocity steam that flows through the narrow gap below the disc. A thermodynamic disc steam trap is shown in Figure 12.7.

12.4 Air Eliminators

Air eliminators are used to eliminate the air released from water within the boiler. These devices are generally installed on hot water boilers. An air eliminator is shown in Figure 12.8.

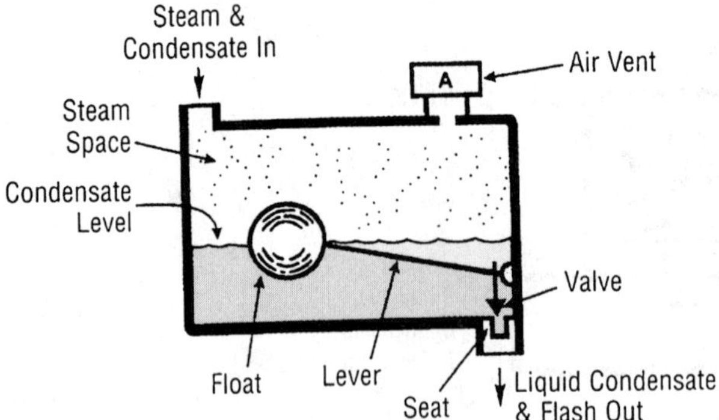

Figure 12.6 Float and thermostatic steam trap. *(Courtesy: Improving Steam System Performance)*

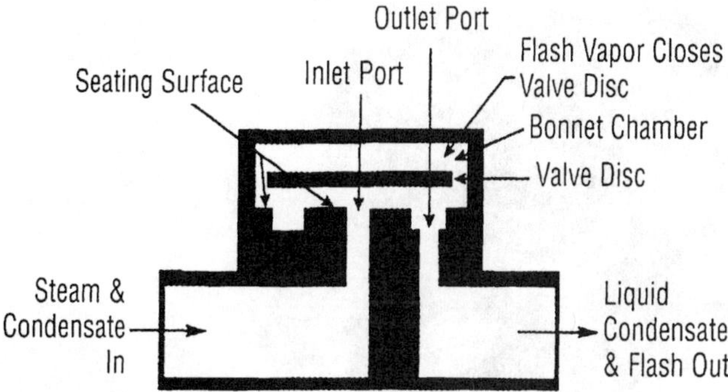

Figure 12.7 Thermodynamic disc steam trap. *(Courtesy: Improving Steam System Performance)*

12.5 Condensate Return Pumps and Return Loop

Condensate is defined as condensed water resulting from the removal of latent heat from steam. Condensate is recovered by a pump where this cannot be done by gravity. This condensate return pump (Figure 12.9) is used with a tank that has a float-operated switch for starting the pump motor. The tank collects the condensate from the heating or process condensate return lines. An atmospheric vent line is installed

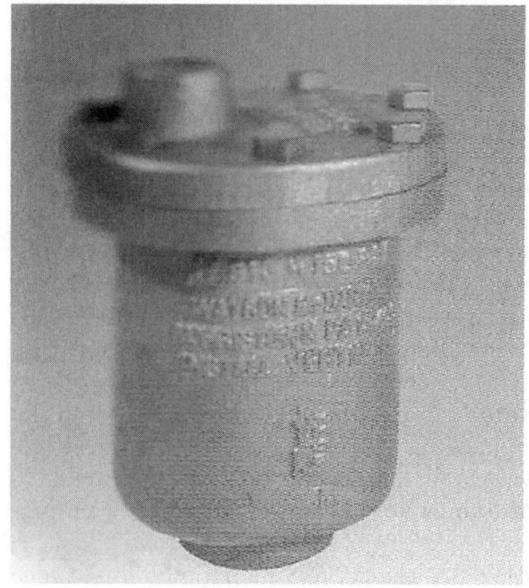

Figure 12.8 Air eliminator. *(Courtesy: Watson McDaniel)*

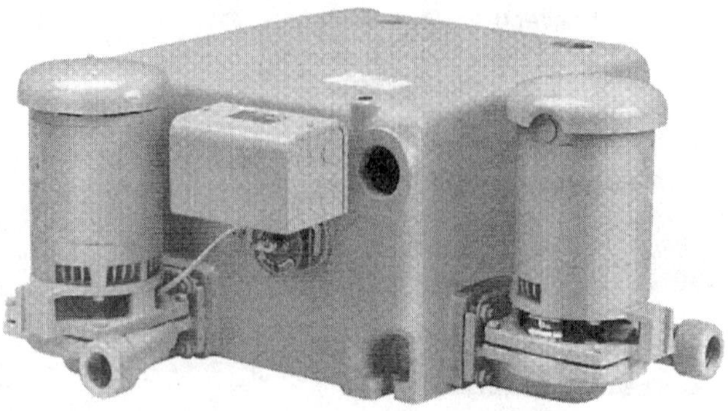

Figure 12.9 Condensate duplex pumps.

to prevent pressure buildup in the tank. A gage glass is attached to the tank for indicating condensate level.

When one condensate return pump serves, two boilers are connected; a vacuum breaker is required on the idle boiler to prevent formation of a vacuum, which may affect functioning of the feed valve. However, the return-pipe connections of each boiler supplying a gravity return of a steam heating system should be arranged like a loop, as shown in Figure 12.10, so that water in a boiler cannot be forced below the safe water level. This loop is required in gravity systems and may also be included in pump-return systems.

12.5.1 Hartford Loop

A Hartford Loop (Figure 12.11) is an arrangement of piping between a steam boiler's header and its gravity-return piping. The end of the header drops vertically below the boiler's waterline and connects into the bottom of the boiler. This pipe is called an "equalizer" because it balances the pressure between the boiler's steam outlet and condensate-return inlet. The "wet" gravity return line, which returns the condensate from the system, rises up from the floor to join the equalizer at a point about two inches below the boiler's lowest operating waterline.

If a return line breaks, water can only back out of the boiler to the point where the wet return connects to the equalizer. The loop works like a siphon that runs out of water. The point where the loop connects to the equalizer is higher than the boiler's crown sheet, and that's what provides the safety.

This loop was introduced by the Hartford Steam Boiler Insurance and Inspection Company in 1919 to reduce the number of boiler failures.

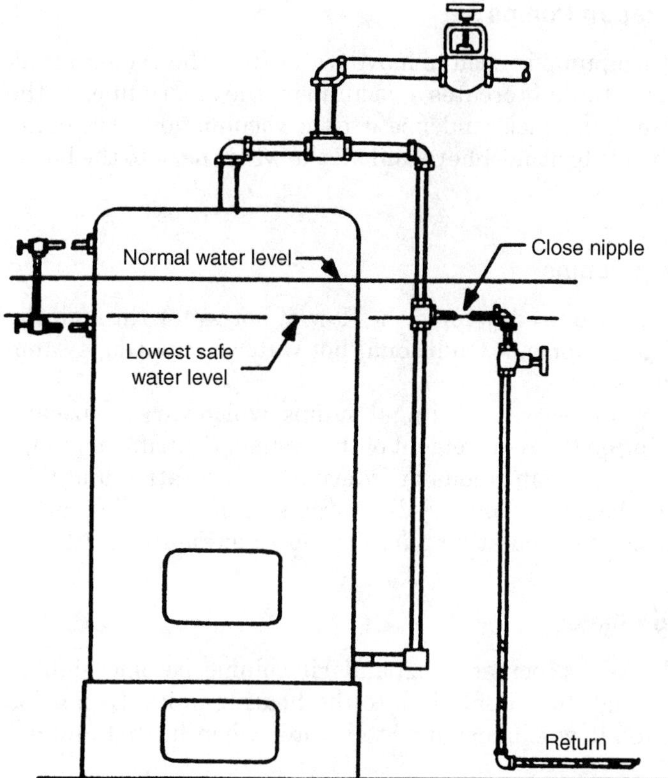

Figure 12.10 Typical return loop. *(Courtesy: ASME Code Section VI)*

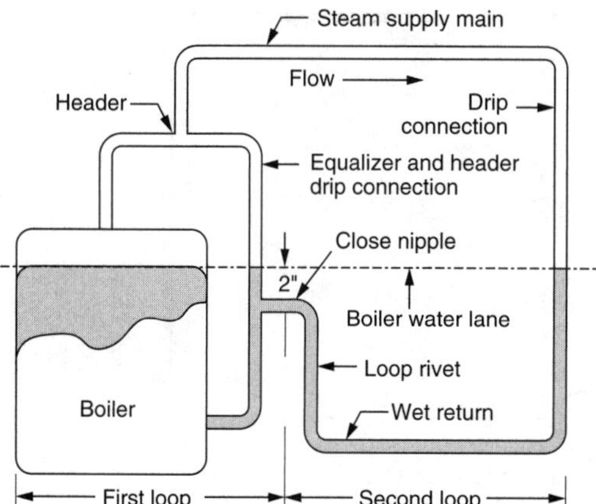

Figure 12.11 Hartford Loop. *(Courtesy: Heatinghelp.com)*

12.6 Vacuum Return Pumps

The vacuum return pump is used to move water from the vacuum tank to the boiler or feed tank. It creates a vacuum on the return lines of the heating system, drawing back condensate to the vacuum tank. The pump discharges all the air to atmosphere and all the water back to the boiler or feed tank.

12.7 Circulating Pumps

Circulating pumps are used to force the flow of hot water through the hot-water heating systems. A traditional hot water circulating system is shown in Figure 12.12.

The circulating pumps are centrifugal pumps, which vary in size and capacity, depending on the requirement of the system. Centrifugal pumps are designed either for continuous or intermittent operation and they preferably should be located on the discharge side of the boiler rather than the return side. A circulating pump is shown in Figure 12.13.

12.8 Expansion Tanks

When water is heated, it increases (expands) in volume, becomes lighter, and flows up through the supply line to the heating units. Expansion tanks are used to allow expansion of the water when heated and are

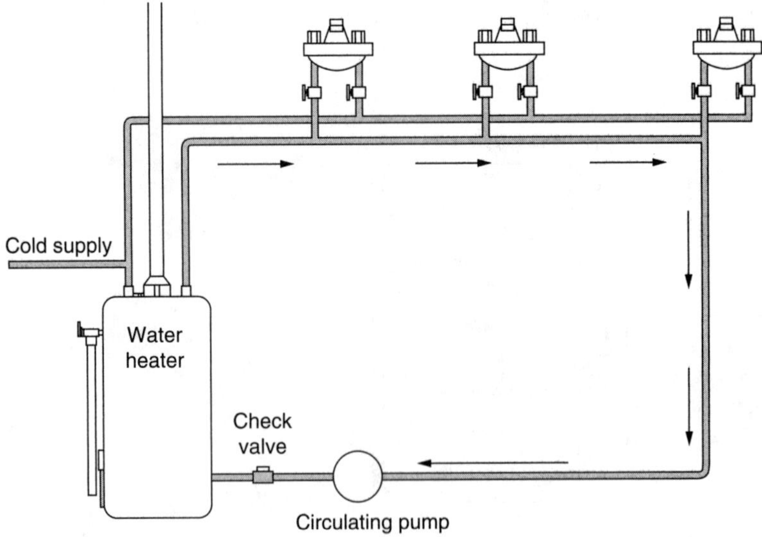

Figure 12.12 Hot water circulating system.

Figure 12.13 Circulating pump. *(Courtesy: Grundfos pump)*

located at the highest point in the system. In fact, the tank functions as a relief valve by collecting excess water from the system as the water is heated and its volume increases. An air cushion in the tank is compressed by the expanding water. Moreover, a vent line, open to the atmosphere, is installed on the top of the expansion tank for releasing air and allowing pressure to vary in the system as required. Alternatively, a precharged bladder-type expansion tank may be mounted on the floor of the boiler room. An expansion tank is shown in Figure 12.14.

12.9 Oil Preheaters

Some heavier grades of fuel must be heated for easy pumping and burning. Oil preheaters are used to condition the heavier grades of fuel oils for easy handling in the oil storage tank or burning at the burner. The oil preheaters are classified as steam, water, or electric oil preheaters. Steam or water oil preheaters are normally used for the initial preheat, with electric preheaters used for start-up and final stages of preheat.

12.10 Fuel Oil Storage and Supply Systems

The fuel oil storage and supply system consists of a fuel oil tank, connecting piping, and necessary strainers. This system may require a fuel

Figure 12.14 Expansion tank. *(Courtesy: Amtrol, Inc.)*

oil transfer pump, depending on the distance of the fuel-oil storage tank from the burner, and the grade of fuel oil. The four fuel oils generally used in low-pressure boilers are No. 2, No. 4, No. 5, and No. 6 fuel oil. Among these, No. 2 is the lightest, and No. 6, also known as "bunker C," is the heaviest fuel oil.

12.10.1 Storage tanks

Hot water or steam heating coils are installed in the storage tanks to heat the fuel oil and the temperature in the tank is controlled between 100 and 120°F. This temperature range allows the fuel oil to be pumped out without giving off vapor that might flash. The Environmental Protection Agency (EPA) requires that the fuel oil tank be located in containment vessel. Also, fuel oils are required to be treated as "hazardous waste" when performing such functions as cleaning out storage tanks.

12.10.2 Fuel strainers

Fuel oil strainers are used to remove foreign matter from fuel oil. Duplex strainers are installed on the suction line before the fuel oil pump. The advantage of duplex strainers is that one strainer can be cleaned, while the other is in use. Fuel oil strainers should be cleaned more frequently when using heavier grades of fuel oil.

12.11 Pressure Gage

Pressure gages are used on both steam and hot water boilers to indicate pressures at various points. Pressure gages allow the operator to maintain pressure within the maximum allowable pressure range. In order to record the pressure correctly, all the pressure gages should be easily visible to the operator. Pressure gages (Figure 12.15) indicate pressure in the boiler in pounds per square inch (psi). *Gage pressure* is the pressure above atmospheric pressure. On the other hand, absolute pressure is the gage pressure plus atmospheric pressure.

A compound gage is a combination of a pressure and a vacuum gage, which is generally used on a low-pressure boiler. The gage reads vacuum in inches of mercury (Hg) on the left side and pressure in pounds per square inch (psi) on the right side.

There are three types of pressure gages, namely, *bourdon, bellows,* and *spiral.* The bourdon tube-type pressure gage has a hollow tube with an oval cross-section called bourdon tube, and it straightens out when pressure builds up. The open end of the tube is connected to the steam, or water side, of the boiler. The closed end is attached to the gage pointer by a mechanical linkage. The bellows-type gage is generally used for pressures below 30 psi (207 kPa).

Figure 12.15 Pressure gages. *(Courtesy: Precision Instrument Company)*

12.11.1 Gage range

The pressure gage range should be 1.5 to 2 times the maximum allowable working pressure of the boiler. This will put gage indicator in about middle position of the gage scale. The gage should be calibrated between the range of 30 psi (207 kPa) to 60 psi (414 kPa) for steam heating boilers and not less than 1.5 or more than 3 times the relief valve setting for hot water boilers.

12.11.2 Accuracy

Pressure gages should have an accuracy of 1 to 1.5 percent of the actual pressure. If a 100-psi (690 kPa) gage is 1.5 percent accurate, then it will be within ± 1.5 psi (10.35 kPa) of the actual pressure.

12.11.3 Calibration

Pressure gages that do not read accurate pressures may be out of calibration due to various reasons. Any pressure gage, which is not accurate to within 1.5 percent of the actual pressure, should be calibrated. The gage used on boilers should be calibrated at least once a year either by using a test gage or dead-weight tester.

12.11.4 Siphon tubes

Siphons are used between the steam boiler and the pressure gage to prevent steam from damaging the bourdon tube. There are two types of siphons, namely, a pigtail siphon and a u-tube siphon. A hand-operated valve is installed to allow changing of the gage in case of failure or malfunction of the pressure gage.

12.12 Pressure or Altitude Gages

A pressure or altitude gage will be installed on a hot water heating or hot water supply boiler, which will be shut off by a cock with a tee or lever handle. The range on the dial of the gage shall be 1.5 to 3.5 times the safety valve set pressure; the connection size will be less than NPS 1 (DN 25), if nonferrous metal is used.

12.13 Water Gage Glasses

One or more water gage glasses are required to be installed to the water column or boiler by means of valve fittings of a minimum size NPS 1/2 (DN 15). The lowest permissible water level in a boiler is recommended by the boiler manufacturer. So that there will be no danger of overheating, the lowest visible part of the gage glass should be a minimum

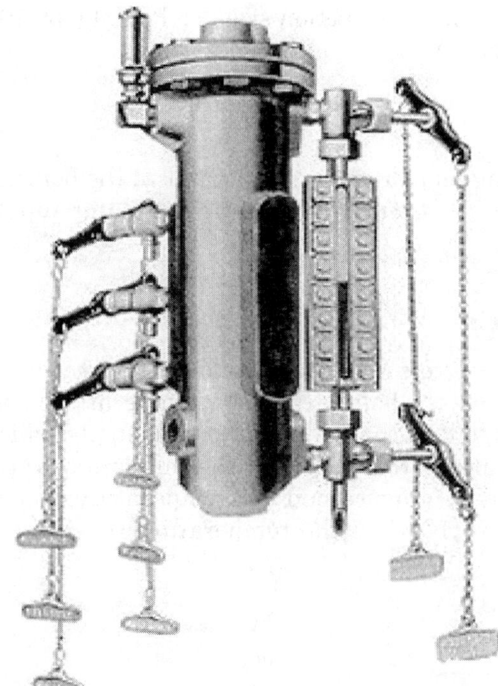

Figure 12.16 Water gage glass. *(Courtesy: Clark Reliance Corporation)*

1 inch (25 mm) above the lowest water level. During fabrication, a boiler is permanently marked indicating the lowest permissible water level.

The water gage glass of an electric boiler, submerged type, shall indicate water levels both at the start-up and under maximum steam load. For an electric boiler, resistance element type, the lowest visible part of the water gage shall be located a minimum of 1 inch above the lowest permissible water level. An automatic low-water cutoff shall be installed on this type of boiler so that power supply is automatically cut off before water level falls below the visible part of the glass. A water gage glass is shown in Figure 12.16.

12.14 Pressure Controls

Two pressure-operated controls are used to protect a steam boiler from overpressure. A safety limit control is used to cut off the fuel supply to prevent steam pressure from exceeding 15 psi (103 kPa) maximum allowable working pressure. A second control will cut off fuel supply if pressure reaches the operating limit, which is less than the maximum allowable pressure. The minimum connection size is NPS 1/4 (DN 10), or for steel it is NPS 1/2 (DN 15).

12.15 Thermometers

A thermometer for indicating temperature of the water in the boiler is required to be installed on a hot water-heating or hot water-supply boiler.

12.16 Temperature Controls

Two temperature-operated controls are used on a hot water heating or hot water supply boiler protect the boiler from over-temperature. One is a high temperature limit control that cuts off the fuel supply, which will prevent water temperature from reaching the maximum water temperature at the boiler outlet. The second is a temperature control that will cut off the fuel supply when the temperature reaches the

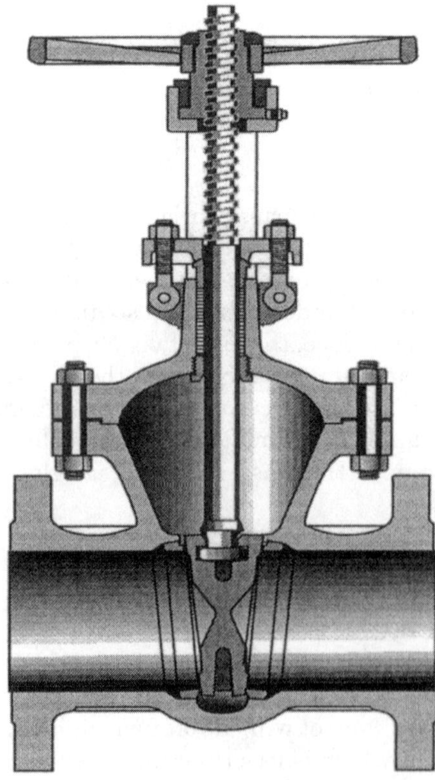

Figure 12.17 Gate valve OS&Y type. *(Courtesy: Crane Valve)*

operating temperature, which is less than the maximum water temperature.

12.17 Stop Valves

Stop valves that are used to isolate a steam boiler from the system should be located on both the supply and return lines as near to the boiler as possible. If the boiler is located above the system and can be drained without draining the system, a stop valve is not required. The minimum pressure rating of a stop valve should be at least equal to the pressure stamped on the boiler and the temperature rating should not be less than 250°F (121°C). This stop valve is properly designated by a metal tag or other material fastened to the valve. A cast steel stop valve, gate OS &Y type is shown in Figure 12.17.

12.18 Bottom Blow-off Valves

Blow-off valves are operated intermittently to remove the accumulated sludge or sediment from the boiler. A bottom blow-off valve is required to be installed on a steam boiler's lowest water space. The minimum pressure rating for blow-off valves and cocks are required to be at least equal to the pressure stamped on the boiler or a minimum of 30 psi (207 kPa). The temperature rating for such valves and cocks should be a minimum of 250°F (121°C). Blow-down valves and piping arrangement for a steam boiler is shown in Figure 12.18.

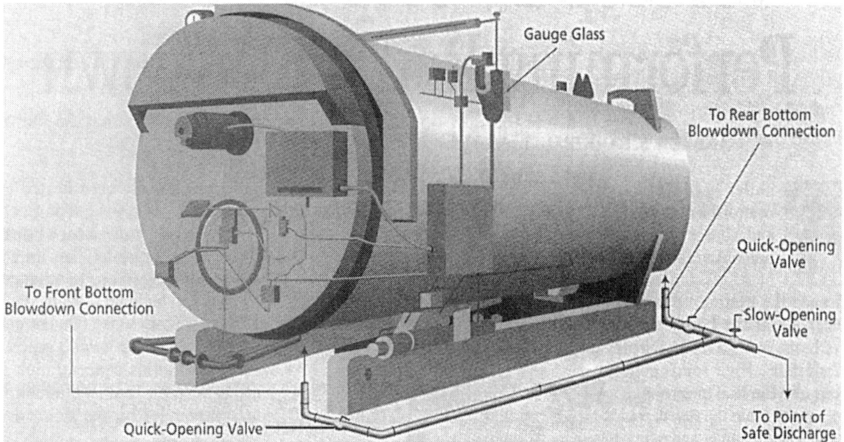

Figure 12.18 Blowdown valves and piping arrangement of a steam boiler. *(Courtesy: Cleaver-Brooks)*

Figure 12.19 Quick-opening valve. *(Courtesy: Everlasting Valve Company, Inc.)*

Two shutoff valves are required for a blow-off system. These two valves may be slow-opening valves that require five full turns of the handwheel to open or close; alternatively, one quick-opening valve (Figure 12.19) and one slow-opening valve (Figure 12.20) may be installed. The quick-opening valve should be installed closest to the boiler shell.

Figure 12.20 Slow-opening valve. *(Courtesy: Everlasting Valve Company, Inc)*

Chapter

13

Instruments and Controls

Each heating boiler requires various instruments and controls for smooth operation and safety of the system. Though some of the instruments and controls have been discussed in other chapters, this chapter describes mandatory requirements in accordance with *ASME Boiler and Pressure Vessel Code Section IV*—Rules for Construction of Heating Boilers, and ASME Code CSD-1—*Control and Safety Devices for Automatically Fire Boilers.*

All instruments and controls are installed in manufacturer's factory before shipment. In a field-erected boiler, instruments and controls may be supplied by the installing contractor. Generally, jurisdictional law requires that these instruments and controls are installed before operation of a boiler.

13.1 Controls for Steam-Heating Boilers

A steam boiler can run into trouble if its controls do not work properly. Some of the troubles caused when the control system malfunctions are:

- Safety valve failure can cause explosion due to over pressure.
- Water-level control failure can cause boiler burn-out.

For an example, Figure 13.1 shows a steam boiler with a feedwater cutoff combination. This control is ideal for use in residential or small commercial applications. The feeder is always ready to add water when a signal is received from the low-water cutoff.

13.1.1 Steam gages

A steam gage or a compound steam gage is required on each steam-heating boiler. The gage should be connected to its steam space or to the

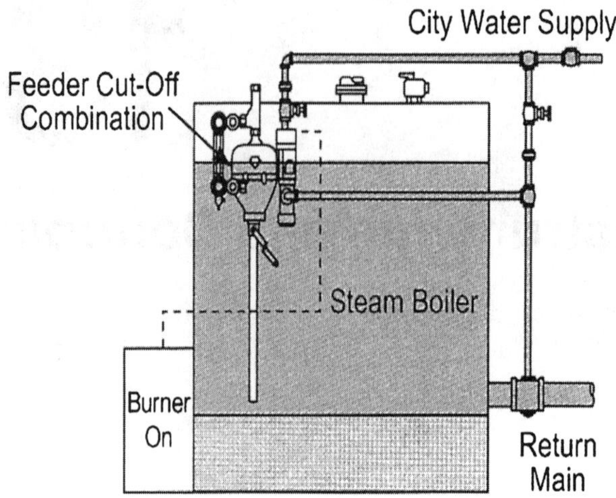

Figure 13.1 Steam boiler. *(Courtesy: McDonnell & Miller)*

water column or to its steam connection. The gage or piping to the gage should contain a siphon or equivalent device to maintain a water seal by preventing steam entering into the gage tube.

The gage connection boiler tapping, external siphon, or piping to the boiler should be minimum NPS 1/4 (DN 8). If steel pipe or tubing is used, the boiler connection and external siphon should not be less than 1/2(DN 15).

The scale on the dial of a gage should be graduated to not less than 30 psi (200 kPa) or nor more than 60 psi (414 kPa). The travel of the pointer from 0 to 30 psi (0 to 200 kPa) pressure should be at least 3 inches (75 mm).

13.1.2 Water gage glasses

One or more water gage glasses are required to be installed to the water column or boiler by means of valve fittings a minimum size NPS 1/2 (DN 15). The lower fitting should be provided with a drain valve with an unrestricted drain opening not less than 1/4 inch (6 mm) in diameter to facilitate cleaning. A boiler mechanic should be able to replace gage glass with the boiler still under pressure. Water gage glass fittings should be attached directly to a boiler.

The lowest permissible water level in a boiler is recommended by the boiler manufacturer. The lowest visible part of the gage glass should be minimum 1 inch (25 mm) above the lowest water level so that there will be no danger of overheating. During manufacture a boiler, is permanently marked, indicating the lowest permissible water level. This marker should be visible at all times.

In a submerged-type, electric boiler, the water gage glass should indicate the water levels both at the start-up and under maximum steam load. The lowest visible part of the water gage for an electric boiler, resistance element type, should be located minimum 1 inch above the lowest permissible water level. An automatic low-water cutoff shall be installed on this type of boiler so that the power supply is automatically cut off before the water level falls below the visible part of the glass. Tubular water glasses on electric boilers with a nominal water content not exceeding 100 gallon (300 L) should be equipped with protective shields.

A water level indicator using an indirect sensing method may be used in lieu of an operating water gage glass. The water level indicator should be attached to a water column or directly to the boiler by means of fittings not less than NPS 1/2 (DN 15). The device should be provided with a drain valve with an unrestricted drain opening not less than 1/4 inch (6 mm) diameter to facilitate cleaning.

13.1.3 Water columns

A water column has two connections. The steam connection to the water column is taken from the top of the shell or upper part of the head, and the water connection is taken from a point not above the centerline of the shell. For a cast-iron boiler, the steam connection is taken from the top of an end section or the top of the steam header, and the water connection is made on the end section, not less than 6 inches (150 mm) below the bottom connection to the water gage glass.

The minimum size of ferrous or nonferrous pipes connecting a water column to a steam boiler should be 1 inch (25 mm). If the water column, gage glass, low-water fuel cutoff, or other water-level control device is connected to the boiler by pipe and fittings, no shutoff valves should be installed in such pipes. The water column drain pipe and valve should be not less than NPS 3/4 (DN 20).

13.1.4 Pressure controls

Two pressure-operated controls are used to protect a steam boiler from overpressure (Figure 13.2). Each automatically fired steam boiler should have a safety limit control that will cut off the fuel supply to prevent steam pressure from exceeding 15 psi (103 kPa) maximum allowable working pressure of the boiler. Each control should be constructed to prevent a pressure setting above 15 psi (100 kPa). Each steam boiler should have a second control that will cut off fuel supply if pressure reaches the operating limit, which is less than the maximum allowable pressure.

Shutoff valves of any type should not be installed in the steam pressure connection between the boiler and the above controls. The control

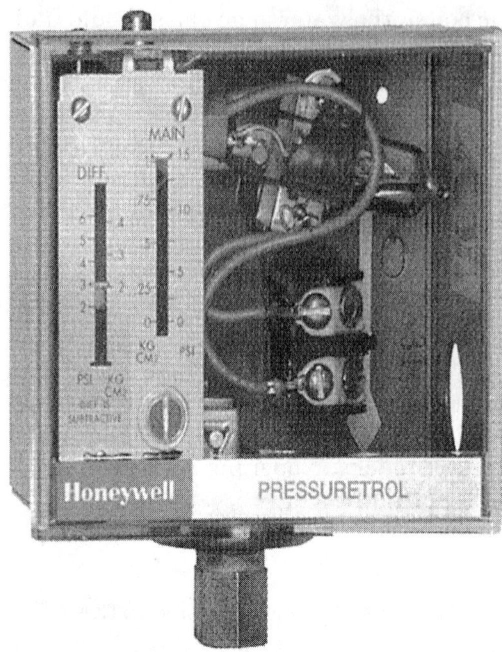

Figure 13.2 Pressure control. *(Courtesy: Honeywell)*

connection piping to the boiler should not be less than NPS 1/4 (DN 8), and steel or wrought iron pipe and tube should not be less than 1/2 (DN 15). The minimum connection size of an external siphon should be NPS 1/4 (DN 10) or 3/8 inch (10 mm) outside diameter, nonferrous tubing.

13.1.5 Automatic low-water fuel cutoffs

An automatic low-water fuel cutoff is required on each boiler to cut off the fuel supply when water level falls below the lowest permissible level. If a water feeding device is installed, it should be constructed in such a way that the water inlet valve cannot feed the water into the boiler through the float chamber.

A low-water cutoff and water-feeding device may be attached to a boiler. A fuel cutoff or water-feeding device may also be installed in the tapped openings available for attaching a water glass directly to a boiler.

Fuel cutoffs and water-feeding devices that embody separate chambers should have vertical drain pipes and blow-off valves not less than NPS 3/4 (DN 20) located at the lowest point in the water-equalizing pipe connection. This is required so that the chambers and equalizing pipes may be flushed, and the device tested.

13.2 Controls for Hot-Water Heating or Hot Water Supply Boilers

It is recognized by the authorities that a water heating system must be protected by basic safety controls. Industry experts recommend that a hot water boiler should be protected by the following basic controls:

- Relief valves
- Low-water cutoffs
- Burner control circuitry
- Makeup water feeders

A hot water boiler with a few controls is shown in Figure 13.3. Hot water boilers are subjected to many of the same problems of steam boilers—one of these is low water. Installation of low-water fuel supply cutoffs and water feeders protects hot water boilers against low water conditions.

13.2.1 Pressure or altitude gages

Each hot-water heating or hot water supply boiler should have a pressure or altitude gage connected to it or its flow connection so that it cannot be shut-off from the boiler except by a cock with a tee or lever handle. The handle of the cock should be parallel to the pipe, in which it is located when the cock is open. The scale on the dial of the gage should be graduated to not less than 1.5 nor more than 3.5 times the safety relief

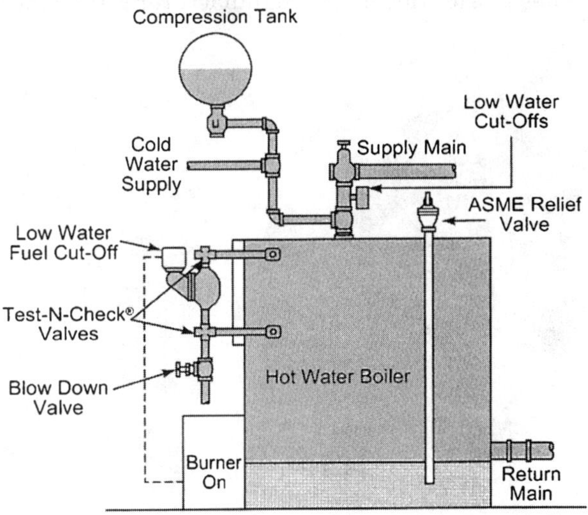

Figure 13.3 Controls on hot water boiler. *(Courtesy: McDonnell & Miller)*

valve set pressure. Piping or tubing connections should be of nonferrous material when their size are smaller than NPS 1 (DN 25).

13.2.2 Thermometers

Each hot-water heating or hot water supply boiler should have a thermometer in a position where it is easily readable (Figure 13.4). The thermometer should be located at a position that indicates the temperature of the water in the boiler at, or near, the outlet.

13.2.3 Temperature controls

Two temperature-operated controls are used on a hot-water heating or hot water supply boiler for protecting the boiler from over-temperature.

A high temperature limit control is used to cut off the fuel supply preventing water temperature from exceeding its marked maximum water temperature at the boiler outlet. This control should be constructed and set to prevent temperatures rising above the maximum allowed.

A second temperature control is used to cut off the fuel supply when system water temperatures reach a preset operating temperature, which is less than the maximum water temperature.

13.2.4 Low-water fuel cutoff

An automatic low-water fuel cutoff designed for hot water service is required to be installed for a hot water boiler with more than 4,000,000 Btu/hr heat input (117 kW). The automatic low-water fuel cutoff should be located so that it automatically cuts off the fuel supply when water levels fall below the safe permissible level. The manufacturer establishes

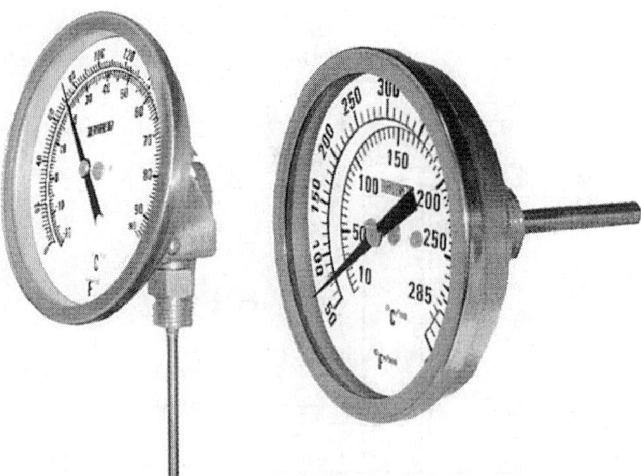

Figure 13.4 Thermometers. *(Courtesy: Precision Instrument Company)*

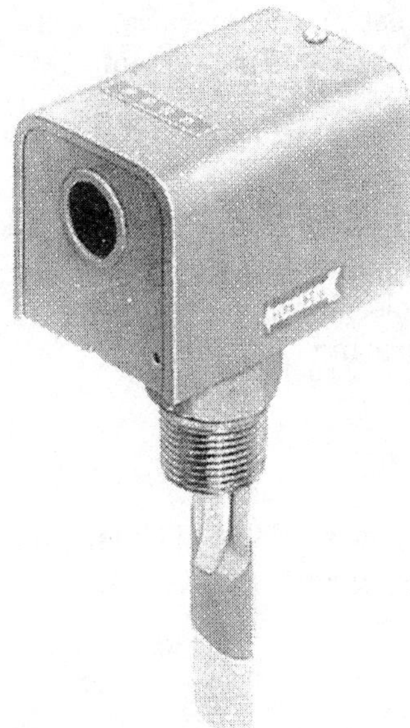

Figure 13.5 Flow switch. *(Courtesy: Johnson Controls, Inc.)*

the safe permissible water level, as there is no standard waterline for all hot water boilers.

A coil-type boiler or a watertube boiler, with heat input more than 4,000,000 Btu/hr (117 kW) forced-circulation type, should have a flow sensing device (Figures 13.5 and 13.6) in lieu of a low-water cutoff. The flow-sensing device cuts off the fuel supply when the circulating flow is

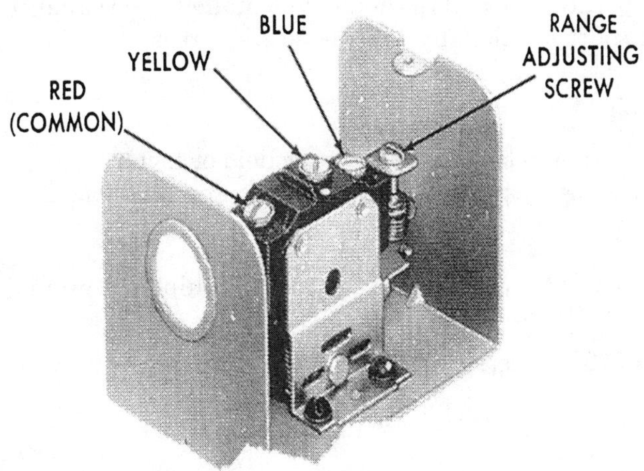

Figure 13.6 Internal view of flow switch. *(Courtesy: Johnson Controls, Inc)*

interrupted. A mean should be provided for testing the operation of the low-water fuel cutoff without needing to drain the entire system.

13.3 Controls for All Boilers

All boilers must be equipped with, at least, minimum controls and safety devices. These are instruments and controls mounted inside jackets, electrical controls, control circuits, limit controls, shutdown switches and circuit breakers, and controls and heat-generating apparatus.

13.3.1 Instruments and control-mounted inside jackets

Instruments and controls may be installed inside the boiler jackets provided the following are always visible through openings:

- Water gage on a steam boiler
- Pressure gage on a steam boiler
- Thermometer and pressure gage on a water boiler

13.3.2 Electrical controls

All electrical wiring, controls, heat-generating apparatus, and the like necessary for the operation of boilers should be installed in accordance with the provisions of the (NFPA)—70 *National Electric Code,* and/or local jurisdictional electrical codes. All boilers supplied with factory-mounted and wired controls, heat-generating apparatuses, and the like, necessary for the operation of boilers should be installed in accordance with the provisions of the nationally recognized standards.

13.3.3 Control circuits

The control circuitry should be 150 V or less for field or factory wiring. One of the following systems may be employed to provide the control circuit:

- Two-wire nominal 120 V system with a separate equipment ground conductor
- Two-wire nominal 120 V system obtained by using an isolation transformer

13.3.4 Limit controls

Limit controls should be wired on the hot, or line side, of the control circuit.

13.3.5 Shutdown switches and circuit breakers

A manually operated remote shutdown switch or circuit breaker should be located just outside the boiler room. A shutdown switch or circuit breaker should be marked for easy identification. If the boiler room door is on the building exterior, the switch should be located inside the door. If there is more than one door to the boiler room, there should be a switch located at each door.

The shutdown switch or the circuit breaker should be installed to perform the following operation:

- The complete burner and controls should be shut off for atmospheric gas burners and oil burners, where a fan is on a common shaft with the oil pump.
- The fuel input supply to the firebox needs to be shut off for power burners with detached auxiliaries.

13.3.6 Controls and heat-generating apparatus

Oil and gas fired and electrically heated boilers should be equipped with primary (flame safeguard) safety controls, safety limit switches, and burners or electric elements as required by a nationally recognized standard.

13.4 Controls and Safety Devices

Many jurisdictions have adopted ASME CSD-1—*Control and Safety Devices for Automatic Controlled Fired Boilers*. These controls, which may be operated electrically or pneumatically shall be provided on a boiler to meet the Code requirements. A boiler owner is required to meet the requirements of the local jurisdiction. A certificate of operation may not be issued for a boiler if controls are not installed according to the local code.

Controls are classified according to the functions they perform for a boiler. Based on functions, controls are classified as operating controls, limit controls, safety controls, and programming controls.

13.4.1 Operating controls

These are the devices that control the operation of a fuel burner to maintain the desired conditions and perform the following functions:

1. Start, stop, and modulate the burner (if designed)
2. Maintain proper water levels in steam boilers
3. Maintain proper water pressure in hot-water heating boilers

Examples of operating controls are flame safeguard control, high limit pressure control, high limit temperature control (hot water), operating limit pressure control (steam), operating limit temperature control (hot water), modulation pressure modulating control steam, modulation temperature control (hot water), low fire hold control, low-water cutoff control, and auxiliary low-water cutoff control.

13.4.2 Limit controls

These are the devices that shut down the boiler when operating limits are reached and they perform the following functions:

1. Stop the burner when steam pressure or hot water temperature exceeds limit control setting (15 psi for steam boilers and 250°F for hot water boilers)
2. Stop the burner when water level falls below the minimum safe level
3. If required, stop the burner in case of unusual conditions such as high stack temperature, high or low gas fuel pressure, and high or low fuel oil temperature.

13.4.3 Safety controls

These are the devices which are incorporated in the burner control circuitry and on the burner to allow the flow of fuel only if required steps and conditions are met. They perform the following functions:

1. Stop the fuel flow in case of ignition failure
2. Stop the fuel flow in case of main flame interruption
3. Stop the fuel flow in case of mechanical draft failure
4. Stop the fuel flow in case of circuit failure

13.4.4 Programming controls

These are the devices which provide proper sequencing of operating controls, limit controls, and safety controls to make sure that all the conditions necessary for proper burner operation are met. Examples of programming controls are prepurge and postpurge cycles to remove gases which might have accumulated in the furnace.

13.4.5 Testing of controls and safety devices

The owner or user of an automatic heating boiler should set up a system of periodic testing of controls and safety devices. While drawing up a detailed check list, manufacturer's instructions and requirements should be followed. ASME CSD-1 recommends the following check list as a minimum:

1. *Daily*
   a. Check gages, monitors, and indicators.
   b. Check instruments and equipment setting.
   c. Test low-water fuel cutoff device and alarm.
   d. Check burner flame.
2. *Weekly*
   a. Check igniter.
   b. Check flame signal strength.
   c. Check flame failure detection system.
   d. Check firing rate control.
   e. Make aural and visual check of pilot and main fuel valves.
3. *Monthly*
   a. Check flue, vent, stack, or outlet dampers.
   b. Test low draft, fan air pressure, and damper position interlock.
   c. Check low fire start interlock.
   d. Test high and low oil pressure and temperature interlocks.
   e. Test high and low gas pressure interlocks.
4. *Semi-annually*
   a. Recalibrate all indicating and recording gages.
   b. Perform a slow drain test of the low-water fuel cutoff device.
   c. Check flame failure detection system components.
   d. Check firing rate control.
   e. Check piping and wiring of all interlocks and shutoff valves.
   f. Inspect burner components.
5. *Annually*
   a. Flame failure detection, pilot turndown test.
   b. Flame failure detection system, test for hot refractory hold-in.
   c. Check dual fuel change over control.
   d. Test high limit and operating temperature or steam pressure controls.
   e. Replace vacuum tubes, scanners, or flame rods in accordance with manufacturer's instructions.
   f. Conduct a combustion test.
   g. Check all coils and diaphragms; test other parts of all safety shutoff and control valves.
   h. Test fuel valve interlock switch in accordance with manufacturer's instructions.
   i. Perform leakage test on pilot and main gas and/or oil fuel valves.
   j. Test purge air switch in accordance with manufacturer's instructions.

k. Test air/steam interlock in accordance with manufacturer's instructions.
l. Test burner position interlock in accordance with manufacturer's instructions.
m. Test rotary cup interlock in accordance with manufacturer's instructions.
n. Test low fire start interlock in accordance with manufacturer's instructions.

6. *As required*
   a. Recondition or replace low-water fuel cutoff device.
   b. For oil burners, clean atomizers and oil strainers.
   c. For gas-fired burners, check drip leg and gas strainers.
   d. Flame failure detection system, pilot turndown test.
   e. Flame detection system, test for hot refractory hold-in.
   f. Test safety/safety relief valves in accordance with *ASME Code* Section VI.

Chapter 14

Installation

Heating boilers are installed by a boiler contractor or an experienced installer. Many jurisdictions have laws and rules for licensing boiler installers. In some jurisdictions, an installation requires a mechanical contractor's license to install boilers. Whether a license is required or not, any individual or company that is engaged in installation, should have experience and knowledge about codes and local jurisdictional requirements. An authorized inspector usually performs the first inspection after the completion of installation for the purpose of issuing an operating certificate.

In addition to the manufacturer's recommendations, the equipment in the boiler room should be installed in accordance with the regulations of the jurisdiction where the boiler room is located. Most jurisdictions in the United States and Canada have adopted the following codes for construction and installation:

- *ASME Code* Section IV—Rules for Construction of Heating Boilers
- *ASME Code* Section VI—Recommended Rules for the Care and Operation of Heating Boilers
- ASME CSD-1—*Control and Safety Devices for Automatically Fired Boilers*
- NFPA 54—*National Fuel Gas Code*
- NFPA 70—*National Electric Code*
- UL 834—*Electric Heating, Water Supply, and Power Boilers*

A contractor should consider five basic factors for boiler installation. They are the boiler room, combustion air, venting, boiler placement, and operation tests.

14.1 Boiler Room

A boiler room is defined as a facility where boiler(s) and associated equipment are installed for supplying steam or hot water. Boiler operators and engineers perform their duties and keep log sheets in a boiler room. All records pertaining to the boiler(s) are maintained in the boiler room. Boiler parts, tools, supplies, and other accessories may be stored in boiler rooms.

The size of a boiler room depends on the physical size and number of boilers, physical dimensions of all the boiler plant-related equipment and water treatment equipment and use of space required for storing boiler related materials. The boiler room should be constructed as per the local codes, and approval should be obtained from the jurisdiction. The owner or user is responsible for meeting all the jurisdictional requirements for drawings, certifications, permits (including environmental permits) licenses, and so forth, before installation of boiler(s) and related equipment in the boiler room.

Exit and egress. All boiler rooms exceeding 500 ft^2 of floor area and containing one or more boilers having a fuel-burning capacity of 1,000,000 Btu should have at least two means of exit. Each exit should have two means of egress, each remotely located from the other.

Ladders and runaways. All walkways, runways, and platforms in the boiler room should be of metal construction, be equipped with handrails 42 inches (107 cm) high with an intermediate rail and 4 inches (10 cm) toeboard, and be provided between or over the top of boilers, which are more than 8 feet (2.8 m) above the operating floor to afford accessibility for operation and maintenance. At least two permanently installed means of egress from walkways, runways, or platforms that exceed 6 feet (183 cm) in length should be provided.

Stairways, which serve as a means of access to the walkways, runways, or platforms, should not exceed an angle of 45 degrees from the horizontal.

Ladders, which serve as a means of access to the walkways, runways, or platforms, should be of metal construction and not less than 18 inches (46 cm) wide and have rungs that extend through the side members and are permanently secured. The ladders should have a clearance width of at least 15 inches (39 cm) from the center of the ladder on either side across the front of the ladder.

14.1.1 Safety

Safe operation of a boiler depends, to a large extent, on the skill and attentiveness of the boiler operator. Boiler room safety rules, which vary depending on the type and size of the plant, should be practiced by the boiler operator at all times. Fuels used in the boiler room are combustible and may cause a fire. The boiler operator is responsible for fire safety in the boiler room. Owners should hire only properly trained boiler operators/engineers to work on or operate mechanical equipment.

14.1.2 Lighting

The boiler room should have sufficient light so that all the instruments can be read clearly and any leakage from the boiler can be detected easily. An emergency light or a flashlight should be kept on standby in case of power failure. Always use low voltage droplights or flashlights when working inside the boiler.

14.1.3 Water and drain connections

Water connections. A means to add water to or fill the boiler, while not under pressure, should be provided. A valve or threaded plug may be used to shut off the fill connection when the boiler is not in service. Provision should be made in every boiler room for proper water supply, which can be used to flush out the boiler and to clean the floor.

Drain connections. There should be at least one drain connection in every boiler room for draining the boiler and cleaning the boiler room. Drains receiving blowdown water should be connected to the sanitary sewer by way of an acceptable blowdown tank or separator or an air gap, which will allow the blowdown water to cool to at least 140°F (60°C) and reduce the pressure to 5 psig (35 kPa) or less.

14.1.4 Fire protection

Fire protection procedures are required because of the combustible materials used in the boiler room. Fire protection procedures and fire protection apparatus for the boiler room should conform to the National Fire Protection Association (NFPA) Standards. The boiler operator should be trained on the procedures to be followed in case of fire in the boiler room.

14.1.5 Housekeeping

The boiler room should be used only for equipment and materials necessary for operation and maintenance of the heating system. A clean boiler room will have fewer accidents because many causes of accidents will be eliminated. Good housekeeping should be practiced to keep the boiler room neat and clean. The boiler room should not be used for storing general goods and materials.

14.1.6 Posting of certificates

Many jurisdictions require that the inspection certificates for the boilers be posted under glass in the boiler room. Generally, the authorized boiler inspectors verify the inspection certificates during boiler inspection. In addition, certificates of boiler operators and/or boiler engineers are required to be posted in the boiler room. Many jurisdictional authorities make sure that these certificates are posted, which show that the boilers have been inspected and qualified personnel are operating boilers. The engineer in charge should make sure that all the certificates are posted in the boiler room.

14.1.7 Recordkeeping

Drawings, diagrams, instruction books. Boiler records such as drawings, wiring diagrams, schematic arrangements, operating and service manuals, spare parts lists, and the like are very important and should be saved during the lifetime of a boiler. If any changes or additions are made, the records should be updated accordingly. These records should be permanently kept either in the boiler room or in another suitable location close to the boiler room.

Log book. A log book containing records of maintenance work, repair work, inspections, tests, and other pertinent data should be maintained permanently in the boiler room. Details of repair and alteration work should be recorded. Any inspection done by the jurisdictional inspector or insurance company inspector should be entered in the log book.

Maintenance schedules and records. A recommended log sheet for maintenance, testing, and inspection is shown in Figure 17.6 (for steam heating boilers) and Figure 17.7 (for hot water heating boiler). This log sheet should be retained in the boiler room for at least 1 year.

14.2 Combustion Air

The boiler room should have an adequate air supply to allow clean, safe combustion, and minimize soot formation. An improperly ventilated room can get excessively hot and cause deterioration of controls and electrical components. A minimum of 19.5 percent oxygen in the air of the boiler room should be maintained at all times. The combustion and ventilation air may be supplied by either an unobstructed opening or by power ventilators or fans.

The boiler room should be provided with suitable-sized openings to ensure adequate combustion air and proper ventilation. In general, two permanent air supply openings shall vent directly through the wall to outside air; one within 12 inches of the ceiling, the other within 12 inches of the floor should be furnished. Each opening should have a minimum free area of 1 in.^{2} (6.5 cm^2) per 2000 Btu/hr input rating of all the burners located in the boiler room or as specified in the National Fire Protection Association Standards.

When a blower is used to supply combustion air, a suitable switch or equivalent should be wired in to the boiler control circuit to prevent the boiler from firing unless the blower is operating.

Combustion air openings are protected with a 1/4-inch mesh screen. Often, louvers are installed on exterior walls for security purposes. Compensation is required for the cross-sectional area that is lost due to the screens or louvers.

The total requirements of the burners for all fired pressure vessels in the boiler room are used to determine the net louvered area in square feet:

Input Btu/hr	Required air ft^3/min	Min net louvered area, ft^2
500,000	125	1.0
1,000,000	250	1.0
2,000,000	500	1.6
3,000,000	750	2.5
4,000,000	1,000	3.3
5,000,000	1,250	4.1
6,000,000	1,500	5.0
7,000,000	1,750	5.8
8,000,000	2,000	6.6
9,000,000	2,250	7.5
10,000,000	2,500	8.3

Boiler rooms should not be under a negative pressure. Where automatic combustion air dampers are used, they must be interlocked with the burner such that the burner will not fire if the damper is not proven open. Limit switches must be on the driven member of the damper.

14.3 Venting

A boiler room requires proper venting to completely remove the products of combustion. The venting in a boiler room requires a vent connector and a chimney, which is shown in Figure 14.1.

An appropriate vent connector is used to connect a chimney. The material for the connector should be suitable for the fuel to be burned. For example, a type B vent is used only for gas and not for oil. When more than one vent is connected to a common stack, the area of stack is required to be equal or greater than the area of the largest vent plus half of the area of the other connected vents. If a boiler is installed with a direct vent, the manufacturer's recommendations should be followed.

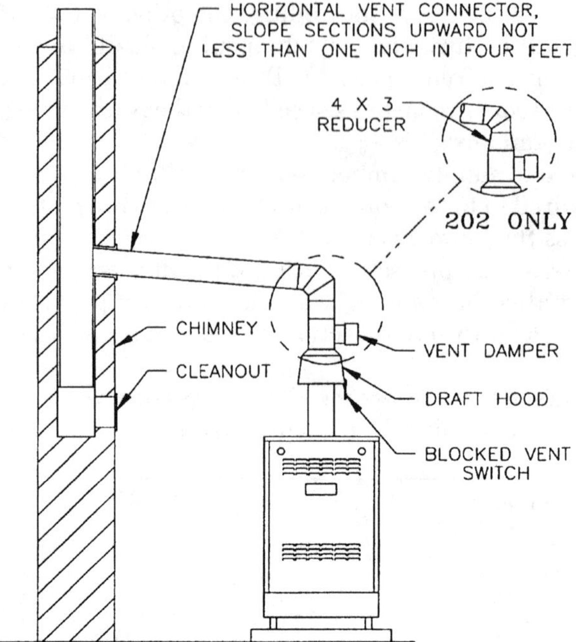

Figure 14.1 Typical vent installation. (*Courtesy: Burnham Corporation*)

Chimneys. Chimneys are used to discharge the products of combustion to the atmosphere. The application of natural draft chimney is limited as modern boilers are designed with a high rate of heat transfer, which results in increased draft loss. This draft loss, when combined with lowering of the gas temperature, required a higher chimney. These conditions can be met by using mechanical draft equipment. The chimneys are installed in connection with both induced-draft and pressurized-furnace units and the products of combustion are discharged at higher elevations.

Chimneys should be installed in accordance with the jurisdictional and environmental requirements, and the manufacturer's recommendations.

Power ventilators. Power ventilators or fans are sized based on 0.2 ft^3/min (0.028 m/m) for each 1000 Btu/hr of maximum fuel input for the total combined burners in the boiler room. The installer should ensure that the blower or fan does not create drafts, which could shutdown the pilot. The velocity of air through the ventilating fan should not exceed 500 ft/min.

14.4 Boiler Placement

Heating boilers should be installed in accordance with the rules of the local jurisdictions. The installing contractor should apply for an installation permit to the local jurisdiction before the boiler is installed. The installation application is submitted with drawings, specifications, and layout plans for the boiler for review by the jurisdictional authority. The boiler can only be installed after the jurisdictional authority grants permission to the installer.

Clearances. The boiler(s) should be placed in a boiler room with at least 36 inches (91.5 cm) of clearance on each side of the boiler(s) for easy operation, maintenance, and inspection. Boilers operated in batteries should not be installed closer than 36 inches (91.5 cm) from each other. A minimum height of at least 36 inches (91.5 cm) should be provided between the top of the heating boiler proper and the ceiling. Heating boilers with manholes should have a minimum 60 inches (153 cm) of clearance between the manhole opening and any wall, piping, or other equipment that may prevent a person from entering the boiler. Modular heating boilers, which require individual units to be installed side by side, front to back, or by stacking, should provide clearances in accordance to the manufacturer's recommendations.

Clearances from combustible surfaces. All boilers should be installed on a noncombustible floor and should not be installed on carpeting. There should be minimum distances maintained from combustible surfaces. Table 14.1 shows the minimum clearance space required to be maintained from combustible surfaces.

14.4.1 Mounting pressure relief valves

Safety and safety relief valves. Safety and safety relief valves should be installed on the top or side of the boiler. They are required to be connected directly to an opening in the boiler or to a header connecting steam or water outlets on the boiler. The valves should be installed with their spindles vertical. When a boiler is fitted with two or more safety valves on one connection, this connection should have a cross-sectional

TABLE 14.1 Minimum Clearances from Combustible Surfaces

Clearance from	Indoor, in.
Top	24
Water Connection Side	24
Opposite Side	24
Front	4
Rear	24
Vent	6

area not less than the combined areas of inlet connections of all the safety valves. When two or more safety valves are used on a boiler, they may be single, directly attached, or mounted on a Y-base.

Safety and safety relief valves should not be connected to an internal pipe in the boiler. No shutoff valve of any description is allowed to be installed between the safety or safety relief valve and the boiler, or on discharge pipes between such valves and the atmosphere.

Discharge piping. A discharge pipe is used at the outlet of the safety or safety relief valve for discharging steam or water. The cross-sectional area of the discharge pipe shall not be less than the full area of the valve outlet. The discharge pipe should be short and straight as possible. The discharge from the safety or safety relief valves should be piped away from the boiler to a safe point and should be drained properly (Figure 14.2).

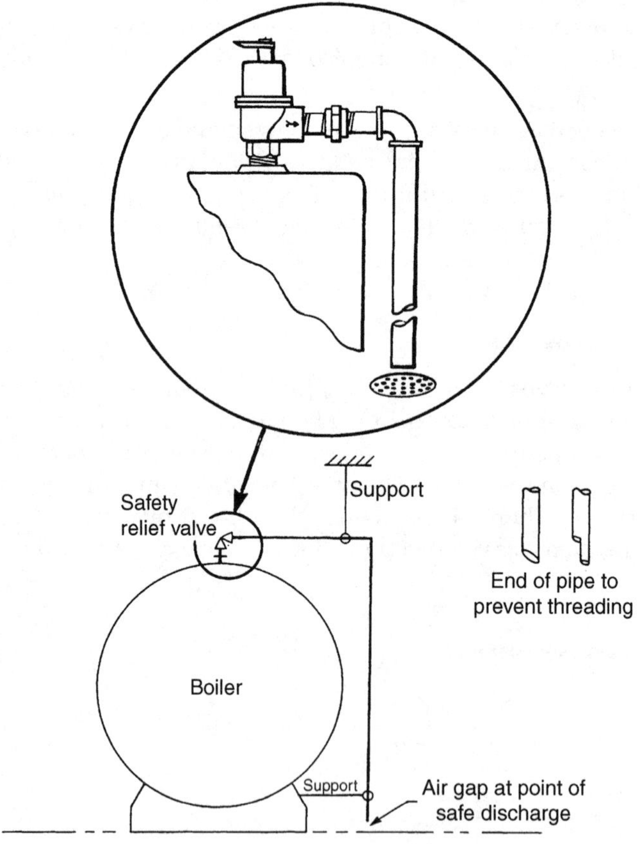

Figure 14.2 Safety relief valve discharge pipe. (*Courtesy: ASME Code Section VI*)

T&P safety relief valves. A temperature and pressure (T&P) relief valve is used on a hot-water heating boiler or supply boilers limited to a water temperature of 210°F (99°C). The T&P valve should be mounted directly on the boiler. The valve may be installed in the horizontal position with the outlet, with the discharge pipe pointed downward.

14.4.2 Feedwater and makeup water connections

Steam boilers. Feedwater for a steam boiler should be introduced through an independent connection or through the return piping system. The feedwater flow from the independent connection should not be discharged directly on any boiler parts exposed to direct radiant heat from the furnace. A check valve should be installed on a feedwater line near the boiler. A stop valve should be installed between the check valve and the boiler, or between the check valve and the return pipe system.

Hot water boilers. Makeup water is normally added for hot water boilers. Makeup water should be introduced through an independent connection or through the piping system. Makeup water should not be discharged directly on boiler parts exposed to direct heat. A check valve should be installed on the makeup water line near the boiler and a stop valve should be installed between the check valve and the boiler, or between the check valve and the piping system.

14.4.3 Oil heaters

An oil heater is used to preheat the oil before it goes to the burner. The *ASME Code* Section IV requires that:

- A heater for oil, harmful for boiler operation, should not be installed directly in the steam or water space within a boiler.

- Means should be provided to prevent introduction of oil harmful for boiler operation.

14.4.4 Storage tanks

Generally a water storage tank is used when capacity of a system exceeds 120 gallons. The storage tank may be designed under *ASME Code* Section VIII, Division 1 or *ASME Code* Section X. For tanks constructed in accordance with *ASME Code* Section X, the maximum allowable temperature marked on the tank should equal or exceed the maximum water temperature marked on the boiler. A typical storage tank for a hot water supply system is shown in Figure 14.3.

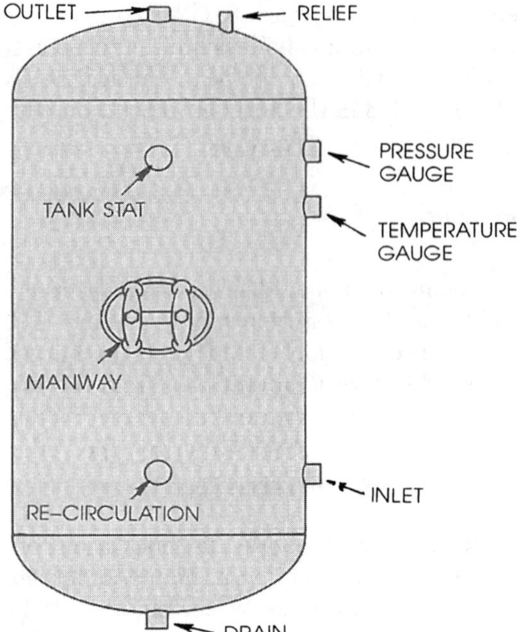

Figure 14.3 Water storage tank.

14.4.5 Expansion tanks

An expansion tank is used to provide thermal expansion in hot water systems. An expansion tank may be an open type or a closed type.

Heating system with open expansion tank. An indoor overflow from the upper part of the open-type expansion tank is provided in addition to an open vent. The overflow is carried within the building to either a suitable plumbing fixture or the basement.

Closed heating system. A closed expansion tank of proper size is installed to handle the capacity of the system. If the system is designed for a working pressure of 30 psi (200 kPa) or less, the tank should be designed for a minimum hydrostatic test pressure of 75 psi (520 kPa). If the system is designed for more than 30 psi (207 kPa), the tank is required to be constructed in accordance with *ASME Code* Section VIII, Division 1 or *ASME Code* Section. There should be provisions for draining the tank without emptying the system except for the pressurized tanks.

The minimum capacity of the closed expansion tank may be determined from Table 14.2:

TABLE 14.2 Expansion Tank Capacities for Forced Hot Water Systems

	Tank capacity, gal (m^3)	
System volume, gal (m^3)	Pressurized diaphragm type	Nonpresurized type
100 (0.38)	9 (0.034)	15 (0.057)
200 (0.76)	17 (0.064)	30 (0.114)
300 (1.14)	25 (0.095)	45 (0.170)
400 (1.51)	33 (0.125)	60 (0.227)
500 (1.89)	42 (0.159)	75 (0.284)
1000 (3.79)	83 (0.314)	150 (0.568)
2000 (7.57)	165 (0.625)	300 (1.136)

NOTES:
1. The capacity is based on an average operating water temperature 195°F (90°C), fill pressure 12 psig (83 kPa), and maximum operating pressure 30 psig (200 kPa).
2. System volume includes volume of water in boiler, radiation, and piping, not including the expansion tank.

Hot water supply system. If the system has a check valve or pressure reducing valve in the cold water line, an airtight expansion tank or other suitable air cushion should be installed. Otherwise, the safety relief valve may lift periodically due to thermal expansion of water.

14.4.6 Stop valves

A stop valve is used on the main feedwater line going from the discharge side of the feedwater pump to the steam boiler. The stop valve could be a gate or globe valve. If a globe valve is used, it should be installed so the pressure from the feedwater pump is under the seat of the valve.

For single steam boiler. If a stop valve is used in the supply pipe connection of a single steam boiler, there should be one used in the return pipe connection.

For single hot water boiler. Stop valves should be located at an accessible point in the supply and return pipe connections close to the boiler nozzle. This permits draining the boiler without emptying the system. A stop valve is not required if the boiler is located above the system and can be drained without draining the entire system.

For multiple boiler installations. A stop valve should be installed in each supply and return pipe connection of two or more boilers connected to a common system.

Type of valves. The minimum pressure rating of all valves should be at least equal to the pressure stamped on the boiler and the temperature rating should not be less than 250°F (121°C). Valves should be flanged, threaded, or have ends suitable for welding or brazing.

Identification of stop valves. The stop valve should be properly designated by a metal tag or other material fastened to the valves:

Supply valve—number ( )

Do not close without "also"

Closing return valve—number ( )

Return valve—number ( )

Do not close without "also"

Closing supply valve—number ( )

14.4.7 Bottom blow-off and drain valves

Bottom blow-off valves. A bottom blow-off connection fitted with a valve or cock should be connected to the lowest water space. The sizes of bottom blow-off piping, valves, and cocks is shown in Table 14.3.

Drain valves. One or more drain connections fitted with valves or cocks are required to be connected to the lowest water containing spaces of a steam or hot water boiler. The minimum size of drain piping, valves, and cocks should be 3/4 inch (19 mm). The discharge piping should be full-sized to the point of discharge.

Minimum pressure rating. The minimum pressure rating for blow-off valves and cocks are required to be at least equal to the pressure stamped on the boiler, but in no case less than 30 psi (207 kPa). The temperature rating for such valves and cocks should be not less than 250°F (120°C).

TABLE 14.3 Size of Bottom Blow-off Piping, Valves, and Cocks

Minimum required safety valve capacity, lb (kg) of steam/hr	Blow-off piping, valves, cocks size, min. in.(mm)
Up to 500 (225)	3/4 (20)
501 to 1250 (225 to 550)	1 (25)
1251 to 2500 (550 to 1200)	1 1/4 (32)
2501 to 6000 (1200 to 2700)	11/2 (40)
6001 (2700) and larger	2 (50)

NOTE:
1. To determine the discharge capacity of safety relief valves in terms of Btu, the relieving capacity in lb of steam/hr is multiplied by 1000.

14.4.8 Methods of support

Loadings. The design and attachment of lugs, hangers, saddles, and other supports should take into account the stresses due to hydrostatic head in calculating the minimum thickness required.

A horizontal-return tubular boiler over 72 inches (1800 mm) in diameter should be supported from steel hangers by the outside suspension-type setting, independent of the furnace wall. The hangers should be so designed that the load is properly distributed.

Lugs or hangers. Lugs, hangers, or brackets made of materials in accordance with Code requirements may be fusion-welded if they are attached by fillet welds. The bracket plates should be spaced at least 2 1/2 inches (64 mm) apart, but this dimension may be increased if necessary to permit access for the welding operation. If it is not practical to attach lugs, hangers, or brackets by welding, studs with not less than 10 threads/in. (approx. 4 threads/cm) may be used.

Setting. Steel boilers of the wet-bottom type having an external width of over 36 inches (900 mm) should not have less than 12 inches (300 mm) between the bottom of the boiler and the floorline. When the width is 36 inches (900 mm) or less, the distance between the bottom of the boiler and the floorline should not be less than 6 inches (150 mm).

14.4.9 Piping

A pipe expands with the application of heat and contracts when in contact with cold fluid. Provisions should be made for handling enough expansion and contraction of steam and hot water piping connected to the boilers by providing sufficient anchorage at suitable points and by providing swing joints when boilers are installed in batteries so that no undue strain is transmitted to the boilers.

Figures 14.4 through 14.8 show typical schematic arrangements of piping for steam-heating boilers and hot water boilers.

Return pipe connections. The return pipe connections of each supplying a gravity-return steam-heating system should be so arranged as to form a loop so that water in each boiler cannot be forced out below the safe water level as shown in Figure 14.6.

Provision should be for cleaning the interior of the return piping at or close to the boiler. Washout openings may be used for return pipe connections, and washout plugs placed in a tee or across so that plug is close to the opening in the boiler.

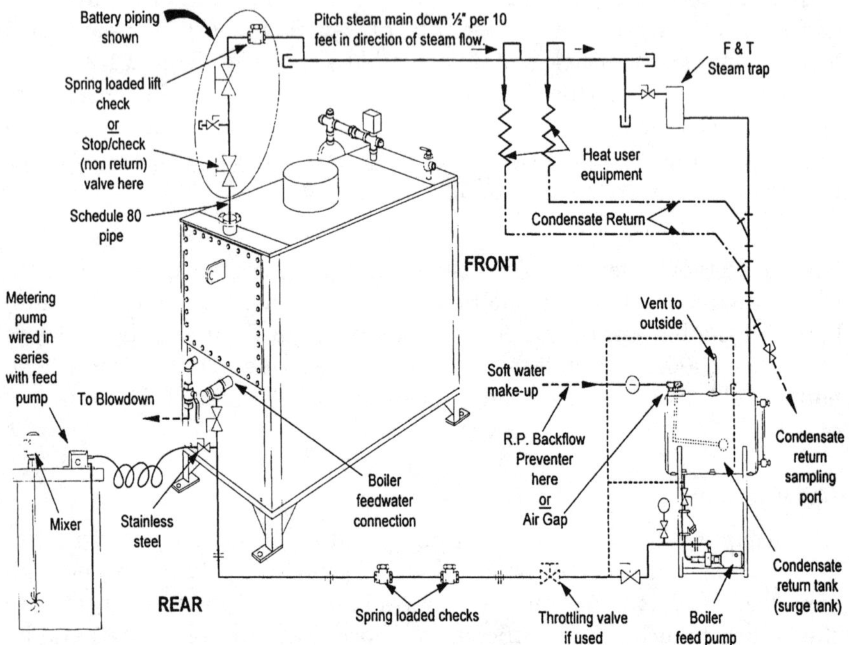

Figure 14.4 Installation of a single steam boiler with condensate tank. *(Courtesy: Rite Engineering & Mfg. Corp.)*

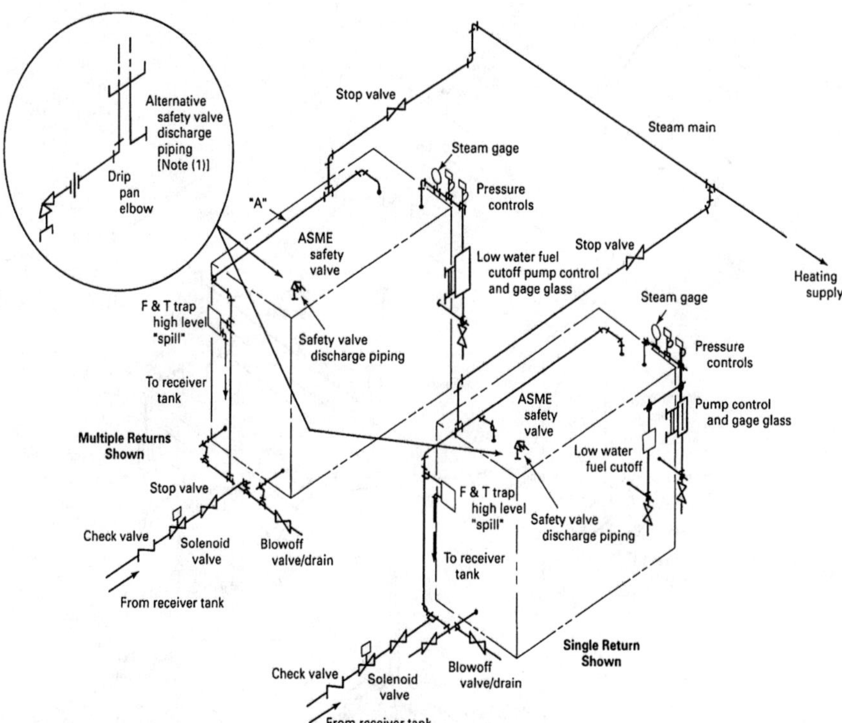

GENERAL NOTE: Return connections shown for a multiple boiler installation may not always insure that the system will operate properly. In order to maintain proper water levels in multiple boiler installations, it may be necessary to install supplementary controls or suitable devices.

NOTE:
(1) Recommended for 1 in. (DN 25) and larger safety valve discharge.

Figure 14.5 Installation of steam boilers in battery-pumped return. *(Courtesy: ASME Code Section IV)*

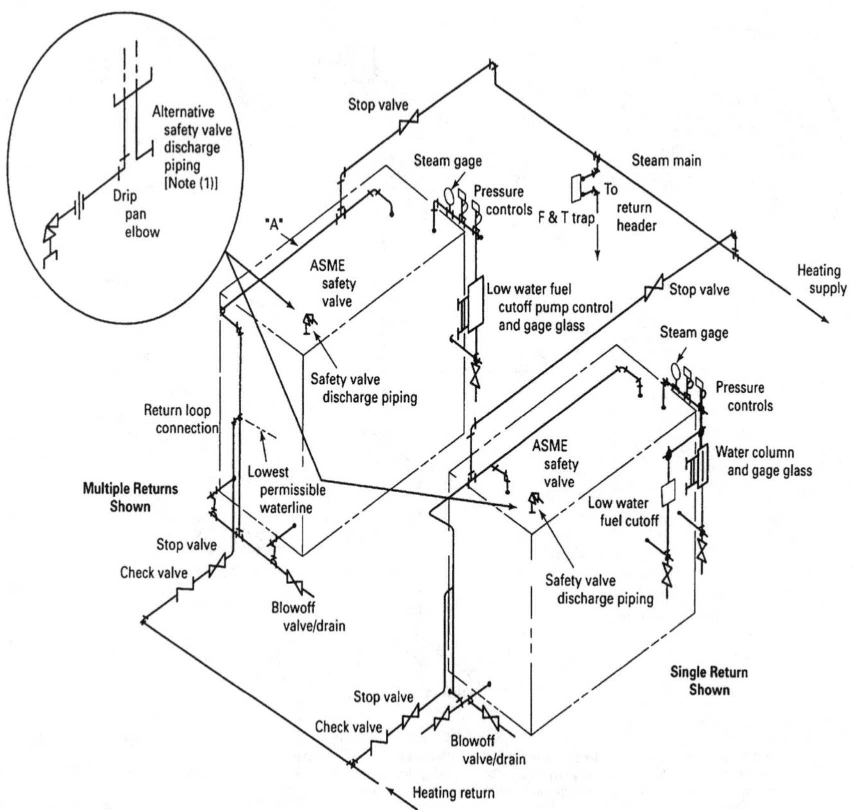

GENERAL NOTE: Return connections shown for a multiple boiler installation may not always insure that the system will operate properly. In order to maintain proper water levels in multiple boiler installations, it may be necessary to install supplementary controls or suitable devices.

NOTE:
(1) Recommended for 1 in. (DN 25) and larger safety valve discharge.

Figure 14.6 Steam boilers in battery-gravity return.

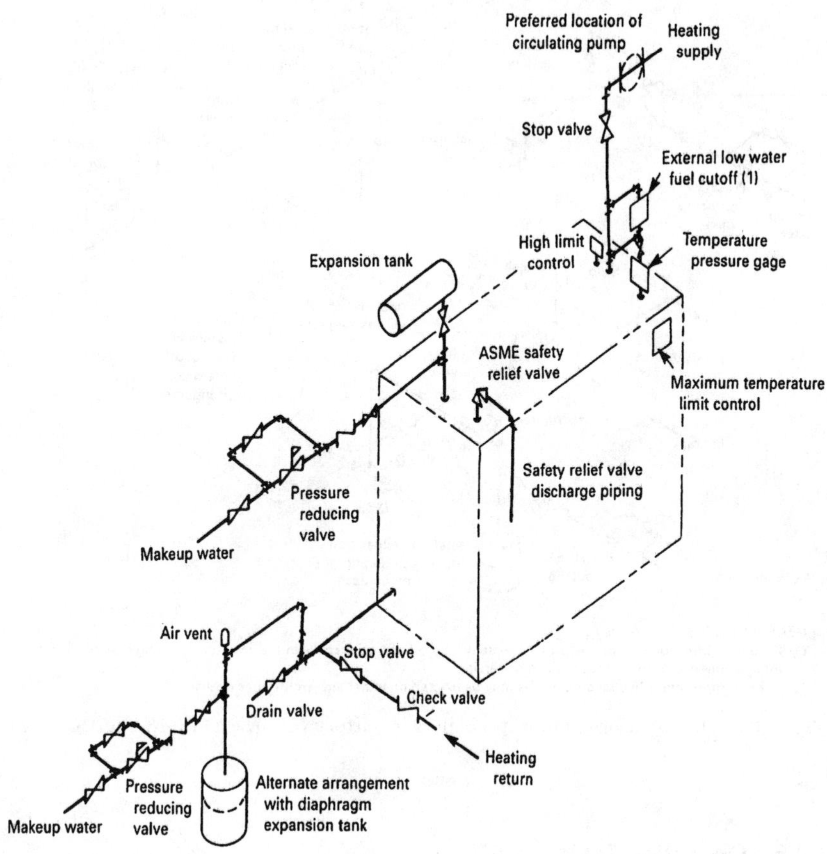

GENERAL NOTE: Plumbing codes may require the installation of a reduced pressure principle backflow preventer on a boiler when the makeup water source is from a potable water supply.

Figure 14.7 Installation of single hot water heating boiler. *(Courtesy: ASME Code Section VI)*

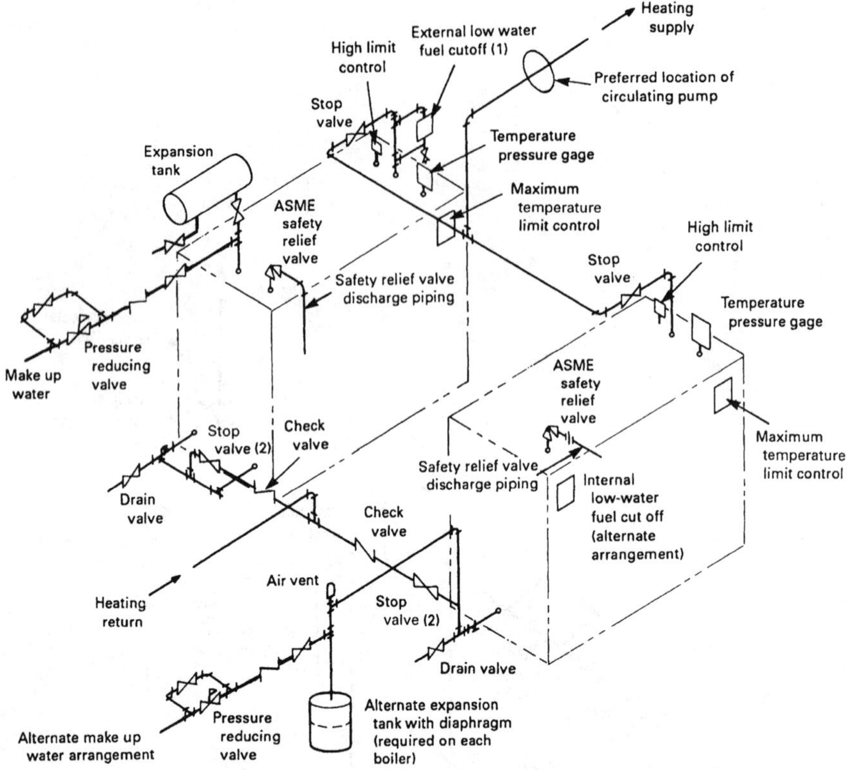

GENERAL NOTES:
(1) Recommended control. See HG-614. Acceptable shutoff valves or cocks in the connecting piping may be installed for convenience of control testing and/or service.
(2) The common return header stop valves may be located on either side of the check valves.

Figure 14.8 Installation of hot water boilers in battery. *(Courtesy: ASME Code Section IV)*

14.5 Operation Tests

The boiler unit is required to be tested after installation is complete. The manufacturer's representative may test the boiler for performance. The owner should not accept the equipment if the unit is not functioning satisfactorily. The manufacturer's representative or the installing contractor must submit an operational test report. This report indicates that all the accessories, auxiliaries, fittings, mountings, controls, and safety devices have been tested, and they are functioning properly. The report may record efficiency test of the boiler, indicating manufacturer's guaranteed efficiency, if any. A sample operation test form for a heating boiler is shown in Figure 14.9.

Job	Contractor	Engineer	Date
Address	Address	Address	Address
City St. Zip	City St. Zip	City St. Zip	Performed By
Phone | Name	Phone | Name	Phone | Name	Authorized By
Model	Serial#	Hydronic ☐ Water Heater ☐	Other ☐

1. Service & installation Manual on job: Yes ☐ No ☐
2. Wiring diagram ☐ Piping diagram ☐

VENTING

3. Atmospheric ☐ Power burner ☐
Power vent ☐ Draft inducer ☐
Other ☐ ______ ☐
4. Stack Material CPVC ☐
Single wall ☐ Stainless ☐
Double wall ☐ Category 1,2,3,4____ ☐
Pressure stack ☐ Other______ ☐
5. Draft diverter Yes ☐ No ☐ Modified ☐
6. Stack temp. at 1 vent diameter above unit top____°
7. Vent termination AGA approved Yes ☐ No☐
8. Combustion analysis: % O_2____ % Co_2____
PPM Co.____ PPM NO_2____% Efficiency ____
9. Exhaust near windows or other openings of building
Yes ☐ How close!________ No☐

COMBUSTION AIR

10. Atmospheric ☐ Power burn ☐
Sealed combustion ☐ Ducted combustion air☐
Meets requirements in manual? Yes ☐ No. ☐
11. Air Filter ☐ Clean ☐ Obstructed ☐
12. Altitude above sea level ______ft.
13. Other equipment in room Yes ☐ No ☐
14. Blockage of louvers in: Doors ☐ Walls ☐
Hi ☐ Low ☐ Only one ☐ Screens ☐
Boarded over ☐ Other ______
15. Powered combustion air fans Interlocked ☐
Intake ☐ Exhaust ☐ Other ____ ☐ None ☐
16. Proximity: Any gases, chemicals, corrosives, cleaning fluids, other volatile material that may contaminate combustion air including dust, lint, or any material that may affect combustion.
Contaminants Yes ☐ No ☐
Chillers - Yes ☐ Freon Leaks Yes ☐ No ☐

CLEARANCES

17. Front ______ Rear ______
Left ______ Right ______ Top ______
Floor: Combustible ☐ Non-Combustible ☐

18. Location of unit: Indoor ☐ Outdoor ☐
Near exterior wall ☐ ____'____" Penthouse ☐
Basement ☐ Interior Room ☐

GAS & PIPING

19. Natural ☐ Propane ☐ High Pressure ☐
20. Supply pressure lock up ______"wc.
21. Supply pressure at full fire ______"wc.
22. Manifold pressure at full fire ______"wc.
23. Gas pipe diameter ______ Orifice size ______
24. Distance from meter ______
25. Meter size: Adequate ☐ No ☐
26. Gas valve size manual ☐ Lubricated valve ☐
27. Gas regulation One stage ☐ Two stage ☐
28. Single stage on/off ______"wc.
Stage two ______"wc.
Stage three ______"wc.
Stage four ______"wc.
29. Gas leaks: Detected Yes ☐ No ☐

WATER & PIPING

30. ASME# ______ Copper ☐ Cupronickel ☐
31. Pipe size ______ = Manual Yes ☐ No ☐
32. P.R.V. in excess of: Input ☐ Output ☐
Pressure ______ PSI. BTU ______
33. P.R.V. plumbed to floor ☐ or drain ☐
34. Heat Exchanger: Clean ☐ Sooted ☐ Corroded ☐
35. Piping: full size ☐ reduced ☐ restricted ☐
36. Cold water inlet to: pressure reducing valve ☐
check valve ☐ to tank ☐ Supply to heater ☐
return from heater to tank ☐
37. Expansion tank Yes ☐ No ☐
Air tank ☐ Precharged bladder ☐ Gravity ☐
38. Temperature rise at full fire ______°
39. Isolation values: Yes ☐ No ☐
40. Water leaks: Yes ☐ No ☐
41. Water hardness: Soft ☐ Normal ☐ Hard ☐
Grains of Hardness 1-7 ☐ 7-17 ☐ >17 ☐
42. Pump: ______ HP Bronze ☐ Cast iron ☐
Factory ☐ Field ☐
Rotation OK ☐ Reversed Rotation ☐

ELECTRICAL WIRING

43. Volts: 110 ☐ 220 ☐ Secondary ______ H_z ☐
44. Unit wired to designated breaker: Yes ☐ No ☐
45. Unit interlocked to pump circuit: Yes ☐ No ☐

CONTROLS

46. Fused Circuit on unit: Yes ☐ No ☐
47. Ignition: Spark ☐ Standing Pilot ☐
100% Lockout ☐ Hot surface ☐
Millivolt ☐ Other ☐
48. Pilot: Clean blue flame ☐ Arcing out of pos. ☐
Corroded/sooted ☐ Cracked ceramic ☐
Broken hot surface ☐ Other ______ ☐
49. Flame rod alignment: Good ☐ Faulty ☐
Millivolt output ______ mv
Flame rod output ______ dc/ma
Hot surface output ______ dc/ma
50. Ignition control: Normal ☐ Faulty ☐ Irratic ☐
51. Alarm package: Yes ☐ No ☐
52. Pump time delay: Yes ☐ No ☐
53. Temp gauge ☐ Press. gauge ☐ Yes ☐ No ☐

54. Firing mode: On/Off ☐ 2 Stage ☐ 4 Stage ☐
Modulation ☐ Mechanical ☐ Motorized ☐
Sequencer ☐ Setpoint ☐ Outdoor Reset ☐
Good Bad
55. Operating control: ______ ☐ ☐

SAFETY CONTROLS

56. High limit: ____ manual ☐ auto ☐ ☐ ☐
57. Flow switch: adjusted out ☐ ☐ ☐
58. High gas manual ☐ auto ☐ ☐ ☐
59. Low gas manual ☐ auto ☐ ☐ ☐
60. Low water cut off: manual ☐ auto ☐ ☐ ☐
61. Terminal strip interfacing energy mgt. ☐ ☐
62. Pressure differential switch:
Adjustable ☐ ______ w/c ☐ ☐
63. Sail switch: manual ☐ auto ☐ ☐ ☐
64. Vent valve: manual ☐ auto ☐ ☐ ☐
65. Codes: IRI ☐ CSDI ☐ California code ☐
FM ☐ Other ☐ ASHRAE90.1 Ill. school ☐
66. Video recording of start-up: Yes ☐ No ☐
Photos of installation: Yes ☐ No ☐
Drawing of system: Yes ☐ No ☐
67. Person in attendance ______
SIGNED

Comments

RBI

Figure 14.9 Operation test form for a heating boiler. *(Courtesy: RBI Boiler Company)*

14.6 Documentation

After installation of the boiler, the contractor must submit a report to the owner about the installation. The contractor should make sure that the boiler unit is operating properly and all instruments and devices are tested and functioning well. In many jurisdictions, where ASME Code CSD-1 has been adopted, the installer must complete a CSD-1 report as shown in Appendix L. The owner should have a copy of this report on file for an authorized inspector's verification at the time of first inspection immediately after installation.

Chapter 15

Operation

Boiler operation is very important as safety, efficiency, and a constant supply of steam or hot water depends on the smooth operation of heating boilers. Boiler operation is a complicated system, especially the modern boilers that are fully automatic. Any mistake in operation can cause plant shutdown, and even explosion of the boilers. On the other hand, a boiler plant can run trouble-free if boilers are operated properly.

A boiler operator or engineer, depending on the size of the boiler, is responsible for its operation. These operators and engineers should be qualified as they make very crucial decisions at times. Many jurisdictions like states, provinces, counties, and cities have laws and rules for qualifying boiler operators and engineers.

There are many textbooks available on boiler operation. It is advisable to follow the manufacturer's operation manuals as they are written for very specific models of boilers. These operation manuals describe in detail the operating procedures, including emergencies in the boiler room. In addition to the manufacturer's operating manuals, *ASME Boiler and Pressure Vessel Code* Section VI—Recommended Rules for the Care and Operation of Heating Boilers may be followed as guidelines for the safe operation of heating boilers.

15.1 Boiler Operators and Engineers

The *ASME Boiler and Pressure Vessel Code* does not cover attendant requirements for heating boilers, as these rules do not address the qualification and experience of boiler operators and engineers. However, it is internationally recognized that a qualified boiler operator or engineer is essential for the safe operation of heating boilers.

The duties of a boiler operator or engineer include observation and manipulation of mechanical, automatic, or remote controls, as well as

The International Society of Boiler and Pressure Vessel Inspectors, Inc.

ISBI

www.boilersociety.org

Safety By Profession

License No. 6178

The International Society of Boiler and Pressure Vessel Inspectors (ISBI) does hereby certify that through education, experience, and written examination,

William P. Smith, BPO

Has duly achieved the standards of competence established by ISBI and shall be forthwith acknowledged by designation of

Boiler Plant Operator (BPO)

The license holder is qualified to take charge of or operate boiler plant(s) up to 2 Million Btu/hour

Certification Date: June 9, 2006

Expiration Date: December 31, 2008

President

Executive Director

Figure 15.1 A boiler plant operator certificate. *(Courtesy: International Society of Boiler and Pressure Vessel Inspectors)*

the testing, clearing, and blowdown or draining of these controls to ensure the proper operation of these devices. A boiler operator or engineer should be able to observe the monitoring equipment or devices of a low-pressure boiler at least once every 8 hours.

A boiler operator or engineer for a low-pressure boiler may be qualified under the laws of a government jurisdiction (state, province, county, or city). However, many governmental jurisdictions require that the owner employ licensed boiler operators and engineers to operate low-pressure boilers.

Many community colleges and technical institutions offer courses for the certification of low-pressure boiler operators. Common titles for these qualified personnel include boiler plant operator, and third, second, and first class boiler engineer. A boiler operator or engineer may take examinations given by the International Society of Boiler and Pressure Vessel Inspectors (ISBI), Inc. (www.boilersociety.org). A Boiler Plant Operator certificate issued by the ISBI is shown in Figure 15.1.

15.2 Steam Boiler Operation

15.2.1 Cleaning and filling a new boiler

A newly installed boiler must be fully cleaned and filled with water before start-up. It is also important to flush fuel systems of construction site debris prior to the first use. The following steps should be taken to prepare the boiler before starting:

1. *Inspection.* Interior should be inspected to make sure that no foreign matter, such as tools, equipment, and rags, is left in the boiler.
2. *Check before filling.* Ensure that the burner is in operating condition, because the water in the boiler should be boiled immediately after it is introduced into the boiler. This boiling, or heating the water to a minimum of 180°F removes dissolved gases, which might otherwise corrode the boiler.
3. *Clean the system.* After the boiler is filled to the proper waterline, the system should be operated with steam in the entire system for a few days to bring oil and dirt back from the system to the boiler.
4. *Boiling out.* Boiler-out the system by the boiling-out method described in Section 15.2.2.

15.2.2 Boiling-out method

Boiling out is a chemical cleaning method used to remove oils and greases from a newly installed boiler or a boiler that has been subjected to major repairs, such as re-tubing. This is accomplished by the following simple boiling-out method:

1. Fill the boiler to the normal waterline.
2. Remove the plug from tapping on the highest point on the boiler.
3. Add an appropriate boilout compound through the opening.
4. Replace the plug or safety valves.
5. Start the firing equipment and check the operating, limit, and safety controls.
6. Boil the water for a minimum of 5 hours.
7. Stop the firing equipment.
8. Drain the boiler.
9. Wash the boiler thoroughly with a high-pressure water stream.
10. Fill the boiler to the normal waterline.
11. Add boiler treatment compound as needed.
12. Boil the water or heat it to a temperature of 180°F.
13. The boiler is ready to put into service.

The boilout compound mentioned in Step 3 is a type of alkaline detergent compound of varying constituents. A qualified water treatment chemical specialist should be consulted regarding the appropriate chemical compound.

15.2.3 Starting a boiler (single boiler installation)

Procedure. The following steps should be taken when a boiler is started up:

1. Review the manufacturer's recommendations for start-up of the burner and boiler.
2. Set control switch in "Off" position.
3. Make sure combustion air to the boiler room is unobstructed.
4. Check availability of fuel.
5. Check water level in gage glass.
6. Use try cocks, if provided, to double-check the water level.
7. Vent combustion chamber to remove unburned gases.
8. Clean glass on flame scanner.
9. Set main steam shutoff valve in the open position.
10. Open the cold water supply valve to water feeder. Open suction and discharge valves on vacuum or condensate pumps, and set electrical switches to the desired position.
11. Check the operating pressure setting of the boiler.
12. Check manual reset, if provided, on the low-water fuel cutoff and high-limit pressure control to determine if they are properly set.
13. Set the manual fuel oil supply or manual gas valve in the "Open" position.
14. Place circuit breaker or fused disconnect switch in "On" position.
15. Place all boiler emergency switches in the "On" position.
16. Place boiler control starting switch in the "On" or "Start" position.
17. Bring pressure and temperature up slowly.
18. During start-up, walk around the boiler frequently to observe that all associated equipment and piping are functioning properly.
19. Immediately after the burner shuts off, inspect the water column, and open each try cock, if provided, individually to determine the true water level.
20. Enter in the log book:
    a. Date and time of start-up.
    b. Any irregularities observed and corrective action taken.

c. Time when controls shut-off the burner at established pressures, tests performed, and the like
d. Signature of operator

21. Check safety valve for evidence of simmering. Perform the try lever test.

Abnormal conditions. If any abnormal conditions occur during light-off or pressure buildup, immediately open the emergency switch. Do not attempt to start the boiler until problems have been identified and corrected.

15.2.4 Condensation

When a gas-fired boiler is started from a cold condition, condensation may occur in the boiler and pipeline. Depending upon the cold condition, quantity of the condensed water may be so much that it appears the boiler is leaking. However, after the boiler is hot, the condensed water dries up and condensation disappears.

15.2.5 Cutting in an additional boiler

When an additional boiler is started and placed on the line with other boilers already in service, the following steps should be taken:

1. Start the additional boiler with the steam stop valve and the return stop valve closed.
2. Open the steam stop valve very slightly when the pressure within the boiler is the same as the pressure in the main steam line.
3. If conditions are normal, continue to open the steam stop valve slowly until it is fully open.
4. Open the stop valve on the return line.
5. *Caution.* When the stop valve at the boiler outlet is closed, the stop valve in the return line of that boiler should also be closed.

15.2.6 Operation

Equipment in the boiler room varies depending on the type and size of the heating plant. A boiler operator or engineer (Figure 15.2) shall be familiar with all the operating conditions and variables for operation of all equipment in the boiler room. It is the operator's or engineer's duty to maintain proper operating conditions and to deliver required steam pressure at the boiler outlet. The following conditions must be checked as a part of routine boiler operation:

Water level. The first thing a boiler operator should do when entering a boiler room is to check the water level on all boilers. This is accomplished by blowing down the gage glass in the following order:

Figure 15.2 A boiler operator learning operation from an engineer. *(Courtesy: Johnston Boiler Company)*

1) Close the lower gage glass valve, then open the drain cock, which is located on the bottom of this valve, and blow the glass clear; 2) close the drain cock and open the lower gage glass valve; 3) at this point, water should return to the glass immediately. If the water is not returned promptly, leave the lower gage glass open and close the upper gage glass valve; 4) open the drain cock and allow water to flow until it runs clear; 5) at this instant, close the drain valve and repeat the first test above, with the lower gage glass valve closed; 6) if water disappears from the gage glass, blow down gage to see if water returns; 7) if water does not appear, shut down the boiler immediately; 8) finally, cool down the boiler and add water to 1 inch in the gage glass.

Steaming pressure. Occasionally, the safety valve(s) fails to open due to the buildup of corrosive deposits between the disc and seat of the safety valve. Also, long periods of leakage can cause the safety valve to stick or freeze. The pressure differential between the safety valve set pressure and the operating pressure should be a minimum of 5 psi; i.e., the boiler operating pressure should not exceed 10 psig. Make sure that the pressure gage indicates correct operating pressure, as the gages cannot be relied upon at the low end of the scale. If operating pressure more than 10 psig is required, it should not exceed 15 psig minus the blowdown pressure of the safety valve.

Blowdown. Blowdown is required to remove precipitated sediments and to maintain chemical concentrates at the desired level. This is accomplished by opening a quick-opening blow-off cock and a slow-opening blow-off valve installed at the bottom of the shell, or mud drum, of the boiler. While blowing down, the quick-opening cock should be opened first, followed by a gradual opening of the slow-opening valve. When the slow-opening valve is shut tight, then close the quick-opening cock.

Appearance of rust. If rust is visible in the water gage glass, then it is an indication of corrosion in the system. First, check the return line and other parts of the system for corrosion. Second, analyze boiler and makeup water, and add water treatment compound accordingly.

Waterline fluctuation. Foaming or priming can occur due to fluctuation of the waterline in the boiler. Foaming is the existence of a layer of foam on the surface of the water in the boiler drum and may be caused by dirt, oil, or other impurities in the boiler water. Priming occurs when small particles of water are carried into steam lines and may be caused by too high of a water level in the boiler. Foaming and priming can be reduced by blowing the boiler down.

Abnormal water losses. An abnormal water loss in a steam boiler is indicated by the requirement of large amounts of makeup water. In case there is abnormal water losses from the boiler, a thorough inspection should be done immediately to determine the cause. As a result of the inspection, proper repair or replacement of the parts should be undertaken to correct the problem(s).

Makeup water. Makeup water is introduced into the steam boiler to replace that lost or removed water from the system. When water makeup is required, it can be added to the boiler either with an automatic water feeder or manually. Only a minimum quantity of makeup water should be added, because as a source of oxygen, it causes corrosion. However, a deaerator may be used to remove dissolved gases, including oxygen, if a large quantity of feed water is required. The use of an oxygen scavenger may be included as a part of the chemical treatment.

Low-water cutoff. Statistics show that most of the accidents to steam boilers occur due to nonfunctioning of the low-water cutoff. For proper operation, check the low-water cutoff, pump control, and the water feeder (if installed). Also, the low-water cutoff should be tested periodically, under actual operating conditions. This test is done by shutting off the supply of feedwater to the boiler. If the boiler is equipped with an individual feed pump, the pump may be cutoff to stop the feedwater supply. With the supply of water shutoff, the water level in the boiler will slowly drop until it falls to the point where the low-water cutoff mechanism should actuate and shut the burner down. This test should be performed when the boiler is operating at low fire conditions.

15.2.7 Removal of boiler from service

Procedure. A boiler may be removed from service either for a short period of time or for an extended period of time. At the same time, a steam boiler may be taken out of service at the end of the heating season or for repair work. The following steps should be taken if a boiler is temporarily removed from service:

1. Start firing and maintain boiler water temperature between 180 to 200°F, and then drain off boiler water from bottom drain until it runs clean.
2. Refill the boiler to top of gage glass and add sufficient water treatment compound.
3. Shut down the firing equipment after all the dissolved gases have been released and disconnect the main switch.
4. Treat the water in the laid-up boilers in accordance with *ASME Code* Section 18.8— Treatment of Laid-Up Boilers.

Cleaning. Tubes and other fire side heating surfaces should be thoroughly cleaned if a boiler is to be removed from service for the nonoperating season or for any extended nonfiring period. Also, clean the flue boxes and other areas where soot or scale may accumulate.

Protection against corrosion. Neutral mineral oil may be applied on fire side equipment to protect against corrosion. Likewise, place a tray of calcium chloride or unslaked lime in the combustion chamber to keep the boiler interior dry.

Water level. The normal water level should be established before putting the boiler back in service. This is accomplished by draining the steam boiler and reducing the water level back to normal water level.

Periodic checks. During an idle period, the boiler should be checked occasionally for corrosion. Any type of situation which may result in boiler corrosion, should be avoided.

15.3 Heating Boiler and Hot Water Supply Boiler Operation

The boilers installed in hot water systems are subjected to thermally induced stress cycling, which is known as thermal shock. This type of failure occurs over a period of time, and is caused by thermally induced stress cycling of the boiler structure. Typically, this is caused by high burner on-off cycling rates (caused by improperly set operating and modulating controls) and/or low water return temperatures. Proper operation procedures should be followed to reduce thermal shock on a hot water boiler.

Boiler manufacturer's guidelines should be adhered to for the following operating parameters:

1. Recommended relationship between operating control and modulating control settings
2. Minimum return water temperature
3. Minimum water flow rate through the boiler
4. Minimum water supply temperature for the system operating pressure
5. Additional protective devices or controls

15.3.1 Cleaning and filling a new boiler

A newly installed hot water supply boiler or a hot water heating boiler must be fully cleaned and filled with water before start-up. The following steps should be taken to prepare the boiler before starting:

1. *Inspection.* Interior should be inspected to make sure that no foreign matter such as tools, equipment, and rags, is left in the boiler.
2. *Checks before filling.* The burner should be in operating condition, as water in the boiler should be boiled immediately after it is introduced into the boiler. When water is boiled or heated to a minimum of 180°F, dissolved gases will be removed. The boiler and complete hot water system should be full of water for smooth operation. And then, water should come out of all the air vents when the system is full.
3. *Operation to clean the system.* Fill the boiler to the proper waterline and operate the boiler with steam in the system for few days to clean out the oil and dirt.
4. *Boiling out.* Boiling out is a chemical cleaning method used to remove oils and greases from a newly installed hot water boiler or a hot water boiler that has undergone major repairs, such as tube replacement. This is accomplished by the following method:
   a. Add an appropriate boilout compound.
   b. Fill the entire system with water.
   c. Start the firing equipment.
   d. Circulate the water through the system.
   e. Vent the system, including the radiation.
   f. Allow boiler water to reach operating temperature, if possible.

g. Continue to circulate the water for a few hours.
h. Stop the firing equipment.
i. Drain the system to a safe location.
j. Wash the water side of the boiler thoroughly, with a high-pressure water stream.
k. Refill the system with fresh water.
l. Bring water temperature to a minimum of 180°F and vent the system at the highest point.
m. Tighten handhole covers, manhole covers, and plugs.
n. The boiler is ready to be put into service.

15.3.2 Starting a new boiler (single boiler installation)

Procedure. The following steps should be taken when a boiler is started up after lay-up:

1. Review the manufacturer's recommendations for start-up of burner and boiler.
2. Fill the boiler and system and vent air at a high point in the system.
3. Check the altitude gage and expansion tank to make sure that the system is filled.
4. Set control switch to the "Off" position.
5. Make sure fresh air to the boiler room is unobstructed and dampers are open.
6. Check availability of the fuel.
7. Vent combustion chamber to remove unburned gases.
8. Clean the glass on fire scanner, if provided.
9. Observe the functioning of water pressure regulator and turn circulator pumps.
10. Check temperature control(s) for proper setting.
11. Check manual reset, if provided, on low-water fuel cutoff and high-limit temperature control.
12. Set manual fuel oil supply or manual gas valve in the "Open" position.
13. Place circuit breaker or fused disconnect switch in the "On" position.
14. Place all boiler emergency switches in the "On" position.
15. Place boiler control starting switch in the "On" or "Start" position.
16. Do not leave the boiler until it reaches the established cutout point to make sure the controls shut off the burner.

17. During start-up, walk around the boiler frequently to observe that all associated equipment and piping are functioning properly. Check the burner visually for proper combustion.
18. Immediately after the burner shuts off, inspect water pressure and open the highest vent to determine if the system is completely full of water.
19. Enter in log book:
    a. Date and time of start-up
    b. Any irregularities observed and corrective action taken
    c. Time when the control shut off the burner at established pressure, test performed, and the like
    d. Signature of operator
20. Check safety relief valve for evidence of simmering. Perform try lever test.

Abnormal conditions. If any abnormal conditions occur during light-off or temperature buildup, immediately open the emergency switch. Do not attempt to start the boiler until problems have been identified and corrected.

15.3.3 Condensation

When a gas-fired boiler is started from cold conditions, condensation may occur in the boiler and pipeline. Depending upon the cold condition, the quantity of the condensed water may be so much that it appears the boiler is leaking. The condensed water dries up and condensation stops after the boiler is hot.

15.3.4 Cutting in an additional boiler

When an additional boiler is started and placed on the line with other boilers already in service, the following steps should be taken:

1. Start the additional boiler with the supply-stop valve and return-stop valve closed.
2. Open the supply stop valve partially when the temperature within the boiler is the same as the operating boiler.
3. If conditions are normal, continue to open the stop valve slowly until it is fully open.
4. Open the stop valve on the return line.
5. *Caution.* When the stop valve at the boiler outlet is closed, the stop valve in the return line of that boiler should also be closed.

15.3.5 Operation

Equipment in the boiler room varies depending on the type and size of the hot water system. A boiler operator or engineer shall be familiar with all the operating conditions and variables on operation of all the equipment in the boiler room. It is the operator's or engineer's duty to maintain proper operating conditions and to deliver the required water pressure and temperature at the boiler outlet. The following conditions must be checked as a part of routine boiler operation:

Check pressure and temperature. A boiler operator should check the pressure and temperature in all water boilers when starting his duty.

Position of hands on combination gage. A combination altitude gage has two hands a stationary hand and a movable hand. These two hands are together when the boiler is cold. However, when the boiler is hot the movable hand should be above the stationary hand. Also, the stationary hand may be set to indicate minimum pressure under which the system can operate.

Operating temperature and pressure. Temperature and pressure of the boiler during operation should be monitored carefully so that their maximum limit is not exceeded.

1. *Operating temperature.* Maximum operating temperature of the boiler should not exceed 250°F (121°C). The boiler should not be operated at a higher temperature as this will accelerate any corrosion process.
2. *Operating pressure.* Due to buildup of corrosive deposits between the disc and the seat, the safety relief valve fails to open at the set pressure. The safety relief valve can stick or freeze if the valve leaks for a long period of time. For these reasons, the pressure differential between the safety relief valve set pressure and the boiler operating pressure should be at least either 10 psi (69 kPa), or 25 percent of the valve set pressure, whichever is greater.
3. *Temperature and pressure safety relief valve.* When boilers limited to a maximum temperature of 210°F (99°C) have a T&P safety relief valve installed, the operating temperature should be low enough to prevent routine operation of the thermal element. When a thermal element opens, it will not close until the temperature is reduced by 25 to 35°F (14 to 19°C). Therefore, the maximum operating temperature should not exceed 160°F (70°C).

It is extremely important that periodic testing of the safety relief valve is performed according to the code adopted by the local jurisdiction.

Example 1: The safety relief valve of a hot-water heating boiler is set to open at 30 psi. What should be the operating pressure of the boiler?

The operating pressure should not exceed either 30 − 10 = 20 psi or 30 − 25% of 30 = 22.50 psi. It is recommended not to exceed 20 psi.

Example 2: The safety relief valve of a hot water heating boiler is set to open at 100 psi. What should be the operating pressure of the boiler?

The operating pressure should not exceed 100 − 10 = 90 psi or 100 −25% of 100 = 75 psi. It is recommended not to exceed 75 psi.

15.3.6 Removal of a boiler from service

Procedure. A hot-water heating boiler may be removed from service at least once a year. The following steps should be taken if at anytime a boiler is removed from the service:

1. Start firing, and maintain boiler water temperature between 180 and 200°F, and then drain off boiler water until the water runs clean.
2. Refill to the normal water fill pressure.
3. Add sufficient treatment compound.

Cleaning. If a boiler is to be removed from service for the nonoperating season or any extended nonfiring period, tubes and other fire side heating surfaces should be thoroughly cleaned. Also, clean the smoke boxes and other areas where soot or scale may accumulate.

Protection against corrosion. Neutral mineral oil may be applied on the fire side equipment to protect against corrosion. By placing a tray of calcium chloride or unslaked lime in the combustion chamber, the boiler interior stays dry.

Periodic checks. During the idle period, the boiler should be checked occasionally for corrosion. This is an opportunity to inspect and service the firing equipment and combustion chamber. Any type of situation which leads to boiler corrosion should be avoided. Exposed metal parts may be repainted.

15.4 Periodic Tests

Periodic tests of all boilers are required to maintain them in good working condition and to ensure safety. Adequate precautions should be taken while tests are being performed to protect personnel making the tests and equipment.

The periodic tests should be performed by an experienced operator or engineer who is familiar with the testing procedures. *ASME Code* Section VI—Recommended Rules for the Care and Operation of Heating Boilers has described procedures for making various tests. The following procedures are recommended:

Flame safeguard device. The following types of tests are recommended to be performed on the flame safeguard device:

- Gas—thermal type
- Oil—thermal type, stack switch
- Gas—electronic flame rod-type with standing pilot
- Gas—electronic flame rod with interrupted ignition
- Gas—electronic flame scanner type
- Oil—electronic flame scanner type
- Oil or gas—electronic type with proven pilot flame detection
- Pilot turndown test

Combustion efficiency tests. A combustion efficiency test should be made on each boiler unit at least once a year. More frequent tests should be made on larger boilers. The efficiency should be checked at the different firing rates if the burners have variable firing rates.

A portable combustion analyzer is used to test combustion efficiency. The analyzer measures and displays flue gas O_2, temperatures, draft, NO_x, and CO. Simultaneously, the instrument calculates and displays combustion efficiency, excess air, CO_2, NO_x-Ref. O_2, and CO air free. A user can store up to 100 tests, generate a personalized printout, and download all of this information to a personal computer for record keeping and trend analysis.

A stack efficiency tester can be used to measure O_2 and CO content, and stack temperature, then calculates CO_2, excess air, and combustion efficiency. A stack efficiency tester is shown in Figure 15.3.

Limit control tests. A boiler operator should test all the limit controls periodically. The manufacturer's recommendations should be followed for complete details of testing. The following controls should be tested periodically:

- High-limit steam pressure control
- Draft-limit control
- Boiler-room temperature limit control

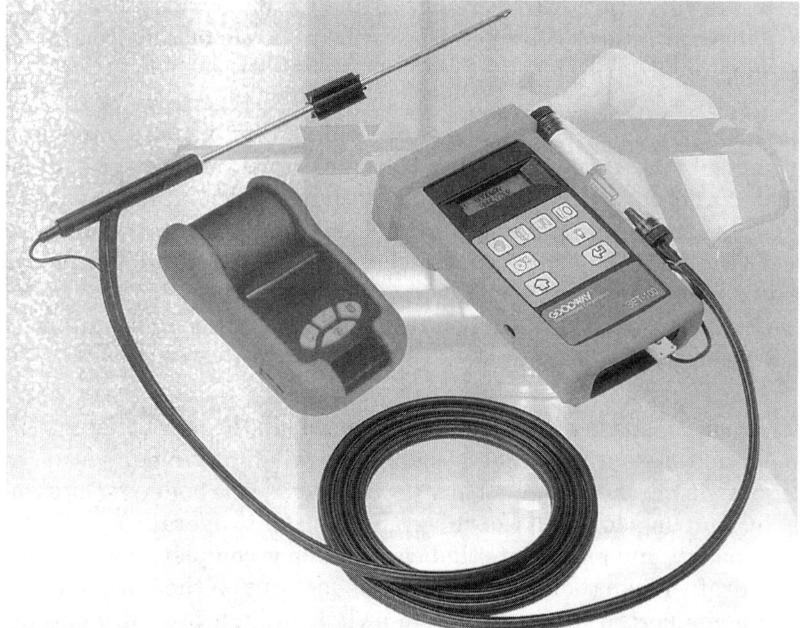

Figure 15.3 Stack efficiency tester. *(Courtesy: Goodway Technologies Corporation)*

- Electrical limit controls
- Low gas pressure control
- High gas pressure control
- Oil pressure supervisory switch
- Air pressure supervisory switch
- Emergency disconnect switch

Safety valve tests. All precautionary measures should be taken while testing safety valves. At least two operators should be present during the safety valve test. Both operators must know the emergency procedures to be taken in case of emergencies such as heavy steam leak, failure of boiler components or heavy water leakage from boiler. The following types of tests are performed on safety valves:

Try lever test. A try lever test should be conducted once a month if the boiler is in operation or when the boiler is started after any period of inactivity. With the boiler under low pressure, lift the try lever on the safety valve to the wide-open position and allow steam to discharge for 5 to 15 seconds. Release the try lever and allow the valve to close.

Operate the valve two or three times to allow the disc to seat properly. If the valve continues to simmer, it must be replaced or repaired by an authorized repairer.

Pop test. A pop test is conducted to determine that a safety valve will open under boiler pressure within allowable tolerances. It should be conducted annually. Isolate the boiler by shutting down all the valves. The test is conducted by using jumper leads across the high pressure control terminals. The safety valve should pop open at an acceptable pressure. A simmering action will be noticed before the valve pops open. The safety valve should stay open until the pressure drops to allow it to close at blowdown pressure.

Capacity test. Capacity tests should be done on all the new boiler installations, and on existing boiler installations when any modification is made that affects the steam-generating capacity of the boiler. Isolate the boiler by shutting down all the valves. Set burner to operate at its maximum capacity, and make sure that combustion is complete with proper overfire draft. When the operating control has shut off the burner, place jumper leads across its terminals to switch to high pressure cutout. When the burner has shut down, place another set of jumper leads across its terminals, and allow the burner to continue without control. The safety valve should pop open at the set pressure.

15.5 Boiler Room Emergencies

Emergencies in the boiler room occur due to many reasons. The emergencies from steam, water, or fuel leakages can be dangerous if not observed and no steps are taken to correct them. Any type of fires, explosions, or rupturing of a boiler and its components can create catastrophic accidents.

A boiler operator or engineer responsible for boiler room operation must follow the procedures in case of emergency. All normal safety precautions should be taken when operating boilers, burners, and fuel systems. The operator or engineer should be knowledgable about the following frequent emergencies in the boiler room as given in Table 15.1.

15.6 Operation Log Sheet

A boiler log sheet is defined as a document wherein all the data pertaining to boiler operation are recorded. The log sheet is signed by the operator or engineer in charge of the shift. All important occurrences in a steam heating boiler or hot water boiler during that shift must be recorded in the log sheet. A typical operation log sheet is shown in Figure 15.4.

TABLE 15.1 Boiler Room Emergencies

Symptoms	Effect	Prevention
1. ***Fuels***		
Gas line leaks	Fire or explosion	Test for leaks before starting
Wet gas	Fire or explosion	Check supply system
High gas pressure	Fireside explosion	Monitor gas pressure
Leaky shutoff valves	Fireside explosion	Monitor shutoff valves
Low oil temperature	Dirty and smoky fire	Check oil temperature
High oil temperature	Oil gasification	Check oil temperature
Damaged atomizers	Incomplete combustion	Replace or repair atomizers
2. ***Steam***		
Steam leaks	Severe burns	Keep all joints tight
Defective safety valves	Rupture boiler	Replace safety valves
Safety valve discharge	Severe burns	Discharge at a safe location
Defective pressure gages	Boiler under excess pressure	Replace defective gages
3. ***Water***		
Low water level	Boiler ruptured	Check normal water level
Scaled internal surfaces	Overheating	Clean internal surfaces
Overpressure in tank	Tank explosion	Check tank pressure
4. ***Air and products of combustion***		
Insufficient air	Furnace explosion	Provide adequate air
Excess negative pressure	Furnace implosion	Maintain proper pressure
Oxygen deficiency	Danger to life	Maintain proper oxygen
Low atomizing air pressure	Poor atomization	Check atomizing air pressure
Carbon monoxide leaks	Death or injury	Check carbon monoxide level
Flue gas leaks	Injury	Stop leakage
5. ***Controls and safety devices***		
Bad combustion controls	Fireside explosion	Check safety controls
Inadequate pilots	Furnace explosion	Check pilots
Delayed ignition	Furnace explosion	Check pilots
Tampering with controls	Improper operation	Prevent tampering
6. ***Electrical***		
Exposed electrical wiring	Electrical shock	Prevent insulation damage
Open electrical boxes	Electrical shock	Cover boxes
Lack of grounding	Electrical shock	Use proper grounding
Dust entering equipment	Fire or explosion	Keep equipment dust proof
7. ***Fire***		
Electrical failure	Fire or explosion	Operate equipment properly
Mechanical failure	Explosion	Operate equipment properly
Coal supply fire	Fire	Practice safe conditions
Fuel stations	Fire	Use fire protection systems

Operation Log Sheet for a Steam or Hot Water Boiler

Shift______________

Date								
Time								
Roller #								
Operator								
Water visible in gauge glass								
Combustion check (visual)								
Steam pressure								
Feedwater pressure								
Feedwater temperature								
Flue gas temperature								
Burner								
Gas pressure								
Gas meter reading								
Oil pressure (regulated)								
Oil tempeature								
Oil meter reading								
Atomizing media pressure								
Amblent air temperature								
Barometric pressure								
Blowdown water column								
Blowdown boiler								
Comments/Observations								

Figure 15.4 Operation log sheet for a steam or hot water boiler.

Chapter 16

Inspection

Inspection is performed on heating boilers after their installation on site. The main objective of in-service inspection is to determine the operational integrity of the boilers, and to issue certificates in accordance with the provisions of the jurisdictional laws and rules. The in-service inspection is conducted by an authorized inspector, who is qualified by the jurisdiction. In-service inspection is a complex technique, and requires experience and knowledge about boiler codes. The *ASME Boiler and Pressure Vessel Code Section VI* provides some guidelines on the care including inspection of heating boilers.

The operating life of a heating boiler begins immediately after installation. An authorized inspector (AI) performs the first inspection of the boiler after completion of installation, and a certificate is issued usually for 2 years. Depending on the jurisdictional laws and rules, the boiler is inspected regularly by an AI, and an operation certificate is issued every time after certificate inspection. The conditions of a boiler are bound to deteriorate over a period of time and safety is in question. The jurisdiction decides, based on the AI's report, whether the old boiler is safe to operate.

16.1 Authorized Inspectors

An authorized inspector is the person responsible for inspection of boilers and their parts when they are placed in service. This AI is also called an in-service inspector to distinguish from the other AI who performs shop inspections for new construction. The inspector conducts in-service inspections according to boiler laws, rules, and regulations of the jurisdiction.

16.1.1 Qualification

An authorized inspector has knowledge, experience, and training in boiler and pressure vessel code and has to qualify a written examination

CERTIFICATE OF COMPETENCY

STATE OF FLORIDA
FLORIDA DEPARTMENT OF FINANCIAL
DIVISION OF STATE FIRE MARSHAL
TALLAHASSEE, FLORIDA 32399-0300

THIS IS TO CERTIFY THAT:
JOHN NORMAN
1055 N OCEAN DR #204
JUNO BEACH, FL 33408

REPRESENTING:
STATE OF FLORIDA
STATE FIRE MARSHAL OFFICE
200 EAST GAINES STREET
TALLAHASSEE, FL 32399-0342

INSPECTOR ID #: 015 ISSUE DATE: 01/01/2005 EXPIRATION DATE: 12/31/2005

IS CERTIFIED PURSUANT TO THE PROVISIONS OF CHAPTER 554, FLORIDA STATUTES, AS A DEPUTY BOILER INSPECTOR OR AS A SPECIAL INSPECTOR.

TOM GALLAGHER
CHIEF FINANCIAL OFFICER

MOHAMMAD A. MALEK, P.E.
CHIEF BOILER INSPECTOR

FORM #DI4-357

Figure 16.1 Certificate of competency. *(Courtesy: State of Florida, Boiler Safety Program)*

given by the jurisdiction. Once a person passes the jurisdictional examination, he or she is issued a Certificate of Competency (Figure 16.1). After receiving further training, an AI is ready for in-service inspection. An AI is employed either by a jurisdiction or by an insurance company licensed to write boiler and machinery insurance in the jurisdiction where the inspector performs in-service inspection. To distinguish between the two types of authorized inspectors, the inspector employed by the jurisdiction is known as a state, county, or city inspector, and the inspector employed by the insurance company is referred as an insurance inspector.

16.1.2 Authorized inspector's responsibilities

An authorized inspector is responsible for certifying a boiler if all the conditions are satisfactory. The authorized inspector is assigned with the following responsibilities:

1. Checking the validity of the last certificate of compliance
2. Inspecting the boilers, externally or internally
3. Reviewing the calculations, drawings, and methods for welded repairs

4. Checking the material for repair to ensure that it complies with the code requirements
5. Checking all the welding procedures to ensure that these are qualified
6. Checking the qualification of welders and welding operators
7. Inspecting repair work after completion
8. Signing the welded repair form
9. Completing first inspection or reinspection reports
10. Submitting inspection reports to the jurisdictional authority in a timely manner

The AI ensures that all the necessary inspections have been performed in accordance with the laws and rules of the jurisdiction.

16.1.3 Inspection tools

An authorized inspector uses a wide variety of tools for in-service inspection of power boilers. The tools used by the AI for shop and in-service inspections are similar. Some of the tools, such as hammers and flash lights, are primitive in nature. Modern electronic instruments, such as ultrasonic thickness detectors and borescopes, are also used for inspection. An inspector's inspection tools may be classified into two categories mechanical tools and electronic instruments.

Mechanical tools are used mostly for visual inspection. Electronic instruments are used for nondestructive examination and finding some of the defects which cannot be determined by visual inspection. Some of the inspection tools are shown in Figures 16.2 through 16.5.

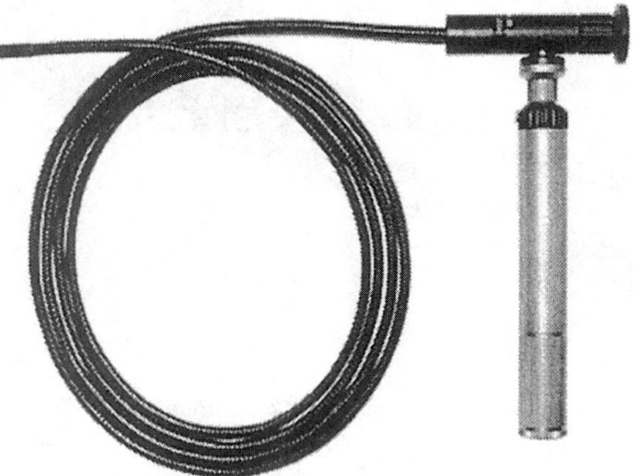

Figure 16.2 Tube scope inspection device. *(Courtesy: Goodway Technologies Corporation)*

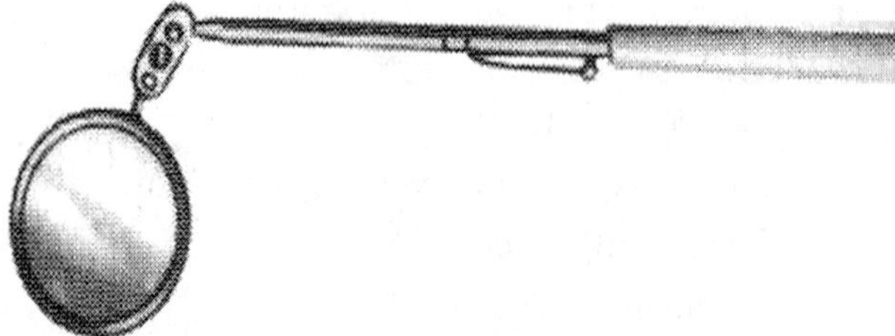

Figure 16.3 Inspection mirror.

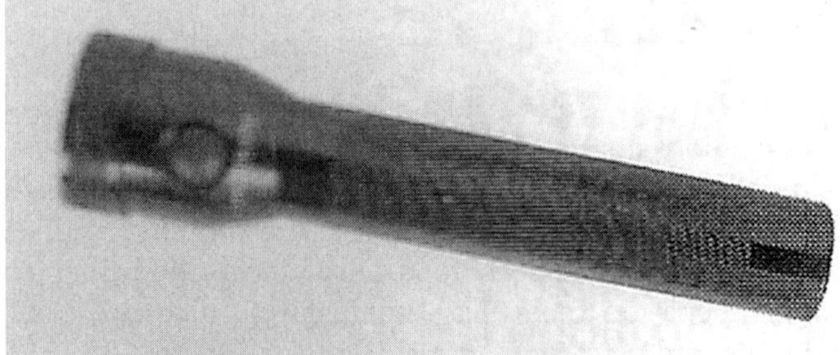

Figure 16.4 Flashlight for inspection.

Figure 16.5 Flexible light. *(Courtesy: Goodway Technologies Corporation)*

16.2 Boiler and Machinery Insurance

Like any other insurance, an owner can have insurance coverage for a boiler plant and accessories. The name of the insurance plan is called "boiler and machinery insurance." Depending on the type of insurance, the owner can submit a claim to the insurance company if the owner suffers damages due to boiler accidents. The insurance company after investigation of accident settles the claim according to the terms of the insurance policy.

Special insurance companies offer boiler and machinery insurance. These companies are licensed by the jurisdiction to perform inspections and write insurance for the boiler plants. These companies are called authorized inspection agencies (AIAs).

These insurance companies employ AIs, who are also licensed by the jurisdiction. The following insurance companies offer boiler and machinery insurance: The Hartford Steam Boiler Inspection and Insurance Company, The Travelers, Arise Incorporated, Chubb & Son, CAN Insurance Companies, Zurich Services Corporation, Seneca Insurance Company, and GCan Insurance Company of Canada.

The AIs of these insurance companies submit inspection reports to the jurisdiction for issuing certificates of compliance. The insurance company must report to the jurisdiction when the company insures any new boilers or discontinues insurance for any boilers.

16.3 Boiler Inspection

Generally, inspection frequency of heating boilers is specified in the boiler laws and rules of the jurisdiction. A heating boiler is inspected once a year or once in every 2 years, or twice a year—depending on the jurisdictional requirements. The inspection that is used to issue a certificate of compliance is called certificate inspection. Either of the external or internal inspections may be used as the certificate inspection. Noncertificate inspection is carried out for checking overall conditions of the boiler in between two certificate inspections.

16.3.1 First inspection

The first inspection of any new boiler is performed by an AI employed either by the local jurisdiction or boiler insurance company. This inspection is also called initial inspection or acceptance inspection. The first inspection ensures that the boiler including boiler supports, piping installations, safety and other valves, water columns, gage cocks, steam gages, and other apparatuses on the boiler have been installed in accordance with the laws and rules of the jurisdiction.

Any violation of the local codes should be corrected before inspection by an authorized inspector. Construction and installation should be done in accordance with the plans and specifications approved by the jurisdiction. An inspection certificate is issued to each boiler if the result of the inspection is satisfactory.

A first inspection report form used by the State of Florida is shown in Appendix F.

16.3.2 Routine inspection

A routine inspection, that is, a reinspection of existing boilers, is performed in accordance with the boiler laws and rules of the jurisdiction. This inspection is also called periodic inspection or annual inspection, if such inspection is performed annually. This inspection indicates that the boilers are in satisfactory and safe conditions, and is performed by an AI employed by the jurisdiction or by an insurance company authorized to issue boiler insurance for the jurisdiction.

It is very important that the inspection be thorough and complete so that all unsafe conditions and code violations are checked. A certificate of inspection, also known as the certificate of compliance, is issued if all conditions of the boiler are satisfactory. A routine inspection report form duly completed is shown in Figure 16.6.

16.4 Inspection of Steam Heating Boilers

A steam heating boiler is required to be prepared for inspection before the AI's arrival. The AI usually notifies the owner or owner's representative about the inspection, especially for internal inspection. The AI may or may not notify the owner for external inspection. The owner is required by law to extend all facilities to the AI during external inspection, though the inspector did not notify the owner.

A periodic inspection, i.e., a reinspection of the existing boilers is performed in accordance with the boiler laws and rules of the jurisdiction. The purpose of this routine inspection is to make sure that boilers are in satisfactory and safe conditions. This inspection is done by an authorized boiler inspector employed by the jurisdiction or an insurance company authorized to write boiler insurance in the jurisdiction. It is very important that the inspections be thorough and complete, so that all unsafe conditions and code violations may be checked. The following guidelines may be followed for routine inspections:

1. *Notification.* The authorized boiler inspector will notify the owner or user to prepare the boiler for inspection. Accordingly, the owner or user should prepare the boiler for internal inspection.

DIVISION OF STATE FIRE MARSHAL
BUREAU OF FIRE PREVENTION
BOILER SAFETY PROGRAM

Boiler-Fired Pressure Vessel
REPORT OF INSPECTION

This inspection is intended for your safety and the safety of the citizens of Florida. Your cooperation is greatly appreciated.

DATE INSPECTED	CERT EXP DATE	CERT POSTED	FOLLOW UP?	JURISDICTION NUMBER	NATL BOARD NO	OTHER NO
06/16/2006	06/16/2008	● YES ○ NO	○ YES ● NO	FL073521		

OWNER	NATURE OF BUSINESS	KIND OF INSP.	CERT INSP?
JACKSONVILLE COUNTRY DAY SCHOOL	PRIMARY & SECONDARY SCHOOL	○ INT ● EXT	● YES ○ NO

OWNER STREET ADDRESS	OWNERS CITY	STATE	ZIP
10063 BAYMEADOWS RD	JACKSONVILLE	FL	32216

USER NAME - OBJECT LOCATION	SPECIFIC LOCATION IN PLANT	OBJECT LOCATION - COUNTY
JACKSONVILLE COUNTRY DAY SCHOOL	E WING BLR R	DUVAL

LOCATION STREET ADDRESS	LOCATION CITY	STATE	ZIP
10063 BAYMEADOWS RD	JACKSONVILLE	FL	32216

TYPE	YEAR BUILT	MANUFACTURER
☐ FT ☐ WT ☑ CI ☐ OTHER	1985	WEIL MCLAI

USE	FUEL	METHOD OF FIRING	PRESSURE TESTED
☐ POWER ☐ PROCESS ☐ STEAM HTG ☑ HWH ☐ HWS ☐ OTHER	Oil	Burner	○ YES ● NO

PRESSURE ALLOWED	SAFETY - RELIEF VALVES	OBJECT CAPACITY
THIS INSPECTION 50 PREV INSP 50	SET AT 30 TOTAL CAPACITY 1649000 BTUs	1419000 BTUs Per Hour

IS CONDITION OF OBJECT SUCH THAT A CERTIFICATE MAY BE ISSUED?	HYDRO TEST
● YES ○ NO (IF NO EXPLAIN FULLY UNDER CONDITIONS)	○ YES ● NO PSI DATE

CONDITIONS: With respect to the internal surface, describe and state location of any scale, oil or other deposits. Give location and extent of any corrosion and state whether active or inactive. State location and extent of any erosion, grooving, bulging, warping, cracking or similar condition. Report on any defective rivets, bowed, loose or broken stays. State condition of all tubes, tube ends, coils, nipples, etc. Describe any adverse conditions with respect to pressure gage, water column, gage glass, gage cocks, safety valves, etc. Report condition of setting, linings, baffles, supports, etc. Describe any major changes or repairs made since last inspection.

NONE

REQUIREMENTS: (List Code Violations)

NCV NOTED.

NAME AND TITLE OF PERSON TO WHOM REQUIREMENTS WERE EXPLAINED
MIKE MORROW

CONTACT PHONE (904) 641-6644

THEREBY CERTIFY THIS IS A TRUE REPORT OF MY INSPECTION
INSPECTOR NAME Roy Williams

SIGNATURE OF INSPECTOR [signature] **IDENT. NO** 278 **EMPLOYED BY** HARTFORD STEAM BOILER INSPECTION AND INSURANCE COMPANY 9265

DI4-379,
10-1-2000

Figure 16.6 Inspection report on hot water heating boiler. *(Courtesy: State of Florida, Boiler Safety Program)*

2. *Opening the boiler.* A boiler should be cooled down slowly and naturally. Every part of the boiler that is accessible should be opened for internal and external examination.
3. *Preparation.* A boiler for internal inspection should be prepared in the following manner:
   a. Drain the water and wash the boiler thoroughly.
   b. Remove all manhole and handhole plates, wash out plugs, and water column connections. Clean the furnace and combustion chamber thoroughly.
   c. Remove all the grates of an internally fired boiler.

   d. Remove brickwork so that the authorized boiler inspector can look at the condition of the furnace, supports, and other parts.
   e. Cutoff any leakage of steam or hot water into the boiler by disconnecting pipe or valve.

4. *Insulation and covering.* Remove insulation material, masonry, or fixed parts if the authorized boiler inspector suspects defects and deterioration. The covering should be removed if there is evidence of leakage through the covering.

5. *Close observation.* The authorized boiler inspector should inspect the parts of the boiler from the closest possible distance. Good artificial light may be used to obtain the best possible vision of any defective surface.

6. *Witnessing the tests.* The authorized boiler inspector will witness the testing of apparatus, controls, and the like, which should be conducted by a boiler engineer/operator.

7. *Scale, oil, and so forth.* Any evidence of oil should be noted by the authorized boiler inspector because any amount of oil present in the boiler is dangerous. Oil and scale act as insulation on the heating surfaces and overheat the pressure parts, weakening to bag or rupture.

8. *Corrosion and grooving.* Corrosion may occur along or adjacent to seams and other points weakening the strength of pressure parts. Grooving along longitudinal seams occurs when the material is stressed. The inspection should reveal any defects due to corrosion and grooving.

9. *Stays.* All stays should be inspected for proper tension. The fastened ends of the stays should be examined for cracks.

10. *Manholes and other openings.* All the openings, including manholes, handholes, and the like should be examined for cracks or deformation. The observation should be made from both inside and outside the boiler to make sure that there are no defects.

11. *Fire surfaces.* Surfaces exposed to fire should be inspected for bulging and blistering. If any pressure part has become deformed due to bulging or blistering, operation of the boiler should be discontinued and defective part should be repaired. Welded seams and tube ends should be inspected for leakage.

12. *Lap joints.* Plates lapped in the longitudinal or straight seam are subject to cracking. If any evidence of leakage is observed at the lap joint, an inspection should be made for cracks in the seam.

13. *Staybolts.* Staybolts should be tested by tapping one end of each bolt with a hammer, and a hammer or other heavy tool at the other end.

14. *Tube defects.* Tubes should be inspected for surface defects such as bulging, flame impingement, steam erosion, cracking, and/or any evidence of defective welds. There should not be a reduction in thickness at any portion of the tube throughout its length.
15. *Ligaments.* A careful examination should be made to detect any cracks of the ligaments between the tube holes in the heads of firetube boilers and in the shells of water tube boilers.
16. *Pipe connections and fittings.* All the pipes, connections, and fittings for steam and water should be thoroughly inspected for leakage, cracks, or any other defect. The piping arrangement should be examined for expansion, contraction, and support. Change of position of the boiler may cause undue strain on the piping. Attention should be given to the blow-off pipes and their connections as expansion and contraction due to water hammer action may impose strain on the blow-off system.
17. *Water column.* The attachments with the water column should be tested for operating conditions. Action of water in the gage glass will prove that the connections of water column and gage glass to the boiler are free. The water column and gage glasses should be blown down to determine that the blow-off piping from the columns and gage glasses are free. Gage glasses should be observed to see that they are clean.
18. *Low-water cutoff and water feeder.* All automatically fired steam boilers should be equipped with an automatic low-water fuel cutoff or water feeding device (Figure 16.7). The device should cut off the fuel supply or deliver requisite feedwater when the water level in the boiler falls below the lowest safe waterline, which should not be lower than the bottom of the water glass.
19. *Baffles.* The baffles in watertube boilers should be inspected for their positions and conditions. Absence of or broken baffles will cause high temperature on some sections of the boiler, which may result in dangerous conditions. Also, broken baffles may cause flame impingement on a particular part of a boiler overheating that section.
20. *Localization of heat.* A defective burner or stoker can cause localization of heat in a particular section in a boiler. In fact, localization of heat is so dangerous that the boiler should be shut down to correct the condition.
21. *Suspended boilers.* If one boiler from a battery of boilers is suspended, attention should be paid to the supports and settings and ensure that another boiler will not produce excessive strains on the structure because of the expansion of the parts under operating conditions.

Figure 16.7 Inspection of low-water cutoff by AI.

22. *Safety valves.* The safety valves should be inspected for evidence of leakage, rust, scale, or other foreign substances. Likewise, opening pressure should be tested by raising the steam pressure to the point of opening. Alternatively, the valve should be tested by opening the try lever. The discharge pipe should have free discharge in accordance with the code requirement. However, no stop valve is permitted between the steam boiler and its safety valve.

23. *Steam gages.* There should be a test gage connection so that the steam gage on the boiler can be tested under operating pressure. The steam gage should be mounted with a siphon or trap between it and the boiler.

24. *Imperfect repairs.* The authorized inspector is responsible for certifying a boiler repair as per the code. The repair work should be done safely and properly by an organization qualified by the National Board of Boiler and Pressure Vessel Inspectors. Imperfect repairs are not acceptable.

25. *Hydrostatic tests.* The authorized inspector should decide whether a hydrostatic test is required to determine the extent of damage. A hydrostatic test should not exceed 1 1/2 times the maximum allowable working pressure. During the test, the temperature of the water should be a minimum of 70°F and a maximum of 160°F. The safety valve and all other controls unable to withstand the test pressure should be removed during the test.

26. *Suggestions.* The authorized boiler inspector should be extremely conscientious and careful during the inspection, observing all the conditions solely and should not accept any statement of other persons. Also, he should make general observation of the condition of the boiler room and apparatus, and make any suggestions for the general care of the steam boiler.

16.5 Inspection of Heating Boilers and Hot Water Supply Boilers

A heating boiler or hot water supply boiler is required to be prepared for inspection before AI's arrival. The AI usually notifies the owner or owner's representative about the inspection, especially for internal inspection. The AI may or may not notify the owner for external inspection. The owner is required by law to extend all facilities to the AI during external inspection, though the inspector did not notify the owner.

A periodic inspection, i.e., a reinspection of the existing boilers is performed in accordance with the boiler laws and rules of the jurisdiction. The purpose of this routine inspection is to make sure that boilers are in satisfactory and safe conditions. This inspection is done by an authorized boiler inspector employed by the jurisdiction or an insurance company authorized to write boiler insurance in the jurisdiction. It is very important that the inspections be thorough and complete, so that all unsafe conditions and code violations may be checked. The following guidelines may be followed for routine inspections:

1. *Notification.* Authorized boiler inspector will notify the owner or user to prepare the boiler for inspection. Accordingly, the owner or user should prepare the boiler for internal inspection.
2. *Opening the boiler.* A boiler should be cooled down slowly and naturally (a forced cooldown is not recommended). Every part of the boiler that is accessible should be opened for internal and external examination.
3. *Preparation.* A boiler for internal inspection should be prepared in the following manner:
   a. Drain the water and wash the boiler thoroughly.
   b. Remove all manhole and handhole plates, wash out plugs, and water column connections. Clean the furnace and combustion chamber thoroughly.
   c. Remove all the grates of an internally fired boiler.
   d. Remove brickwork so that the authorized boiler inspector can look at the condition of the furnace, supports, and other parts.
   e. Cut off any leakage of hot water into the boiler by disconnecting pipes or valves.

4. *Insulation and covering.* Remove insulation material, masonry, or fixed parts if the authorized boiler inspector suspects defects and deterioration. The covering should be removed if there is evidence of leakage through the covering.
5. *Close observation.* The authorized boiler inspector should inspect the parts of the boiler from the closest possible distance. Good artificial light may be used to obtain the best possible vision of any defective surface.
6. *Witnessing the tests.* The authorized boiler inspector will witness the testing of apparatus, controls, and the like, which should be conducted by a boiler engineer/operator.
7. *Scale, oil, and so forth.* Any evidence of oil should be noted by the authorized inspector because any amount of oil present in the boiler is dangerous. Oil and scale act as insulation on the heating surfaces and overheat the pressure parts, weakening to bag or rupture.
8. *Corrosion and grooving.* Corrosion may occur along or adjacent to seam and other points weakening the strength of pressure parts. Grooving along longitudinal seams occurs when the material is stressed. The inspection should reveal any defects due to corrosion and grooving.
9. *Stays.* All stays should be inspected for proper tension. The fastened ends of the stays should be examined for cracks.
10. *Manholes and other openings.* All the openings, including manholes, handholes, and the like should be examined for crack or deformation. The observation should be made from both inside and outside the boiler to make sure that there are no defects.
11. *Fire surfaces.* Surfaces exposed to fire should be inspected for bulging and blistering. If any pressure part has become deformed due to bulging or blistering, operation of the boiler should be discontinued and defective part should be repaired. Welded seams and tube ends should be inspected for leakage.
12. *Lap joints.* Plates lapped in the longitudinal or straight seam are subject to cracking. If any evidence of leakage is observed at the lap joint, an inspection should be made for cracks in the seam.
13. *Staybolts.* Staybolts should be tested by tapping one end of each bolt with a hammer, and a hammer or other heavy tool at the other end.
14. *Tube defects.* Tubes should be inspected for surface defects such as bulging, flame impingement, steam erosion, cracking, or and any evidence of defective welds. There should not be a reduction in thickness at any portion of the tube throughout its length.

15. *Ligaments.* Careful observation should be made to detect any crack of the ligaments between the tube holes in the heads of firetube boilers and in the shells of watertube boilers.
16. *Pipe connections and fittings.* All the pipes, connections, and fittings for steam and water should be thoroughly inspected for leakage, crack, or any other defect. The piping arrangement should be examined for expansion, contraction, and support. Change of position of the boiler may cause undue strain on the piping. Attention should be paid to the blow-off pipes and their connections as expansion and contraction due to water hammer action may impose strain on the blow-off system.
17. *Low-water cutoff.* All automatically fired hot water heating or supply boilers should be equipped with an automatic low-water fuel cutoff device. The device should cut off the fuel supply when the water level in the boiler falls below the lowest safe waterline, which should not be lower than the bottom of the water glass.
18. *Equalizing pipe.* Designs embodying a float or a float bowl should have a vertical straightway-valve drain pipe at the lowest point in the water-equalizing pipe connection, by which the bowl and equalizing pipe can be flushed and the device can be tested.
19. *Baffles.* The baffles in watertube boilers should be inspected for their positions and conditions. Absence of, or broken baffles will cause high temperatures on some sections of the boiler, which may result in dangerous conditions. Also, broken baffles may cause flame impingement on a particular part of a boiler, overheating that section.
20. *Localization of heat.* Defective burner or stoker can cause localization of heat in a particular section in a boiler. In fact, localization of heat is so dangerous that the boiler should be shut down to correct the condition.
21. *Suspended boilers.* If one boiler from a battery of boilers is suspended, attention should be paid to the supports and settings to ensure that another boiler will not produce excessive strains on the structure because of the expansion of the parts under operating conditions.
22. *Safety relief valves.* The safety relief valves should be inspected for evidence of leakage, rust, scale, or other foreign substances. Likewise, opening pressure should be tested by raising the water pressure to the point of opening. Alternatively, the valve should be tested by opening the try lever. The discharge pipe should have free discharge in accordance with the code requirement. However, no stop valve is permitted between a boiler and its safety valve.

23. *Combination temperature and pressure gages.* There should be a test gage connection so that the gage on the boiler can be tested under operating pressure. It is important not to expose the gage to high ambient temperatures.

24. *Imperfect repairs.* The authorized inspector is responsible for certifying boiler repair as per the code. This repair work should be done safely and properly by an organization qualified by The National Board of Boiler and Pressure Vessel Inspectors. Imperfect repairs are not acceptable.

25. *Hydrostatic tests.* An authorized inspector should decide whether a hydrostatic test is required to determine the extent of damage. A hydrostatic test should not exceed 1 1/2 times the maximum allowable working pressure. During the test, the temperature of the water should be a minimum of 70°F and a maximum of 160°F. The safety relief valve and all other controls unable to withstand the test pressure should be removed during the test.

CERTIFICATE OF COMPLIANCE

STATE OF FLORIDA
FLORIDA DEPARTMENT OF FINANCIAL SERVICES
DIVISION OF STATE FIRE MARSHAL
TALLAHASSEE, FLORIDA 32399-0300

THIS IS TO CERTIFY THAT:
RAMADA INN
GENERAL MANAGER
3130 HARLEY ROAD
JACKSONVILLE, FL 32257

LOCATED AT:
RAMADA INN
3130 HARLEY ROAD/MECH RM
JACKSONVILLE, FL 32257

HAS COMPLIED WITH STATUTE 554.1101 OF THE STATE OF FLORIDA AND IS AUTHORIZED TO OPERATE THIS BOILER BEARING THE BELOW STATE ID NUMBER.

MANUFACTURED BY: LOCHINVAR YEAR BUILT: 1998
MAXIMUM ALLOWABLE WORKING PRESSURE: 160 PSI
INSPECTED BY: ROGER GUIST/ZURICH SERVICES CORPORATION

INSPECTION DATE	INVOICE NUMBER	FEES	STATE ID NUMBER	NAT'L BOARD #	EXPIRATION DATE
04/18/2006	B06016982	30.00	FL087927	95161	04/18/2008

THIS CERTIFICATE MUST BE POSTED IN A CONSPICUOUS LOCATION IN THE ROOM CONTAINING THE BOILER AND FRAMED WITH A TRANSPARENT COVER.

Tom Gallagher
TOM GALLAGHER
CHIEF FINANCIAL OFFICER

Mohammad Malek
MOHAMMAD A. MALEK, P.E.
CHIEF BOILER INSPECTOR

FORM #DI4-402

Figure 16.8 Certificate of compliance for a hot water supply boiler. *(Courtesy: State of Florida, Boiler Safety Program)*

26. *Suggestions.* The authorized inspector should be extremely conscientious and careful during the inspection, observing all the conditions solely, and should not accept any statement of other person. Also, he should make general observations of the condition of the boiler room and apparatus, and, make any suggestions for the general care of the hot water heating and supply boilers.

Certificate of compliance. The local jurisdiction issues a certificate of compliance (certificate of operation) if the results of inspection were found satisfactor by the AI. A certificate of compliance for a hot water supply boiler is shown in Figure 16.8.

Chapter

17

Maintenance and Repairs

Maintenance is performed on heating boiler plants to increase efficiency, ensure safety, and prevent unscheduled shutdowns. The maintenance of boiler plants should be controlled by a planned program, which should be established based on the type of equipment, history, and age of equipment. A preventive maintenance schedule, which is a part of the maintenance program, can prevent power boilers from costly shutdowns. The *ASME Boiler and Pressure Vessel Code Section VI* provides guidelines for the care of boilers including recommendations for maintenance and repair to improve boiler availability and keep heating boilers in a safe operating condition.

Boiler maintenance is done to prevent heating boilers from breakdown. Any breakdown of boilers can be very costly for a plant in terms of loss of production, and repair costs. If the boilers are old, more maintenance is required than the maintenance on the boilers when they are new. Regular maintenance is the key to efficient continuous boiler operation.

Generally, maintenance personnel are mechanics, technicians, and engineers who have experience in maintenance of boilers and their related equipment. The maintenance personnel should be very familiar with the equipment; otherwise the boiler plant may not be maintained properly.

If any outside contractor is hired for maintenance, the contractor should be licensed and bonded. Many government jurisdictions issue licenses to mechanical contractors who undertake boiler maintenance. Before engaging any contractor, ensure that the organization has experience in maintenance of the particular type of boiler(s) at the boiler plant.

17.1 Maintenance Tools

A variety of tools and equipment is required to perform maintenance on boilers. Some of these tools and equipment are specialized and designed especially for boiler applications. The type of maintenance work also decides the type of tools required. Here are examples of major and specialized tools and equipment required for boiler maintenance:

1. *General tools.* Flashlights, general purpose tools, tapes, carts, glue guns, tool boxes, test pumps, and test instruments.
2. *Hand tools.* Chisels, pliers, wrenches, sockets, screw drivers, keys, cutters, crimpers, punches, benders, pipe wrenches, pipe threaders, pipe stand, spinners, sheet metal tools, trimmers, shears, knives, hammers, steel files, vises, clamps, and masonry tools.
3. *Power tools.* Drills, hammer drills, saws, screwdrivers, impact wrenches, chipping hammers, cutters, band saws, routers, trimmers, shears, sanders, planners, engravers, grinders, polishers, drill presses, and buffers.
4. *Tube cleaning tools.* Firetube boiler cleaning kits, watertube boiler cleaning kits, air-driven tube cleaning systems, air/water-actuated tube cleaning systems, electric tube cleaning systems, brushes, expanding scraper, buffing tools, cutter heads, drill heads, cutter bits, flexible shafts, and tube cleaning extension rods.
5. *Tube expansion tools.* Electronic tube expansion systems, air-driven tube expansion systems, tube expanders, combination beading expanders, tube wall reducing tools, tube pullers, tube sheet hole brushes, tube plugs, and expander lube.
6. *Tube removal tools.* Collapsing tools, chisels, knock-out tools, hydraulic tube cutters, hydraulic cutter, super-heater tube cutters, and chipping hammers.
7. *Tube finishing tools.* Beading tools, weld remover, and tube/pipe-end cutter.
8. *Pneumatic tools.* Wheel grinders, right angle grinders, offset handle drills, handhole seat cleaning kits, handhole seat grinding kits, tube-end polishing kits, and boiler maker's grinding kits.
9. *Vacuum tools.* Electric-powered vacuums, air-powered vacuums, walk-behind vacuums, soot/powder vacuums, continuous duty vacuums, and vacuum accessories.
10. *Pressure washers.* Electric- and air-powered washers, gasoline-powered washers, hot water pressure washers, pressure washer accessories, and vapor-steam cleaners.
11. *Welding equipment.* Welding machines, electrodes, welder/generators, soldering equipment, solders, fluxes, portable torch kits, regulators, and welding gages.

TABLE 17.1 Effect of Deposits in Tubes on Heat Transfer

Scale in boiler tubes	Efficiency loss	Average fuel loss
1/32 in. (0.8 mm)	8%	2%
1/16 in. (1.6 mm)	12%	2 1/2%
1/8 in. (3.2 mm)	20%	4%

17.2 Tube Cleaning

The surfaces of the tubes should be clean and scale-free for a boiler to function efficiently. The formation of mud deposits, and hard or soft scales on the tube surfaces, reduces the flow through the watertubes. These fouling agents also have some chemical characteristics that are corrosive to the tube metal. All these deposits act as insulators and reduce over-all heat transfer efficiency. The industrial study given in Table 17.1 shows how deposits in tubes can affect heat transfer.

Soot accumulation in firetubes wastes fuel and reduces the boiler efficiency. Soot has five times the insulating value of asbestos. The industrial study shown in Table 17.2 indicates loss of efficiency and fuel loss as a result of soot accumulation.

A tube cleaning is a regularly scheduled maintenance job. The best time to clean tubes is when a boiler is shut down due to technical trouble or for annual inspection. Boiler tubes may be cleaned either by the mechanical method or the chemical method discussed in the next two sections, 17.2.1, and 17.2.2.

17.2.1 Mechanical cleaning

Mechanical tube cleaning is done with tube-cleaning tools. Tube-cleaning tools cover a wide variety of devices including a number of air, water, and electrically driven motors designed to rotate drills, cutter heads, scraper heads, and brushes through tubes or pipes. One has to be knowledgeable and experienced in selecting proper tools for tube cleaning. Boiler tubes may be damaged during cleaning if proper tools are not selected.

Examples of mechanical tube-cleaning tools are shown in Figure 17.1. In Figure 17.2, a maintenance technician is cleaning firetubes. This cleaning device has been specially designed for the tubes of firetube boilers.

TABLE 17.2 Loss of Efficiency Due to Soot Accumulation

Soot in boiler tubes	Efficiency loss	Average fuel loss
1/32 in. (0.8 mm)	12%	2 1/2%
1/16 in. (1.6 mm)	24%	4 1/2%
1/8 in. (3.2 mm)	47%	8 1/2%

Figure 17.1 Tube-cleaning tools. *(Courtesy: Goodway Technologies Corporation)*

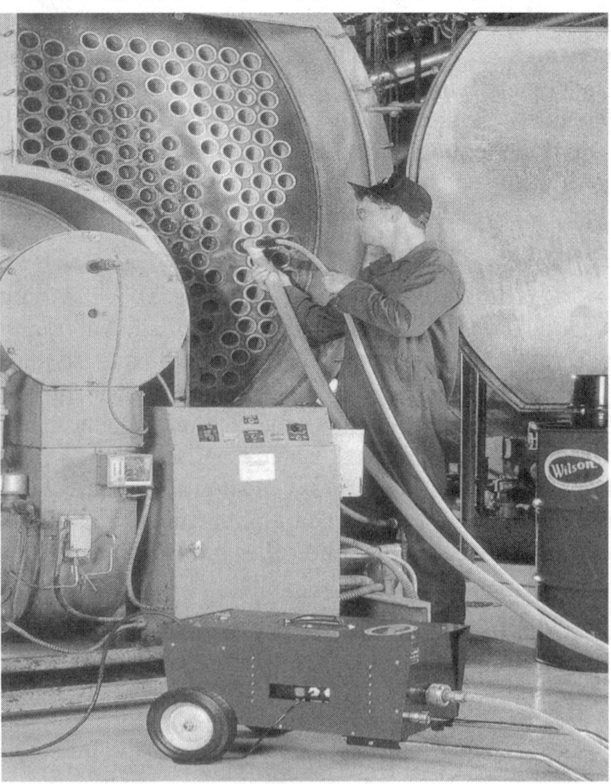

Figure 17.2 Cleaning of firetubes. *(Courtesy: Goodway Technologies Corporation)*

Selection of tube-cleaning tools. In selecting tube-cleaning tools for a particular boiler, the mechanic should determine whether an internal or external cleaning unit is going to be used.

An internal cleaner is one in which both the drive motor and the cleaning tool enter the tube and is generally used for cleaning straight or curved tubes found in boilers. External cleaners are those with drive motors external to tubes and driving tools with a length of rigid, hollow shafting or flexing shafting. The external cleaners are commonly used for cleaning small, straight tubes up to 2 inches of outside diameter found in condensers and heat exchangers.

A mechanic should check the typical tube deposit to be removed: whether thick, thin, hard, soft, powdery, wet, or the like. The type of deposit determines the type of drill, cutter, or brush that will be used. A manufacturer's catalog should be consulted for selecting the correct sizes of tools based on the tube sizes and the types of deposits.

It is recommended that a complete tube cleaner kit should be ordered based on the tube cleaner selection. A complete air-driven cleaner kit for tube inside diameter sizes from 2.125 to 2.250 inches consists of: 1 motor, 1 wrench, 1 cutter head assembled, 1 head coupling, 2 sets of cutters, 2 sets of cutter pins, 1 set of alarm pin, 1 set of keeper pins, 1 brush and 1 set of refills, 1 brush coupling, 1 drill, 1 expanding brush, 1 air valve, 1 tool box, 1 operating hose, and 1 lubricator.

17.2.2 Chemical cleaning

Chemical cleaning of the boilers is achieved by using acid. This method is used especially when the scales are thick and hard, and it is practically impossible to remove them by mechanical methods. Most hard scales are of the calcium or magnesium type, with calcium being the most prevalent. The acid recommended for elimination of these types of scale is muriatic acid. Only a specialist experienced in chemical cleaning should perform chemical cleaning. The tubes may be damaged by acid if proper chemicals and methods are not used.

For chemical cleaning, the quantity of chemical solution needed has to be determined first. The boiler catalog will indicate the dry and wet weight of the unit. Subtract the dry weight from the wet weight and divide the result by 8.33 to obtain the total number of gallons.

Mix one unit of muriatic acid to three units of fresh water. After the acid has been diluted, add to this solution pure sodium dichromate—1 ounce per 1 gallon of the above solution. Pour this solution in the boiler and fill to the top of the manhole. Bring the boiler temperature to just below the boiling temperature and allow the boiler to remain at this temperature for 8 to 12 hours.

Open the manhole and observe the appearance of the waterside, which had hard scale prior to cleaning. After the solution has removed the scale

(scale will fall to bottom), it is necessary to neutralize the acid solution to prevent corrosion or thinning of the tubes. For every 150 pounds of muriatic acid and water solution, add one pound of soda ash. Bring water to just below the boiling point and allow the solution to remain for approximately 1 hour.

After the boiler water has been neutralized, drain the unit and flush the boiler two times with fresh water. When this is done, add sufficient quantities of sodium sulfite content to the boiler within 25 to 50 ppm. Treat it for proper pH and carry out any other treatment recommended.

17.3 Tube Expansion

Tube expansion is defined as the technology of reducing a tube wall by compressing the outside diameter of the tube against a fixed container such as expanding tubes into tube sheets or drums. This is a method of attaching tube to the tube sheet. Tube expansion is also known as tube rolling.

Tube-joint leakage is caused by one of the following: cracked tube sheet holes, under-rolling, over-rolling, improper preparation of tube sheets, and differential thermal expansion. When a minor water leakage is found from the tube joint at the tube sheet, expansion of that particular tube can stop the leakage. A boiler mechanic can do this tube expansion job by using the pneumatic tube tools. Under-rolling occurs when the tube is not expanded to fill the tube sheet hole and the proper amount of wall reduction is not obtained. Over-rolling happens when the expansion of the inside diameter of the tube surpasses the ultimate limit of the tube material required for proper wall reduction. It is better to perform under-rolling than over-rolling, as over-rolling can damage the tube sheet.

Selection of tube expansion tools. Proper tools are required for expansion of tubes used for different types of boilers. Watertube boilers require different tools than firetube boilers. Tools for super heaters are different from the tools for economizers. Selection of expansion tools depends on the type of applications and tube sizes.

A manufacturer's catalog should be consulted for ordering proper expansion tools. The selection process begins with the tube size (outside diameter and thickness) and tube sheet thickness. The type of mandrel required should be decided based on the selected type of expander. Different types of mandrels are: standard mandrels, header mandrels, reverse mandrels, drum mandrels, and short mandrels. It is recommended to order one spare mandrel, roll set for each expander, and other accessories.

Like tube cleaning kits, tube expansion tools are also available as a complete kit. A typical electronic kit for expansion of tube sizes 3/4 to 2 1/2 inches consists of: an electric motor, mountings and cables, telescopic

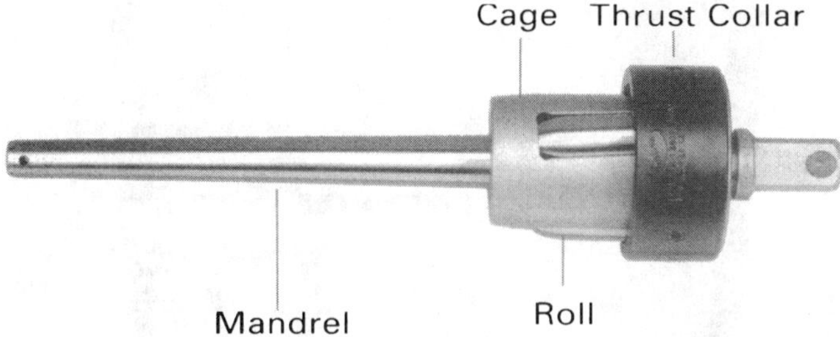

Figure 17.3 Tube expander with drum mandrel. *(Courtesy: Thomas C. Wilson, Inc.)*

shaft, electronic control unit, mobile stand, stationary stand, adapters (3/8, 1/2, 3/4, and 1 in.), 1/2 inch chuck, expanders, expander lube, accessories, and service manual.

A tube expander with a drum mandrel is shown in Figure 17.3. The tube expansion operation inside the drum is shown in Figure 17.4. The

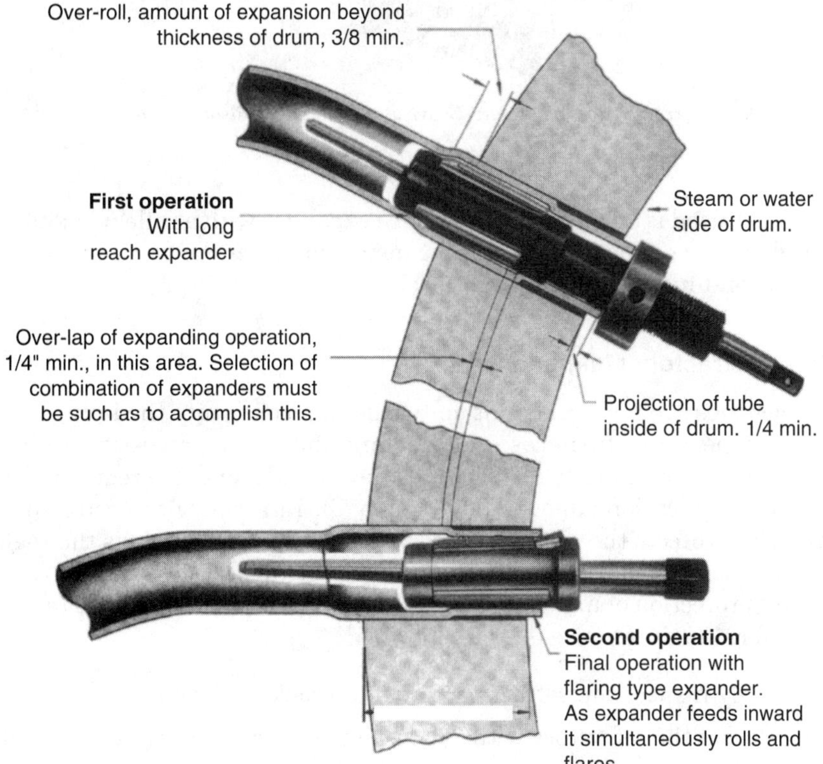

Figure 17.4 Tube expansion operation. *(Courtesy: Thomas C. Wilson, Inc.)*

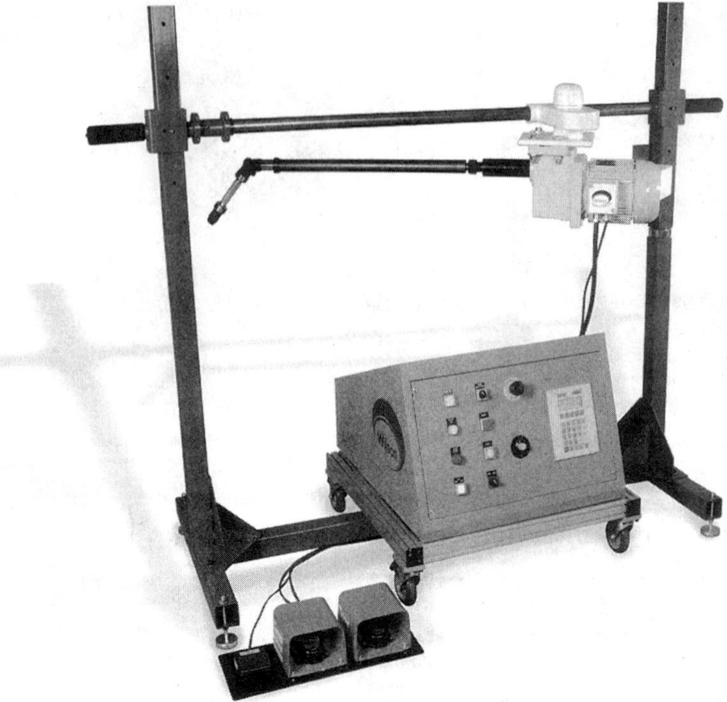

Figure 17.5 Electronic tube expander. *(Courtesy: Thomas C. Wilson, Inc.)*

expander used is an electronic-type tube expander with a roller expander suitable for a particular size of tube inside diameter. An electronic tube expansion unit is shown in Figure 17.5.

17.4 Refractory Maintenance

Refractories are heat-resistant materials that constitute the linings for high-temperature furnaces and reactors and other processing units. Refractories are more heat-resistant than metals, and are required for heating applications above 1000°F (538°C). In a boiler the refractory material protects the metal surfaces at critical points, such as the rear door and the furnace. The refractory should be inspected periodically to ensure protection of metal surfaces. The following steps should be taken to perform maintenance on refractory:

- Visually inspect the refractory. Look for cracks or broken pieces.
- Wash coat the refractory with a high-temperature bonding cement and air dry mortar.

- Face all cracks and joints with high-temperature bonding cement.
- If any bricks have fallen out or show signs of excessive wear, replace them.
- Follow the manufacturer's recommendation for curing the refractory.

17.5 Preparing for Inspection

A heating boiler has to be prepared for the AI when the inspection is due. Generally, the maintenance program receives a notice from the AI with the time and date of inspection. The maintenance department is responsible for preparing the boiler for inspection.

In a small organization, where no maintenance department exists, the boiler operator or engineer is responsible for preparing the boiler. Sometimes, a small organization engages an outside boiler mechanic or contractor to prepare the boiler for inspection, especially for internal inspection.

A boiler has to be prepared according to the type of inspection. The preparation for external inspection is comparatively easy. The preparation for an internal inspection can be a big job depending on the size of the boiler.

Use of a flashlight. A light—either a flashlight or an extension light—is used for preparation for an internal inspection. A flashlight is preferred, however, for safety reasons. When an extension light is used, it should be of low voltage and be properly grounded with a nonconducting holder and handle.

17.6 Maintenance of Steam Heating Boilers

Cleaning. A boiler is properly maintained when a boiler is "clean and tight," a motto for the boiler room. *Clean* means cleaned tubes, other heating surfaces, flue boxes, combustion chambers, and the like; *tight* refers to settings, baffles, and all the pressure parts. The boiler should be in excellent condition for higher efficiency, safety, and reliability.

Draining. A steam-heating boiler should be drained only if freezing conditions are expected, and/or heavy sludge and dirt have accumulated on the waterside, and/or repair work is necessary. Draining is required if a small amount of makeup water has been added, or if water treatment is maintained regularly.

Protection against freezing. Antifreeze, if used, should be tested as per the recommendations of the antifreeze manufacturer. The antifreeze solution should be heated in an indirect heat exchanger and should not be circulated through the boiler proper.

Fireside corrosion. Boilers can corrode on the fireside. This corrosion may occur from the presence of sulfur, vanadium, or sodium in fuels, and of course, the intensity of corrosion depends on the amount of substances contained in the fuels, burning conditions, and the like. To avoid corrosion, fireside surfaces should be cleaned thoroughly at the end of the firing session and the heating surfaces should be kept dry when the boiler is out of service.

Safety valves. In case of replacement, ASME-rated safety valves of the required capacity should be used. The safety valves should be tested at regular intervals in accordance with the *ASME Code Section* and local jurisdictional rules.

Burner maintenance. Burners should be maintained in accordance with the manufacturer's recommendations. Also, the type of maintenance depends on the type of burners. Periodic maintenance is done on oil burners to clean burner nozzles, oil line strainers, ignition electrodes, air intake screens, blowers, and air passages. From time-to-time, oil levels in the gear cases, belts, and all linkages are required to be checked. In the case of gas burners, the presence of dirt, lint, or foreign matter has to be checked. Also, pilot burners, linkages, belts, and moving parts on power burners are required to be checked and adjusted.

Low-water fuel cutoff and water feeder maintenance. These devices should be dismantled at least once a year for proper functioning of the working parts. All the components should be inspected for deposits of mud, scale, and the like, and then cleaned. Also, check floats and mercury bulbs to make sure that they are in satisfactory condition.

Flame safeguard maintenance. All boilers are not equipped with automatic flame failure controls. Small boilers are equipped with a flame scanner that shuts the main fuel valve if pilot failure or loss of main flame occurs. In all cases, flame safeguards should be maintained in accordance with the manufacturer's recommendations.

1. *Thermal-type detection device.* Check the device for electrical continuity and current generation. On completion of the maintenance, test the detection device.
2. *Electronic-type detection device.* Replace vacuum tubes on older units annually, and also check operation of the unit and examine for damaged or worn parts. Replace any defective device, do not repair it.

Limit-control maintenance. Inspect the device visually for wear, corrosion, and the like. Next check for mercury separation and discoloration of the bulb, if the control is a mercury bulb-type. Do not repair any defective pressure-limiting control—just replace it.

Steel boiler maintenance. Clean-heating surfaces by using appropriate methods. Remove all accumulations of soot, carbon, dirt from the fireside, breeching, and stack. Also, inspect refractory, and repair if required. The internal surfaces of the boiler should be cleaned either by high-pressure water stream or by chemical method. First, allow the boiler to cool, drain, and then remove all handholes and manhole covers. Never open manholes or handholes before the boiler is fully cooled and drained. Next, run a high-pressure water stream inside the boiler to remove any solidified sludge and dirt. Finally, flush thoroughly after cleaning. Chemical cleaning is required if the scale is hard or scale buildup is difficult to remove.

Use of flashlight for internal inspection. A light, either a flashlight or extension light, is required for an internal inspection. Because of safety reasons, a flashlight is preferred. However, when an extension light is used, it should be of low voltage, with a nonconducting holder and handle, and properly grounded.

Leaking tubes. Check tubes for corrosion. If one tube is leaking due to corrosion, probably some other tubes are corroded also. A boiler inspector should be called to inspect all the tubes before a decision is made for replacement of corroded tubes.

Use of sealant. Sealant should not be used for sealing any leak on a steam boiler. Instead, proper repair should be done.

Maintenance of condensate return. Drain and flush the condensate tank. Check strainer, pump packing, float switches, and vacuum switches and then follow the manufacturer's recommendations for maintenance and repair.

Maintenance schedule. Routine maintenance and tests should be conducted on boilers to prevent any possible accidents. The frequency of maintenance and testing depends on the type of boilers, type of service, and duration of operation. In conjunction with inspection and operation, the following maintenance frequencies are suggested:

1. *Daily (boilers in service).* Observe operating pressures, water level, and general conditions. Check for any unusual noises or conditions and then correct.

2. *Weekly (boilers in service)*
   a. Test low-water fuel cutoff and/or water feeder. Blow down the boiler if considerable makeup water is used.
   b. Test water column or gage glass.
   c. Observe condition of flame; correct if flame is smoky or if burner starts with a puff.
   d. Check fuel supply (oil only).
   e. Observe operation of condensate or vacuum pump.
3. *Monthly (boilers in service)*
   a. Safety valve—try lever test.
   b. Test flame detection devices.
   c. Test limit controls.
   d. Test operating controls.
   e. Sludge blowdown where required.
   f. Check boiler room floor drains for proper functioning.
   g. Inspect fuel supply systems in boiler room area.
   h. Check condition of heating surfaces.
   i. Check combustion air supply opening to ensure that it is not closed or stopped up.
4. *Annually*
   a. Internal and external inspection after thorough cleaning.
   b. Routine burner maintenance.
   c. Routine maintenance of condensate or vacuum return equipment.
   d. Routine maintenance of combustion control equipment.
   e. Combustion and draft tests.
   f. Safety valve pop test.
   g. Slow drain test of low-water cutoff.
   h. Inspect gas piping for proper support and tightness.
   i. Inspect boiler room ventilation louvers and intake.

Maintenance log sheet. Every steam-heating boiler should have a maintenance log sheet. Any maintenance done on the boiler should be recorded in detail. The log sheet should be signed by a mechanic, and countersigned by a maintenance engineer or supervisor.

A sample of a maintenance log sheet is shown in Figure 17.6. The maintenance log sheet should also show the name of the person or contractor who completed the job.

17.7 Maintenance of Hot Water Heating and Hot Water Supply Boilers

Cleaning. A boiler is properly maintained when a boiler is "clean and tight," a motto in the boiler room.

Maintenance, Testing, and Inspection Log Steam Heating Boilers	Building:	Month:	Year:
	Address:	Fuel Type:	
Person(s) to be Notified in Emergency (Name and Telephone No.)		Boiler No.:	

DAILY CHECKS

	1	2	3	4	5	6	7	8	9	10	11	12	13	14	15	16	17	18	19	20	21	22	23	24	25	26	27	28	29	30	31
(1) Observe Water Level																															
(2) Record Pressure																															
(3) Record Flue Gas Temperature																															

WEEKLY CHECKS (Enter Date)

	WEEK 1	WEEK 2	WEEK 3	WEEK 4
(1) Test Low Water Cutoff				
(2) Test Gage Glass				
(3) Observe Flame Condition				

MONTHLY CHECKS (Enter Date)

(1) Manual Lift Safety Valve				
(2) Review Condition of or Test Each Item	(A) Linkages		(F) Floor Drains	
	(B) Damper Controls		(G) Flame Detection Device	
	(C) Stop Valves		(H) Limit Controls	
	(D) Refractory		(I) Operating Controls	
	(E) Flue-Chimney Breeching			
(3) Inspect Fuel Piping				
(4) Combustion Air Adequate/Unobstructed				

General Comments:

(Instructions on Reverse)

Figure 17.6 Maintenance log sheet for steams-heating boilers. *(Courtesy: ASME Code Section VI)*

1. *General.* Clean the boiler tubes, other heating surfaces, smoke boxes, combustion chambers, and the like, and all the pressure parts. The frequency of cleaning may be determined by trial to get the best result. A boiler should be in excellent condition for higher efficiency, safety, and reliability.
2. *Backwashing.* A water heater connected to a boiler should be backwashed regularly. The purpose of backwashing is to reduce the amount of scale, which is accumulated at the outlet side of the heater. This is done by using the valves to reverse the direction of flow through the heater. It is important that backwashing continue until the water runs clear.

Draining. A heating boiler should be drained only if freezing conditions are expected, and/or heavy sludge and dirt have accumulated on the waterside, and/or if repair work is necessary. Draining is required if a some makeup water has been added, and water treatment has been maintained regularly.

Protection against freezing. If used, antifreeze solutions should be an ethylene-glycol type with an inhibitor added; then it should be tested as per the recommendations of the antifreeze manufacturer. Concentration of antifreeze solution should be between 33 and 66 percent. Indeed, metal temperature in contact with the solution should be kept below 350°F (177°C), as high metal temperatures accelerate depletion of the antifreeze inhibitors. Note that the use of antifreeze in hot water heating systems, in cold climates, will reduce the boiler output—consult the boiler manufacturer for recommendations.

Fireside corrosion. Boilers can corrode on the fireside, and that corrosion may result from the presence of sulfur, vanadium, and sodium compounds in fuels. The intensity of corrosion, of course, depends on the amount of foreign substances contained in the fuels, burning conditions, and the like. Fireside surfaces should be cleaned thoroughly at the end of the firing session to avoid corrosion. Also, the heating surfaces should be kept dry when the boiler is out of service.

Safety relief valves. ASME-rated safety relief valves should be installed on hotwater heating and hotwater supply boilers. In case of replacement, ASME-rated safety relief valves of the required capacity should be used. The safety relief valves should be tested at regular intervals in accordance with *ASME Code Section IV* and local jurisdictional requirements.

Burner maintenance. Burners should be maintained in accordance with the manufacturer's recommendations. Also, the type of maintenance depends on the type of burners.

1. *Oil burners.* Periodic maintenance is done on oil burners to clean burner nozzles, oil line strainers, ignition electrodes, air intake screens, blowers, and air passages. From time to time, the oil levels in the gear cases, belts, and all linkages are required to be checked.
2. *Gas burners.* Check burners for presence of dirt, lint, or foreign matter. In addition, linkages, belts, and moving parts on power burners are required to be checked and adjusted. For proper flame adjustment, check pilot burners and ignition equipment.

Low-water fuel cutoff and water feeder maintenance. These devices should be dismantled at least once a year for proper functioning of the working parts. All the components should be inspected for deposit of mud, scale, and so forth, and then cleaned. Also, check floats and mercury bulbs to make sure that they are in satisfactory condition. Do not repair, but replace complete mechanisms in the field.

Flame safeguard maintenance. All boilers are not equipped with automatic flame failure controls. Small boilers are equipped with a flame scanner that shuts the main fuel valve if pilot failure or loss of main flame occurs. In all cases, flame safeguards should be maintained in accordance with the manufacturer's recommendations.

1. *Thermal-type detection device.* Check the device for electrical continuity and current generation. On completion of the maintenance, test the detection unit.
2. *Electronic-type detection device.* Replace vacuum tubes on older units annually, and also check operation of the unit and examine for damaged or worn parts. Replace any defective device—do not repair it. On completion of the maintenance, test the detection unit.

Limit-control maintenance. Inspect temperature limit-controls visually for evidence of wear, corrosion, and the like. Next, check for mercury separation and discoloration of the bulb, if the control is a mercury bulb-type. Do not repair any defective pressure limiting control—just replace it.

Steel boiler maintenance. Clean heating surfaces by using appropriate methods. Remove all accumulations of soot, carbon, and dirt from the fireside, breeching, and stack. Also, inspect refractory and repair if required. The internal surfaces of the boiler should be cleaned either by high-pressure water stream or by chemical methods. First, allow the boiler to cool and drain, and then remove all handholes and manhole covers. Next, run a high-pressure water stream inside the boiler to remove any solidified sludge and dirt. Finally, flush thoroughly after cleaning. Chemical cleaning is required if the scale is hard or scale buildup is difficult to remove.

Use of flashlight for internal inspection. A light, either a flashlight or extension light, is required for internal inspection. Because of safety reasons, a flashlight is preferred. However, when any extension light is used, it should be of low voltage, with a nonconducting holder and handle, and properly grounded.

Leaking tubes. Check tubes for corrosion. If one tube is leaking due to corrosion, probably some other tubes are corroded also. A boiler inspector should be called to inspect all the tubes before a decision is made to replace corroded tubes.

Use of sealant. Sealant should not be used for sealing any leak on hot-water heating or hot water supply boilers. Instead, proper repair should be done.

Maintenance of circulating pumps and expansion tanks. Check pumps, their switches, and controls. Lubricate the pumps in accordance with the manufacturer's recommendations. Check expansion tanks for dirt, tightness, and evidence of corrosion.

Maintenance schedule. Routine maintenance and tests should be conducted on boilers to prevent any possible accidents. The frequency of maintenance and testing depends on the type of boilers, type of service, and duration of operation. The following frequencies are suggested for inspection and maintenance of boilers:

1. *Daily (boilers in service).* Observe operating pressures, water level, and general conditions. Check for any unusual noises or conditions and correct.
2. *Weekly (boilers in service)*
   a. Observe condition of flame; correct if flame is smoky or if burner starts with a puff.
   b. Check fuel supply (oil only).
   c. Observe operation of circulating pump(s).
3. *Monthly (boilers in service)*
   a. Safety relief valve—try lever test.
   b. Test flame detection devices.
   c. Test limit controls.
   d. Test operating controls.
   e. Check boiler room floor drains for proper functioning.
   f. Inspect fuel supply systems in boiler room area.
   g. Check condition of heating surfaces.
   h. Perform combustion and draft tests (preheated oil only).
   i. Test low-water fuel cutoff and/or water feeder if piping arrangement permits without draining considerable water from the boiler.
4. *Annually*
   a. Internal and external inspection after through cleaning.
   b. Routine burner maintenance.
   c. Routine maintenance of circulating pump and expansion tank equipment.
   d. Routine maintenance of combustion control equipment.
   e. Combustion and draft tests.
   f. Safety relief valve(s)—pop test.
   g. Slow drain test of low-water cutoff.
   h. Inspect gas piping for proper support and tightness.
   i. Inspect boiler room ventilation louvers and intake.

Maintenance log sheet. Every hot water boiler should have a maintenance log sheet. Any maintenance done on the boiler should be recorded

Maintenance, Testing, and Inspection Log Hot Water Heating Boilers	Building:	Month:	Year:
	Address:	Fuel Type:	
Person(s) to be Notified in Emergency (Name and Telephone No.)		Boiler No.:	

DAILY CHECKS

	1	2	3	4	5	6	7	8	9	10	11	12	13	14	15	16	17	18	19	20	21	22	23	24	25	26	27	28	29	30	31
(1) Record Pressure																															
(2) Record Boiler Water Temperature																															
(3) Record Flue Gas Temperature																															

WEEKLY CHECKS (Enter Date)

	WEEK 1	WEEK 2	WEEK 3	WEEK 4
(1) Observe Flame Condition				
(2) Observe Circulating Pumps				

MONTHLY CHECKS (Enter Date)

(1) Manual Lift Relief Valve				
(2) Review Condition of or Test Each Item	(A) Flame Detection Devices		(F) Refractory	
	(B) Limit Controls		(G) Stop Valves	
	(C) Operating Controls		(H) Check Valves	
	(D) Floor Drains		(I) Drain Valves	
	(E) Fuel Piping		(J) Linkages	
(3) Observe Gage Glass on Expansion Tank				
(4) Combustion Air Adequate/Unobstructed				

General Comments:

(Instructions on Reverse)

Figure 17.7 Maintenance log sheet for hot water heating boilers. *(Courtesy: ASME Section VI)*

in detail. The log sheet should be signed by a mechanic, and countersigned by maintenance engineer or supervisor.

A sample of a maintenance log sheet is shown in Figure 17.7. The maintenance log sheet should also show the name of the person or contractor who completed the job.

17.8 Boiler Repairs

Repair work is necessary to restore steam boilers to a safe and satisfactory operating condition. The repair method should be conformed to a national standard repair code and work should be done by the qualified mechanics.

The *National Board Inspection Code* (NBIC) is used as guidance for performing repairs and alterations to boilers and pressure vessels. Also, the National Board of Boiler and Pressure Vessel Inspectors issues a repair symbol stamp, "R", and a Certificate of Authorization to qualified boiler repair companies. Many jurisdictions require that the welding repair on a boiler should be completed by a repair company that possesses an "R" Certificate of Authorization (Figure 17.8).

Precaution. Repair work should be performed only when steam boilers are out of service and under zero pressure. The repair work shall not be undertaken without the approval of an authorized boiler inspector employed by the jurisdiction or insurance company that is licensed to

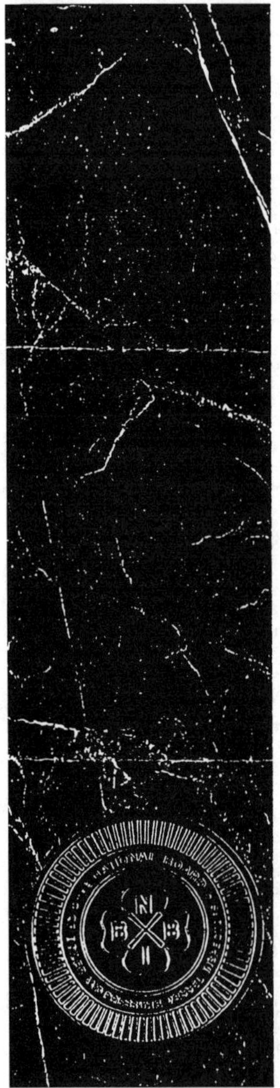

THE NATIONAL BOARD
OF
BOILER & PRESSURE VESSEL INSPECTORS

Certificate of Authorization

R

This is to certify that

ALPHA BOILERS, INCORPORATED
780 N. W. LEJEUNE ROAD, SUITE 417
MIAMI, FLORIDA 33126

is authorized to use the "R" SYMBOL in accordance with the provisions of the National Board.

The scope of Authorization is limited as follows:

REPAIRS AND/OR ALTERATIONS AT FIELD SITES ONLY AS CONTROLLED BY THE ABOVE LOCATION

CERTIFICATE NUMBER: R-2584
ISSUE DATE: NOVEMBER 30, 2000
EXPIRATION DATE: DECEMBER 18, 2003

Executive Director

NB 243 Rev. 1

Figure 17.8 "R" Certificate of Authorization. *(Courtesy: Alpha Boilers, Inc.)*

write boiler insurance as per jurisdictional law. In the absence of a jurisdiction, the work should be performed under the supervision of an engineer responsible for the repair work.

Notification. When repair work is completed, notify the authorized boiler inspector for inspection of the repair work. The boiler can be put back into operation only if the repair work is certified as satisfactory after being duly inspected and tested by the authorized boiler inspector.

Welding requirements. All repair methods shall be approved by an authorized boiler inspector before undertaking any welding repair. Welding should be done by qualified welders using procedures properly qualified in accordance with *ASME Code Section IX.*

Safety. Necessary precautions should be taken to ensure the safety of the people working in the boiler room. The main burner switch must be turned off and locked out. Tags should be placed showing that the burner is out of service. Always, one person should stand outside when another person is working inside the boiler.

Repair nameplate. When a welding repair is completed on a boiler, a repair nameplate should be attached to the boiler. As required by many jurisdictions, the repair nameplate should be attached to the original manufacturer's stamping, or nameplate. An example of a repair nameplate is shown in Figure 17.9.

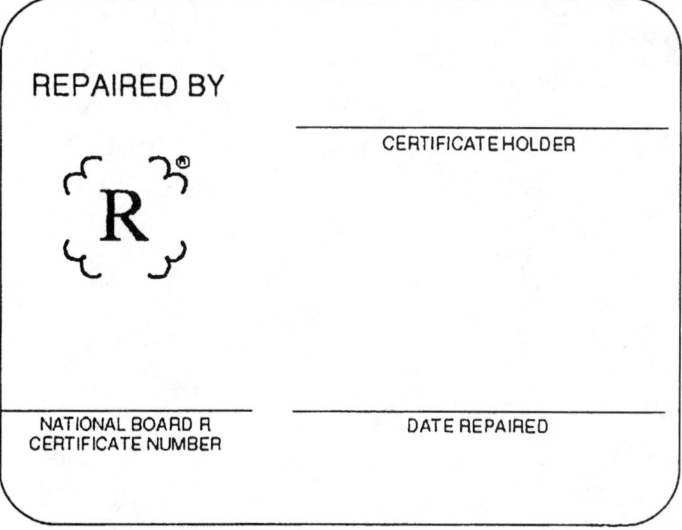

Figure 17.9 Repair nameplate. *(Courtesy: National Board of Boilers and Pressure Vessel Inspectors)*

Chapter

18

Water Treatment

Water contains various amounts of impurities like solids, gases, industrial pollutants, and other wastes. The solid impurities in untreated water can settle on the heating surfaces of the boiler in the form of scale. Gas impurities in water can lead to corrosion and pitting of metal in the boiler. Untreated water, if used in a boiler, can cause a lot of problems like caustic embrittlement, scale, corrosion, carryover, overheating, and the like, resulting in the failure of the boiler or system components. Water used in a boiler should be treated for safety and maximum efficiency.

This chapter covers considerations for water treatment, boiler water troubles, chemicals used for water treatment, functions of chemicals, treatment alternatives, procedures, and methods of treatment of laid-up boilers. These recommended rules are applicable to the treatment of water in steam and hot water heating boilers.

The management of a boiler plant should make a decision in advance whether water treatment is required for the boilers or not. Once a decision is finalized in favor of water treatment, then the type of treatment has to be considered very carefully. The following factors should be considered in determining the type of water treatment:

1. The type of boiler, i.e., cast iron or steel, steam or hot water; different types of water treatment are used for different types of boilers operating in the same system. In such cases it is necessary to inject the water treatment into the individual boilers, rather than into a central location such as a boiler feed tank. In this manner the water chemistry of each boiler can be properly controlled.
2. The nature of raw water, i.e., hard or soft, corrosive or scale forming.

3. Preliminary treatment of the water, i.e., soft waters, preheaters, deaerators.
4. The amount of makeup water and blowdown required.
5. The use of steam, i.e., for heating only or for other purposes.
6. The amount of supervision and control testing available.

Water treatment is a specialized field and only *a specialist* can recommend water treatment procedures suitable for a particular boiler installation. There are reputable water treatment companies who provide expert services, and/or chemicals for boiler treatment. A boiler engineer should consult a water treatment specialist for recommendations based on plant water conditions and specific boiler installation. The specialist can furnish test kits to the boiler operators for simple day-to-day water analysis. In addition, the specialist will arrange to take samples from the boiler water, send samples to a laboratory for analysis, and adjust the chemicals to obtain proper results.

A boiler operator should check with the local jurisdiction for any ordinances about disposal of blowdowns and draining of boilers. A boiler owner must comply with the local ordinances with respect to disposal of boiler compounds or chemicals.

18.1 Boiler Water Troubles

Raw water can cause a lot of trouble to the steam and hot water heating boilers. Some of the problems generated by untreated water are as follows:

Corrosion. Raw water contains impurities including dissolved gases such as oxygen and carbon dioxide. These gases make the water corrosive, causing and corrosion localized pitting of boiler metal. If unchecked, serious corrosion and pitting can result in the thinning of boiler plates and tubes, which can lead to rupture.

Scale deposits. Raw water contains dissolved salts, mainly calcium and magnesium compounds. When a boiler is under operation, calcium bicarbonate and magnesium bicarbonate is deposited on hot metal surfaces, forming hard scale deposits. The scale acts like insulation on the boiler tubes, resulting in heating losses and loss in efficiency. These deposits can also cause overheating, resulting in tube or furnace failure. Scale deposits on various parts of boilers are shown in Figure 18.1.

Metal (caustic) embrittlement. Any high alkaline solution that works into cracks in seams of improperly caulked joints or rivets may cause

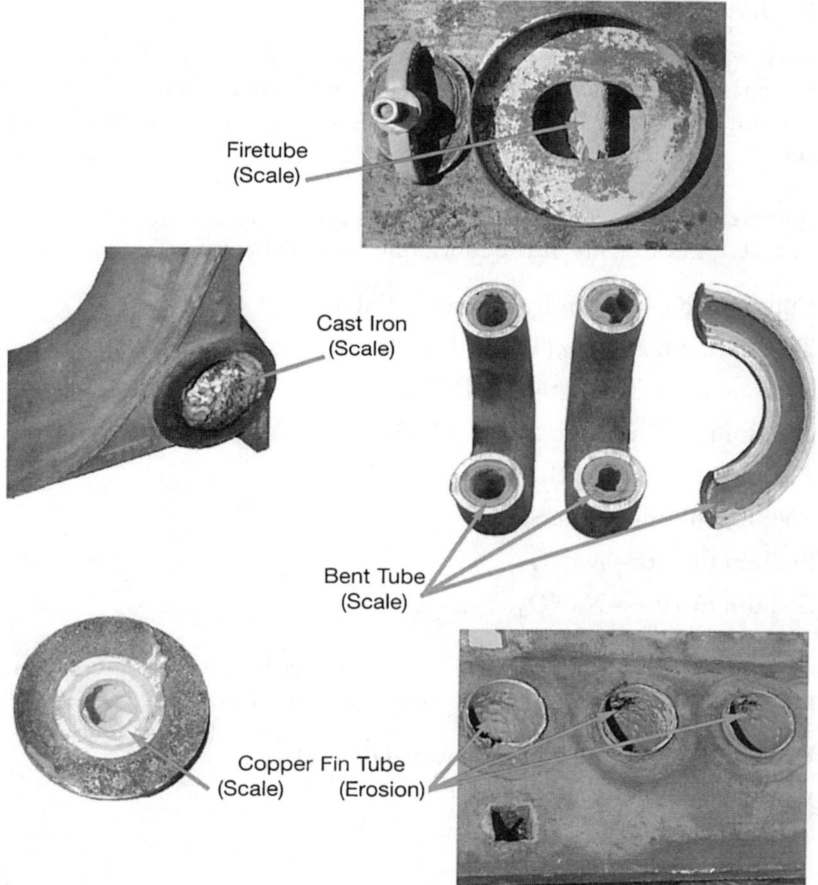

Figure 18.1 Scale and erosion on water surfaces. *(Courtesy: Rite Engineering & Mfg. Corp.)*

caustic embrittlement. The alkalinity of a solution is affected by the amount of hydroxides, such as caustic soda, present in the solution. Caustic embrittlement can cause metal to crack below the waterline and under rivets, welds, longitudinal seams, and at the ends of tubes in the boiler. This type of failure is limited to steel boilers.

Foaming, priming, and carryover. Foaming is the rapid fluctuation of the water level due to impurities on the surface of water. Priming is the carrying over of small water particles. Carryover is the carrying over of water into steam lines caused by high alkalinity, dissolved solids, oil, and sludge. Carry over can cause water hammer in the steam lines, which can result in rupturing the steam line or header, damage to system components, or an inability to meet load demand.

18.2 Water Treatment Chemicals

There are many chemicals used for treating of boiler water. The purpose of using the chemicals should be determined before applying any chemical to the boiler. The following chemicals are generally used for boiler water treatment:

Inorganic chemicals. Inorganic chemicals are used for precipitating the hard water salts. Some of the inorganic chemicals are:

1. Caustic soda (sodium hydroxide)—NaOH
2. Trisodium phosphate (TSP)—Na_3PO_4
3. Sodium acid phosphate—NaH_2PO_4
4. Sodium tripolyphosphate—$Na_5P_3O_{10}$
5. Sodium borate—$Na_2B_4O_7$
6. Sodium sulphite—Na_2SO_3
7. Sodium nitrate—$NaNO_3$
8. Sodium nitrite —$NaNO_2$

Organic chemicals. The organic chemicals act as protective colloids. Some of the organic chemicals used in water treatment are:

1. Sodium alginate and other seaweed derivatives
2. Quebrancho tannin
3. Lignin sulfonate
4. Starch

18.3 Functions of Chemicals

Various chemicals are used for treatment of boiler water. These chemicals, which may be classified in groups, perform certain functions and treat the raw water making it useable for the boiler.

Caustic soda. Sodium hydroxide (caustic soda) is used to control alkalinity of boiler water and precipitation of the magnesium salts. The pH value of boiler water should be maintained between 7 and 11 to prevent scale formation. Boiler water with a pH value of 7 is neutral. Boiler water with a pH value above 7 is alkaline.

Chromates and sulphites. Sodium chromate and sodium sulphite are added to the boiler to control corrosion. Sodium chromate acts as an inhibitor for the protection of boiler metal, whereas sodium sulphite is an oxygen scavenger, which removes oxygen from the boiler water. Oxygen present in the boiler water changes sodium sulphite to sodium sulphate.

Phosphates. The various phosphates of sodium are used to control precipitation of the hard water salts as insoluble lime and magnesia phosphates.

Nitrates and nitrites. Nitrates are used to control metal embrittlement. Nitrites act in a similar way to sulphites in controlling oxygen corrosion, but under certain conditions where dissimilar metals, particularly copper, brass, or soft solder, are immersed in the boiler water, they can cause very severe localized corrosion, unless suitable inhibiting agents are used. Use is generally confined to hot water systems.

Organic agents. Organic chemicals are used to control insoluble matter in suspension as sludge, and prevent the formation of dense adherent scale on the heat transfer surfaces.

Boiler compounds. Commercial boiler compounds are used for treatment of water in small boilers. These compounds, which may be solid or liquid, are mixtures of different water treatment chemicals. The manufacturer keeps the mixing formula a trade secret. There are two types of boiler compounds generally used:

1. Boiler compounds based on chromates
2. Boiler compounds based on alkaline salt combinations plus sodium sulphite

18.4 Treatment Alternatives

External or internal treatment. Boiler water can be treated before it is introduced in the boiler (external treatment) or it can be treated in the boiler (internal treatment). However, boiler water is best treated by a combination of external and internal treatment to provide necessary protection against problems arising from untreated water.

External treatment. External treatment is used to remove the scale-forming salts and dissolved gases. This type of treatment includes zeolite softener, lime soda process, deaerator, and demineralization.

Internal treatment. In internal treatment, chemicals are added directly into the boiler. This treatment method depends on the condition of the water in the steam and water drum. Sodium sulphite, an oxygen scavanger, may be injected into the boiler water, but it reacts with oxygen to form sodium sulphate, which accumulates at the bottom of the boiler. The sodium sulphate is then removed through the bottom blowdown line or continuous blowdown line.

Seasonal or continuous treatment. Boiler installations, for which water treatment is considered, may be classified into three categories:

Class 1—No treatment

Class 2—Seasonal or semiseasonal treatment with limited chemical control

Class 3—Complete treatment with continuous chemical control

Treatment of cast iron boilers. Since cast iron boilers are less subject to corrosion than steel boilers, water treatment may not be required if annual cleanout on boilers that are used as low-pressure heating units is done properly. If the steam system is a closed system, i.e., steam is returned to the boiler as condensate and the makeup water is small, there may negligible effects from boiler water. It may be necessary after a few years to use acid cleaning for removal of scale, if water treatment has not been used.

18.5 Blowdown

Because an excess total of dissolved solids may cause foaming, priming, and carryover, blowdown is used to control the amount of total dissolved solids and sludge in the boiler. As a rule of thumb, about 1000 ppm is considered as a safe maximum limit for dissolved solids. The frequency of blowing down is determined by boiler water analysis; yet this frequency should be kept at a minimum because it involves loss of heat and waste of treatment chemicals.

18.6 Feeders

Feeders are used to add treatment chemicals to the boiler. The type of feeder required depends on the frequency of adding treatment, amount of blowdown, and loss of condensate. The following types of feeders are commonly used:

Open-type gravity feeders—where the chemical treatment is fed manually in one slug or in periodic small shots.

Closed-type gravity drip and bypass feeders—where the treatment is fed in proportion to the amount of makeup water.

Pot-type proportional feeders—where slowly dissolving treatment crystals or briquettes are used.

18.7 Procedures

Determination of water-holding capacity. Total water-holding capacity of the boiler is required for calculating the amount of water treatment compound. This capacity should be available from the manufacturer's catalogs; if it is not available, a water meter should be used at the time of initial filling to record the capacity.

Making a pH or alkalinity test. The pH value, which determines the alkalinity of boiler water, can be quickly tested with hydrion paper. A color chart on the side of the hydrion dispenser indicates the reading in pH. If the precise measurement of pH is required, a color slide comparison kit may be used instead of hydrion paper.

Mixing and handling chemicals. Chemicals, solid or liquid, should be handled carefully as they may be highly alkaline or skin irritatants. Always remember that liquid chemicals are diluted before use; solid chemicals are dissolved in accordance with the supplier's instructions, but make sure that the latter are dissolved before use. If the chemicals are slow to dissolve, use a spatula to stir the solution or heat the water by steam and agitate the mixture.

18.8 Treatment of Laid-Up Boilers

Boilers used on a seasonal basis that will be idle for a length of time (in excess of 30 days) should be laid-up by a proper method, which is determined by the length of the out of service period and the plant conditions. Improper lay-up methods can result in damage caused by oxygen corrosion and pitting. Regardless of the method used, the boiler should first be thoroughly cleaned on both the fire and watersides, and all repairs be made. During the inactivity period, the boiler is protected from corrosion either by draining it and keeping the surfaces thoroughly dry or by completely filling the boiler with properly treated water.

Dry method. In the event that the boiler could be subjected to freezing temperatures or if it is to be idle for an excessive period of time, the following steps should be taken:

1. Drain and clean the boiler thoroughly.
2. Dry the boiler out by means of hot air.
3. Place lime, silica gel, or suitable moisture absorbent in open trays inside the boiler shell.
4. Close the boiler tightly to exclude all moisture and air.
5. All the allied equipment, such as condensate tanks and pumps, should be thoroughly drained.

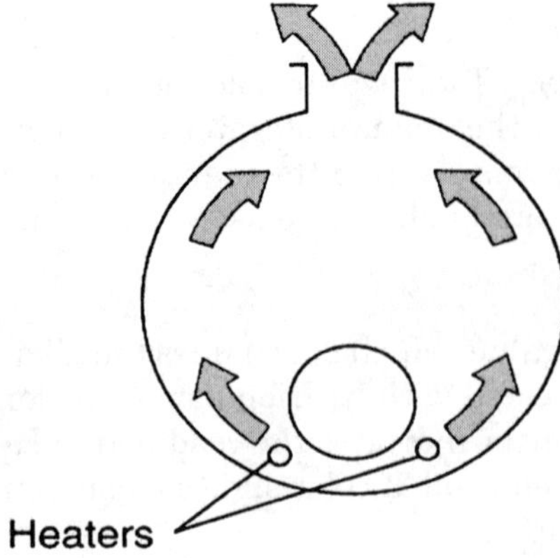

Figure 18.2 Dry storage method.

6. Check the boiler every 2 to 3 months and replace the lime or silica gel if necessary.
7. Do not use this method for cast-iron boilers.

A dry storage method is shown in Figure 18.2.

Wet method. In order to protect the boiler during any short period of time, the following steps should be taken:

1. Drain and clean the boiler thoroughly.
2. Fill the boiler to the top of the drum and steam for a short time with the vent open to release dissolved gases.

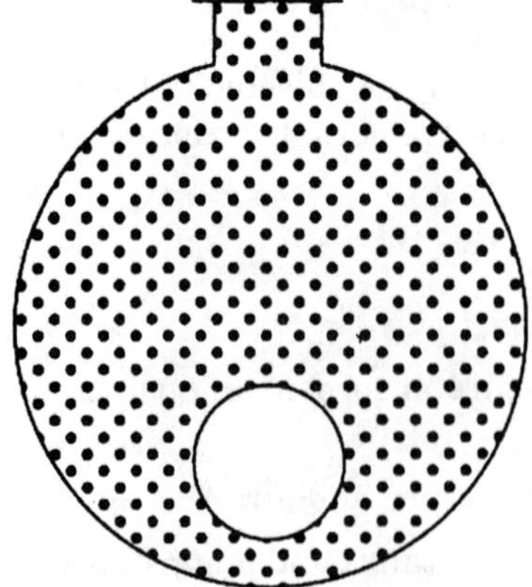

Figure 18.3 Wet storage method.

3. Add caustic soda (400 ppm) and sodium sulphite (100 ppm) to the hot water. Maintain sodium sulphite concentration at a minimum of 100 ppm on steam boilers and 300 ppm on hot water boilers.
4. Close all the valves including the boiler vent.
5. Check all boiler connections for leaks and take a weekly sample to make sure that the alkalinity and sulphite are stable.
6. Occasionally circulate the water with a pump.
7. This method is recommended for cast-iron boilers.

A wet storage method is shown in Figure 18.3.

Chapter 19

Glossary of Terms

The common terms relating to boilers, fuels, and fuel-burning equipment, and water treatment have been defined in this section.

19.1 Boilers and General Terms

Absolute pressure—pressure above zero pressure, the sum of the gage and atmospheric pressures.

Accumulator (steam)—a pressure vessel containing water and steam, which is used to store the heat of steam for use at a later period and at some lower pressure.

Air purge—the removal of undesired matter by replacement with air.

Air vent—a valve opening in the top of the highest drum of a boiler or pressure vessel for venting air. Also a device, manual or automatic, which will effect the removal of air from a steam or hot water heating system. It is usually located at the highest point in the system.

Altitude gage—a pressure gage calibrated in feet, indicating the height of the water in the system above the gage; usually found as combination pressure, altitude, and temperature gage.

Ambient temperature—the temperature of the air surrounding the equipment.

Baffle—a plate or wall for deflecting gases or liquids.

Blowdown—the difference between the opening and closing pressures of a safety or safety relief valve.

Boiler—a closed pressure vessel in which a liquid, usually water, is vaporized by the application of heat.

Watertube—a boiler in which the tubes contain water and steam; heat is applied to the outside surface.

Bent tube—a watertube boiler consisting of two or more drums connected by tubes, practically all of which are bent near the ends to permit attachment to the drum shell on radial lines.

Horizontal—a watertube boiler in which the main bank of tubes is straight and on a slope of 5 to 15 degrees from the horizontal.

Sectional header—a horizontal boiler of the longitudinal or cross drum type, with the tube bank composed of multiple parallel sections, each section made up of a front and rear header connected by one or more vertical rows of generating tubes and with the sections or groups of sections having a common steam drum.

Box header—a horizontal boiler of the longitudinal or cross drum type consisting of a front and a rear inclined rectangular header connected by tubes.

Cross drum—a sectional header or box header boiler in which the axis of the horizontal drum is at right angles to the center lines of the tubes in the main bank.

Longitudinal drum—a sectional header or box header boiler in which the axis of the horizontal drum or drums is parallel to the tubes in a vertical plane.

Low head—a bent tube boiler having three drums with relatively short tubes in the rear bank.

Firetube—a boiler with straight tubes, which are surrounded by water and steam and through which the products of combustion pass.

Horizontal return tubular—a firetube boiler consisting of a cylindrical shell, with tubes inside the shell attached to both end closures. The products of combustion pass under the bottom half of the shell and return through the tubes.

Locomotive—a horizontal firetube boiler with an internal furnace, the rear of which is a tube sheet directly attached to a shell containing tubes through which the products of combustion leave the furnace.

Horizontal firebox—a firetube boiler with an internal furnace, the rear of which is a tube sheet directly attached to a shell containing tubes. The first-pass bank of tubes is connected between the furnace tube sheet and the rear head. The second-pass bank of tubes, passing over the crown sheet, is connected between the front and rear end closures.

Refractory lined firebox—a horizontal firetube boiler, the front portion of which is set over a refractory or water-cooled refractory furnace. The rear of the boiler shell has an integral or separately connected section containing the first-pass tubes through which the products of combustion leave the furnace, then return through the second-pass upper bank of tubes.

Vertical—a firetube boiler consisting of a cylindrical shell, with tubes connected between the top head and the tube sheet, which forms the top of the internal furnace. The products of combustion pass from the furnace directly through the vertical tubes.

Submerged vertical—the same as the plain type above, except that by use of a waterleg construction as a part of the upper tube sheet, it is possible to carry the waterline at a point above the top ends of the tubes.

Scotch—in stationary service, a firetube boiler consisting of a cylindrical shell, with one or more cylindrical internal furnaces in the lower portion and a bank of tubes attached to both end closures. The fuel is burned in the furnace, the products of combustion leaving the rear end to return through the tubes to an uptake at the front head—known as dry-back type.

In the marine service, this boiler has an internal combustion chamber of waterleg construction covering the rear end of the furnace and tubes, in which the products of combustion turn and enter the tubes—known as water-back type.

Modular—a steam or hot water heating assembly consisting of a grouping of individual boilers called modules, intended to be installed as a unit. Modules may be under one jacket or may be individually jacketed.

Vacuum—a factory sealed steam boiler that is operated below the atmospheric pressure.

Boiler horsepower—the evaporation of 34.5 lb (15.6 kg) of water per hour from temperature of 212°F (100°C) into dry saturated steam at the same temperature equivalent to 33,472 Btu (9.81 kJ)

Boiler layup—any extended period of time during which the boiler is not expected to operate and suitable precautions are made to protect it against corrosion, scaling, pitting,and the like on the water and firesides.

Boiler trim—piping on or near the boiler, which is used for safety, limit, operating controls, gages, water column, and so forth.

Breeching—a duct for the transport of the products of combustion between the boiler and the stack.

Casing—a covering of sheets of metal or other material such as fire-resistant composition board used to enclose all or a portion of a steam generating unit.

Cleanout door—a door placed so that accumulated refuse may be removed from a boiler setting.

Cock—a plug- or ball-type valve in which a 90 degree turn of the handle will move the valve to full-open or closed position.

Combustion chamber—that part of a boiler where combustion of the fuel takes place.

Condensate—condensed water resulting from the removal of latent heat from steam.

Conductivity—the amount of heat (Btu) transmitted in 1 hour through 1 ft^2 (645 mm^2) of a homogeneous material 1 inch (25 mm) thick for a difference in temperature of 1°F (0.556°C) between the two surfaces of the material.

Crown sheet—in a firebox boiler, the plate forming the top of the furnace.

Damper—a device for introducing a variable resistance for regulating the volumetric flow of gas or air.

Butterfly type—a blade damper pivoted about its center.

Curtain type—a damper, composed of flexible material, moving in a vertical plane as it is rolled.

Flap type—a damper consisting of one or more blades each pivoted about one edge.

Louver type—a damper consisting of several blades, each pivoted about its center and linked together for simultaneous operation.

Slide type—a damper consisting of a single blade, which moves substantially normal to the flow.

Design pressure—the maximum allowable working pressure permitted under the rules of *ASME Code Section IV*.

Downcomer—a tube in a boiler or waterwall system through which fluid falls downward.

Drain—a valved connection at the lowest point for the removal of all water from the pressure parts.

Drum—a cylindrical shell closed at both ends designed to withstand internal pressure.

Efficiency—the ratio of output to the input. The efficiency of a steam generating unit is the ratio of the heat absorbed by water and steam to the heat in the fuel fired.

Electric boiler—a boiler in which electric heating means serve as the source of heat.

Equalizer—connections between parts of a boiler to equalize pressures.

Equivalent direct radiation (EDR)—the amount of heating surface that will give off 240 Btu/hr (0.070 kW) for steam and 150 Btu/hr

(0.044 kW) for hot water. EDR may have no direct relation to actual surface area.

Equivalent evaporation—evaporation expressed in pounds of water evaporated from a temperature of 212°F (100°C) to dry saturated steam at 212°F (100°C).

Externally fired boiler—a boiler in which the furnace is essentially surrounded by refractory or water-cooled tubes.

Firebox—the equivalent of a furnace, a term usually used for the furnaces of locomotives and similar types of boilers.

Firetube—a tube in a boiler having water outside and carrying the products of combustion inside of it.

Fired pressure vessel—a vessel containing a fluid under pressure exposed to heat from the combustion of fuel.

Furnace—an enclosed space provided for the combustion of fuel.

Fusible plug—a hollow threaded plug having the hollowed portion filled with a low melting-point material, usually located at the lowest permissible water level.

Gage glass—the transparent part of a water gage assembly connected directly or through a water column to the boiler, below and above the waterline, to allow for visual inspection of the water level in the boiler.

Handhole—an opening in a pressure part for access, usually not exceeding 6 inches (152 mm) in the longest dimension.

Hydrostatic test—a strength and tightness test of a closed pressure vessel by water pressure.

Impingement—the striking of moving matter, such as the flow of steam, water, gas, or solids, against similar or other matter.

Inspector, authorized—a boiler inspector who, according to the local requirements, is authorized to inspect the boilers. He may be a city, state, province, or insurance company employee.

Internally fired boiler—a firetube boiler having an internal furnace such as a scotch, locomotive firebox, vertical tubular, or other type having a water-cooled plate–type furnace.

Low-water fuel cutoff—a float-operated device, which shuts down the fuel burner when the water level in the boiler drops below its operating level.

Lowest safe waterline—that water level in the boiler below which the burner is not allowed to operate.

Makeup water—water introduced into the boiler to replace that lost or removed from the system.

Miniature boiler—fired pressure vessels which do not exceed the following limits: 16 inches (406 mm) inside diameter of shell, 42 inches (1066 mm) overall length to outside of heads at center, 20 ft^2 (1.9 m^2) water heating surface, or 100 psi (690 kPa) maximum allowable working pressure.

Packaged steam generator—a boiler equipped and shipped complete with fuel burning equipment, mechanical draft equipment, automatic controls, and accessories.

Pressure, accumulation test—that steam pressure at which the capacity of a safety, safety relief, or relief valve is determined.

Pressure vessel—a closed vessel or container designed to confine a fluid at a pressure above the atmospheric pressure.

Receiver—the tank portion of a condensate or vacuum return pump where condensate accumulates.

Safety valve—an automatic pressure-relieving device actuated by the static pressure upstream of the valve and characterized by full-opening pop action. It is used for gas or vapor service.

Safety relief valve—an automatic pressure-relieving device actuated by the pressure upstream of the valve and characterized by opening pop action with further increase in pressure over popping pressure.

Smokebox—an external compartment on a boiler to catch unburned products of combustion.

Stop valve—a valve (usually gage type) which is used to isolate a part of a heating system or a boiler from the other parts.

Strainer, condensate—mechanical means (screen) for removing solid material from the condensate before it reaches the pump.

Trap—a device installed in steam piping, which is designed to prohibit the passage of steam but allow the passage of condensate and air.

Unfired pressure vessel—a vessel designed to withstand internal pressure, neither subjected to heat from products of combustion nor an integral part of a fired pressure vessel system.

Vent—an opening in a vessel or other enclosed space for the removal of gas or vapor.

Water column—a vertical tubular member connected at its top and bottom to the steam and water space, respectively, of a boiler, to which the water gage, gage cocks, and high- and low-level alarms may be connected.

19.2 Fuels, Fire Burning Equipment, and Combustion

Air-atomizing oil burner—a burner for firing oil in which the oil is atomized by compressed air, which is forced into and through one or more streams of oil, breaking the oil into a fine spray.

Air purge—the removal of undesired matter by replacement with air.

Ambient air—the air that surrounds the equipment. The standard ambient air for performance calculations is air at 80°F, 60 percent relative humidity, and a barometric pressure of 29.92 in. Hg.

Aspirating burner—a burner in which the fuel in a gaseous or finely divided form is burned in suspension, the air for combustion being supplied by bringing it into contact with the fuel. Air is drawn through one or more openings by the lower static pressure created by the velocity of the fuel stream.

Atmospheric pressure—the barometric reading of pressure exerted by the atmosphere—at sea level, 14.7 lbs/in.^{2} (101 kPa) or 29.92 in. Hg (101 kPa).

Auxiliary air—additional air, either hot or cold, which may be introduced into the exhaust inlet or burner lines, to increase the primary air at the burners.

Balanced draft—the maintenance of a fixed value of draft in a furnace at all combustion rates by controling incoming air and outgoing products of combustion.

Blower—a fan used to force air under pressure.

British thermal unit—the mean British thermal unit is 1/180 of the heat required to raise the temperature of 1 lb (0.45 kg) of water from 32°F (0°C) to 212°F (100°C) at a constant atmospheric pressure. It is about equal to the quantity of heat required to raise 1 lb (0.45 kg) of water to 1°F or (0.556°C).

Bunker C oil—residual fuel oil of high viscosity commonly used in marine and stationary steam power plants (No. 6 fuel oil).

Burner—a device for the introduction of fuel and air into a furnace at the desired velocities, turbulence, and concentration to establish and maintain proper ignition and combustion of the fuel.

Burner windbox—a plenum chamber around a burner, in which an air pressure is maintained to insure proper distribution and discharge of secondary air.

Calorific value—the number of heat units liberated per unit of quantity of fuel burned in a calorimeter under prescribed conditions.

Calorimeter—apparatus for determining the calorific value of a fuel.

Centrifugal fan—consists of a fan rotor or wheel within a scroll type of housing which discharges air at a right angle to the axis of the wheel.

Chimney—a brick, metal, or concrete stack.

Combustion—the rapid chemical combination of oxygen with combustible elements of a fuel resulting in the production of heat.

Combustion rate—the quantity of fuel fired per unit of time, as pounds of coal per hour, or cubic feet of gas per minute.

Complete combustion—the complete oxidation of all the combustible constituents of a fuel.

Control valve—a valve used to control the flow of air or gas.

Damper—a device for introducing a variable resistance for regulating the volumetric flow of gas or air.

Butterfly type—a single blade damper pivoted about its center.
Curtain type—a damper composed of flexible material, moving in a vertical plane as it is rolled.
Flap type—a damper consisting of one or more blades, each pivoted about one edge.
Louver type—a damper consisting of several blades, each pivoted about its center and linked together for simultaneous operation.
Slide type—a damper consisting of a single blade, which moves substantially normal to the flow.

Dew point—the temperature at which condensation starts.

Draft—the difference between atmospheric pressure and some lower pressure existing in the furnace or gas passages of a steam-generating unit.

Draft differential—the difference in static pressure between two points in a system.

Draft gage—a device for measuring draft, usually in inches of water.

Electric ignition—ignition of a pilot or main flame by the use of an electric arc or glow plug.

Excess air—air supplied for combustion in excess of that theoretically required for complete oxidation.

Explosion—combustion, which proceeds so rapidly that high pressure is generated suddenly.

Fan—a machine consisting of a rotor and housing for moving air or gases at relatively low-pressure differentials.

Flame detector—a device which indicates if fuel, such as liquid, gaseous, or pulverized, is burning, or if ignition has been lost. The indication may be transmitted to a signal or to a control system.

Flame safeguard

Thermal—bimetallic strip thermocouple, which is located in the pilot flame. If the pilot goes out, a circuit is broken and the fuel valve is shut. Response time is 1 to 3 min. Suitable for small installations.

Electronic—electrode is used in a flame rectification system, which detects pilot and main flame, and prevents fuel flow if pilot is not detected or stops fuel flow if main flame is not detected. Response time is 1 to 4 seconds. Suitable for large programmed installations.

Flareback—a burst of flame from a furnace in a direction opposed to the normal flow; usually caused by the ignition of an accumulation of combustible gases.

Flash point—the lowest temperature at which, under specified conditions, fuel oil gives off enough vapor to flash into momentary flame when ignited.

Flue gas—the gaseous products of combustion in the flue to the stack.

Forced draft fan—a fan supplying air under pressure to the fuel-burning equipment.

Fuel-air ratio—the ratio of the weight or volume of fuel to air.

Furnace draft—the draft in a furnace, measured at a point immediately in front of the highest point at which the combustion gases leave the furnace.

Grate—the surface on which fuel is supported and burned, and through which air is passed for combustion.

Hand-fired grate—a grate on which fuel is placed manually, usually by means of a shovel.

Ignition—the initiation of combustion.

Ignition temperature—the lowest temperature of a fuel at which combustion becomes self-sustaining.

Incomplete combustion—the partial oxidation of the combustible constituent of a fuel.

Induced draft fan—a fan exhausting hot gases from the heat-absorbing equipment.

Intermittent firing—a method of firing by which fuel and air are introduced into and burned in a furnace for a short period, after which the flow is stopped, this succession occurring in a sequence of frequent cycles.

Lignite—a consolidated coal of low classification according to ran moist (bed moisture only) Btu less than 8300 (MJ less than 89.76).

Low heat value—the high heating value minus the latent heat of vaporization of the water formed by burning the hydrogen in the fuel.

Manometer—device used to detect small changes in pressure, usually a tube with water, the pressure variations measured in inches of water.

Mechanical draft—the negative pressure created by mechanical means.

Mechanical stoker—a device consisting of a mechanically operated fuel-feeding mechanism and a grate, used for the purpose of feeding solid fuel into a furnace, distributing it over the grate, admitting air to the fuel for the purpose of combustion, and providing a means for removal or discharge of refuse.

Overfeed stoker—a stoker in which fuel is fed onto grates above the point of air admission to the fuel bed. Overfeed stokers are divided into four classes, as follows:

1. A front-feed inclined grate stoker is an overfeed stoker, in which fuel is fed from the front onto a grate inclined downward toward the rear of the stoker.
2. A double-inclined-side feed stoker is an overfeed stoker, in which the fuel is fed from both sides onto grates inclined downward toward the centerline of the stoker.
3. A chain or traveling grate is an overfeed stoker having a moving endless grate, which conveys fuel into and through the furnace where it is burned, after which it discharges the refuse.
4. A spreader stoker is an overfeed stoker that discharges fuel into the furnace from a location above the fuel bed and distributes the fuel onto the grate.

Underfeed stoker—a stoker in which fuel is introduced through retorts at a level below the location of air admission to the fuel bed. Underfeed stokers are divided into three general classes, as follows:

1. A side-ash-discharge underfeed stoker is a stoker having one or more retorts, which feed and distribute solid fuel onto side tuyeres, or a grate, through which air is admitted for combustion and over which the ash is discharged at the side parallel to the retorts.
2. A rear-ash-discharge underfeed stoker is a stoker having a grate composed of transversely spaced underfeed retorts, which feed and distribute solid fuel to intermediate rows of tuyeres through which air is admitted for combustion. The ash is discharged from the stoker across the rear end.

3. A continuous-ash-discharge underfeed stoker is one in which the refuse is discharged continuously from the normally stationary stoker ash tray to the ash pit without the use of mechanical means other than the normal action of the coal feeding and agitating mechanism.

Modulation of burner—control of fuel and air to a burner to match fluctuations of the load on the boiler.

Multifuel burner—a burner by means of which more than one fuel can be burned, either separately or simultaneously, such as pulverized fuel, oil, or gas.

Natural draft boiler—a boiler that operates with the furnace pressure less than atmospheric pressure due to buoyant action of the venting system.

Natural draft stoker—a stoker in which the flow of air through the grate is caused by difference of pressure between the furnace and the atmosphere.

Oil burner—a burner for firing oil.

Orifice—the opening from the whirling chamber of a mechanical atomizer or the mixing chamber of a steam atomizer through which the liquid fuel is discharged.

Orsat—a gas-analysis apparatus in which certain gaseous constituents are measured by absorption in separate chemical solutions.

Oxidation—chemical combination with oxygen.

Perfect combustion—the complete oxidation of all the combustible constituents of a fuel, utilizing all the oxygen supplied.

Pilot flame—the flame, usually gas or light oil, which ignites the main flame.

Pour point—the temperature at which the oil flows.

Primary air—air introduced with the fuel at the burners.

Products of combustion—the gases, vapors, and solids resulting from the combustion of fuel.

Pulverized fuel—solid fuel reduced to a fine size.

Radiation loss—a comprehensive term used in a boiler-unit heat balance to account for the conduction, radiation, and convection heat losses from the settings to the ambient air.

Ringleman chart—a series of four rectangular grids of black lines of varying widths printed on a white background, and used as a criteria of blackness for determining smoke density.

Rotary oil burner—a burner in which atomization is accomplished by feeding oil to the inside of a rapidly rotating cup.

Secondary air—air for combustion supplied to the furnace to supplement the primary air.

Secondary combustion—combustion which occurs as a result of ignition at a point beyond the furnace (see also Delayed combustion).

Smoke—small gasborne particles of carbon or soot, less then 1 μin. (0.025 μm) size, resulting from incomplete combustion of carbonaceous material and of sufficient number to be observable.

Soot—unburned particles of carbon derived from hydrocarbons.

Spray nozzle—from which a fuel is discharged in the form of a spray.

Stack—a vertical conduit, which due to the difference in density between internal and external gases, creates a draft at its base.

Standard air—dry air weighing 0.075 lb/ft^3 (0.001 kg/m^3) at sea level [29.92 in. (736.6 mm) barometric pressure] and 70°F (21°C).

Steam atomizing oil burner—a burner for firing oil, which is atomized by steam. It may be of the inside- or outside-mixing type.

Strainer, fuel oils—metal screen with small openings to retain solids and particles in fuel oil, which could detrimentally affect the operation of the oil burner.

Tertiary air—air for combustion supplied to the furnace to supplement the primary and secondary air.

Unburned combustible—the combustible portion of the fuel which is not completely oxidized.

Viscosity—measure of the internal friction of a fluid or its resistance to flow.

Volatility—measurement of a fuel oil's ability to vaporize.

Windbox—a chamber below the grate or surrounding a burner, through which air under pressure is supplied for combustion of the fuel.

19.3 Terms Relating to Water Treatment

Acid—any chemical compound containing hydrogen that dissociates to produce hydrogen ions when dissolved in water.

Alkali—any chemical compound of a basic nature that dissociates to produce hydroxyl ions when dissolved in water.

Amines—a class of organic compounds that may be considered as derived from ammonia by replacing one or more of the hydrogen ions with organic radicals.

Chelating—the property of a chemical, when dissolved in water, that keeps the hard water salts in solution and thus prevents the formation of scale.

Colloid—a fine dispersion in water that does not settle out, and which is not a true solution.

Hydrazine—a strong reducing agent having the formula N_2NNH_2 in the form of a colorless hygroscopic liquid. Used as an oxygen scavenger.

Inhibitor—a compound that slows down or stops an undesired chemical reaction such as corrosion or oxidation.

Muriatic acid—commercial hydrochloric acid.

Neutralize—the counteraction of acidity with an alkali or of alkalinity with an acid to form neutral salts.

Parts per million (ppm)—the most commonly used method of expressing the quantity of a substance present in water.

pH—a scale used to measure the degree of acidity or alkalinity of a solution. The scale runs from 1 (strong acid) to 14 (strong alkali) with 7 (distilled water) as the neutral point.

Precipitation—the formation and settling out of solid particles in a solution.

Zeolite—originally a group of natural minerals capable of removing calcium and magnesium ions from water and replacing them with sodium. The term has been broadened to include synthetic resins, which similarly soften water by ion exchange.

Appendix A

RULES FOR THE CONSTRUCTION

OF

LOW-PRESSURE HEATING BOILERS

SECTION IV

A.S.M.E. BOILER CONSTRUCTION CODE

REPORT OF SUB-COMMITTEE OF THE BOILER CODE COMMITTEE ON HEATING BOILERS

THE AMERICAN SOCIETY OF MECHANICAL ENGINEERS
29 WEST THIRTY-NINTH STREET, NEW YORK
1923

To the Boiler Code Committee of the American Society of Mechanical Engineers:

Gentlemen:

Your Sub-Committee on Heating Boilers respectfully submits the following revised Code of Rules for the Construction of Low-Pressure Steam Heating, Hot-Water Heating and Hot-Water Supply Boilers.

In organizing the work of revision of the Heating Boiler Section of the Code, it was your decision to invite the co-operation of the American Society of Heating and Ventilating Engineers which was requested to appoint a conferring committee composed of members having engineering and technical knowledge of the subject with two representatives of the following branches: Consulting Engineers; Engineering Education; Designers; Manufacturers; and Contractors. It was pointed out that the conferring committee would co-operate with your Sub-Committee on Heating Boilers, with report to be made to the Boiler Code Committee, which in turn would report to the Council of the American Society of Mechanical Engineers.

The Conference Committee appointed by the American Society of Heating and Ventilating Engineers was constituted as follows:

L. A. Harding W. S. Timmis	Consulting Engineers.	P. J. Dougherty J. F. McIntire	Designers.
F. P. Anderson L. P. Breckenridge	University Professors.	Homer Addams A. A. Landon	Manufacturers.
H. M. Hart W. H. Driscoll	Contractors.		

Upon careful review of the Heating Boiler Section of the 1918 Edition of the A. S. M. E. Boiler Code, it was determined by your Sub-Committee that an entire rewriting of this section would be necessary, involving a complete revision of the rules, dividing them into separate sections for steel and cast-iron heating boilers. Intensive work on the revisions began in 1921 and was completed in the early part of 1922, the proposed revisions of this section having been published in November, 1922, issue of Mechanical Engineering (the Journal of the American Society of Mechanical Engineers). Following this publication, a Public Hearing on the proposed revisions was held on December 5, 1922, in connection with the Annual Meeting of the Society, at which a number of criticisms and suggestions for changes were received. Your Sub-Committee having co-ordinated the suggestions offered, respectfully submits the accompanying revised code for your final approval.

Respectfully submitted:

S. F. Jeter, *Chairman* M. F. Moore
R. W. Birchfield O. T. Nelson
Wm. H. Boehm Wm. B. Reed
Chas. E. Gorton

The revised code thus submitted was approved by the Boiler Code Committee and upon its recommendation, by the Council of the Society on May 28, 1923.

C. W. Obert, *Secretary.*

2M-1-25

A.S.M.E. BOILER CONSTRUCTION CODE

SECTION IV

Boilers used Exclusively for Low-Pressure Steam Heating, Hot-Water Heating, and Hot-Water Supply

These rules are divided into two sections: Section 1 applying to steel-plate boilers, and Section 2 applying to cast-iron boilers. They do not apply to economizers or feedwater heaters. These Rules were not formulated to cover apparatus such as ordinary range water backs, range boiler and gas water heaters for the production of domestic hot-water supply.

I—STEEL-PLATE BOILERS

H-1. These rules for steel-plate boilers shall apply:

a To all steam boilers for operation at pressures not exceeding 15 lb. per sq. in.

b To steel-plate hot-water boilers not exceeding 60 in. in diameter, or 160 lb. working pressure, or temperatures not exceeding 250 deg. fahr.

c For conditions exceeding those specified above, the rules for power boilers shall apply.

H-2. Wherever the term maximum allowable working pressure is used herein, it refers to gage pressure or the pressure above the atmosphere in pounds per square inch.

H-3. The maximum allowable working pressure shall not exceed 15 lb. per sq. in. on a steel-plate boiler built under these rules to be used for low-pressure steam heating.

The maximum allowable working temperature at or near the outlet of a hot-water steel-plate boiler shall not exceed 250 deg. fahr., or the maximum allowable working pressure 160 lb. per sq. in.

H-4. The maximum allowable working pressure on the shell or drum of steel plate steam heating and hot-water boilers shall be determined by the strength of the weakest course, computed from the thickness of the plate, the tensile strength stamped thereon, as provided for in Par. S-17 of Section II of the Code, the efficiency of the longitudinal joint, or of the ligament between the tube holes in shell or drum (whichever is the least), the inside diameter of the course, and the factor of safety, but in no case shall the pressure on which the factor of safety is based be considered less than 30 lb.

$$\frac{TS \times t \times E}{R \times FS} = \text{maximum allowable working pressure, lb. per sq. in.}$$

where TS = ultimate tensile strength stamped on shell plates, as provided for in Par. S-17 of Section II of the Code, lb. per sq. in.

t = minimum thickness of shell plates in weakest course, in.

E = efficiency of longitudinal joint or of ligaments between tube holes (whichever is the least)

R = inside radius of the weakest course of the shell or drum, in.

FS = factor of safety, or the ratio of the ultimate strength of the material to the allowable stress.

For new constructions, FS in the above formula = 5.

Materials

H-5. Specifications are given in Pars. S-1 to S-213 of Section II of the Code, for the important materials used in the construction of boilers, and where so given the materials herein mentioned for boiler parts required to resist internal pressure shall conform thereto except as specified herein for autogenously welded boilers.

H-6. Steel plates for any part of a boiler where under pressure, also manhole and handhole covers and other parts subjected to pressure, and braces and lugs when made of steel plate, shall be of firebox or flange quality as designated in the Specifications for Steel Boiler Plate, except for base metal as specified for autogenous welding.

H-7. Braces when made of parts welded together shall be of wrought iron of the quality designated in the Specifications for Extra-Refined Bar Iron.

Ultimate Strength of Material

H-8. Tensile Strength of Steel Plate. In determining the maximum allowable working pressure, the tensile strength used in the computations for steel plates shall be that stamped on the plates as herein provided, which is the minimum of the stipulated range, or 55,000 lb. per sq. in. for all steel plates except for special grades having a lower tensile strength.

H-9. Crushing Strength of Steel Plate. The resistance to crushing of steel plate shall be taken at 95,000 lb. per sq. in. of cross-sectional area.

H-10. Strength of Rivets in Shear. In computing the ultimate strength of rivets in shear, the following values in pounds per square inch of the cross-sectional area of the rivet shank shall be used:

Iron rivets in single shear	38,000
Iron rivets in double shear	76,000
Steel rivets in single shear	44,000
Steel rivets in double shear	88,000

The cross-sectional area used in the computations shall be that of the rivet shank after driving.

Minimum Thickness of Plates and Tubes

H-11. The minimum thickness of any boiler plate under pressure shall be ¼ in.

H-12. The minimum thickness of shell plates, heads and tube sheets for various shell diameters of steel-plate heating boilers shall be as shown in Table H-1.

TABLE H-1 MINIMUM ALLOWABLE THICKNESS OF SHELL PLATES

Diameter of Shell, Tube Sheet or Head	Minimum Thickness Allowable Under Rules: Shell, in.	Minimum Thickness Allowable Under Rules: Tube Sheet or Head, in.
42 in. or under	$\frac{1}{4}$	$\frac{5}{16}$
Over 42 in. to 60 in.	$\frac{5}{16}$	$\frac{3}{8}$
Over 60 in. to 78 in.	$\frac{3}{8}$	$\frac{7}{16}$
Over 78 in.	$\frac{7}{16}$	$\frac{1}{2}$

H-13. The minimum thickness of butt straps for double-strap joints shall be as given in Table H-2. For plate thickness exceeding $\frac{3}{4}$ in., the thickness of butt straps shall be not less than two-thirds of the thickness of the plate.

TABLE H-2 MINIMUM THICKNESS OF BUTT STRAPS

Thickness of Shell Plates, in.	Minimum Thickness of Butt Straps, in.
$\frac{1}{4}$	$\frac{1}{4}$
$\frac{9}{32}$	$\frac{1}{4}$
$\frac{5}{16}$	$\frac{1}{4}$
$\frac{11}{32}$	$\frac{1}{4}$
$\frac{3}{8}$	$\frac{5}{16}$
$\frac{13}{32}$	$\frac{5}{16}$
$\frac{7}{16}$	$\frac{3}{8}$
$\frac{15}{32}$	$\frac{3}{8}$
$\frac{1}{2}$	$\frac{7}{16}$
$\frac{17}{32}$	$\frac{7}{16}$
$\frac{9}{16}$	$\frac{7}{16}$
$\frac{5}{8}$	$\frac{1}{2}$
$\frac{3}{4}$	$\frac{1}{2}$

H-14. The minimum thickness of tubes used in water-tube or fire-tube boilers measured by Birmingham wire gage, shall be as given in Table H-3.

TABLE H-3 TUBES FOR WATER-TUBE AND FIRE-TUBE BOILERS

	Minimum Thickness of Tubes
Diameters 1 in. or over but less than $2\frac{1}{2}$ in.	No. 13 B.W.G.
Diameters $2\frac{1}{2}$ in. or over but less than $3\frac{1}{4}$ in.	No. 12 B.W.G.
Diameters $3\frac{1}{4}$ in or over but less than 4 in.	No. 11 B.W.G.
Diameters 4 in. or over but less than 5 in.	No. 10 B.W.G.
Diameters 5 in.	No. 9 B.W.G.

Hydrostatic Test

H-65. All hot-water boilers, the maximum allowable working pressure of which is not in excess of 30 lb. per sq. in., and steam-heating boilers, shall be subjected to a hydrostatic test of 60 lb. per sq. in. at the shop where constructed. Hot-water boilers, the maximum allowable working pressure of which exceeds 30 lb. per sq. in., shall be subjected to a hydrostatic test of 1½ times the maximum allowable working pressure both at the shop where constructed and in the field when erected and ready for service.

Any hydrostatic pressure test to be made on either a steam-heating boiler or hot-water boiler, after the boiler has been in service, shall be at a pressure 1½ times the maximum allowable working pressure.

In making hydrostatic pressure tests the pressure shall be under such control that in no case shall the required test pressure be exceeded by more than 6 per cent.

H-66. Individual shop inspection shall not be required for boilers which come under the rules of this section, except for boilers constructed by autogenous welding (see Par. H-81).

H-67. Each plate of a completed boiler shall bear the plate maker's name with brand and tensile strength, except that these marks need not appear on the butt straps after completion of boiler.

H-68. All boilers built according to these rules and no other boilers shall be marked A.S.M.E. Std.—Heat., and with the manufacturer's name and maximum allowable working pressure. These markings shall be stamped with letters and figures at least $\frac{5}{16}$ in. high on some conspicuous portion of the boiler proper, preferably over or near the fire door. Boilers suitable for use for both steam and water shall have the stamps arranged substantially as follows:

A.S.M.E. STD.—HEAT
. . . .
(Manufacturer's Name)
Max. W. P., Steam 15 lb.
Water . . . lb

Boilers suitable for use for water only shall have the stamps arranged substantially as follows:

A.S.M.E. STD.—HEAT
. . . .
(Manufacturer's Name)
Max. W. P., Water . . lb.

These stamps shall not be covered with insulating or other material. The symbol authorized for use on power boilers shall not be used on heating boilers, nor shall any accessory or part of the boiler be marked A.S.M.E. or A.S.M.E. Std. unless so specified in the Code.

H—CAST-IRON BOILERS

H-84. These Rules for cast-iron boilers shall apply:

a To all steam boilers for operation at pressures not exceeding 15 lb. per sq. in.

b To hot-water boilers to be operated at pressures not exceeding 160 lb. per sq. in., or temperatures not exceeding 250 deg. fahr.

c For conditions exceeding those specified above, cast-iron construction is not permitted.

H-85. Wherever the term maximum allowable working pressure is used herein, it refers to gage pressure or the pressure above the atmosphere in pounds per square inch.

H-86. The maximum allowable working pressure shall not exceed 15 lb. per sq. in. on a cast-iron boiler built under these rules to be used exclusively for low-pressure steam heating.

The maximum allowable working temperature at or near the outlet of a hot-water cast-iron boiler shall not exceed 250 deg. fahr.

Boiler Openings

H-87. Washout Openings. All cast-iron steam and hot-water boilers shall be provided with suitable washout openings to permit the removal of any sediment that may accumulate therein. Washout openings may be used for return pipe connections and the washout plug placed in a tee so that the plug is directly opposite and as close as possible to the opening in the boiler.

H-88. Flanged Connections. Flanged pipe connection openings in boilers shall conform to the American Standard given in Tables A-5 or A-6 of the Appendix to the Code, for the corresponding pipe size, and shall have the corresponding drilling for bolts or studs.

H-89. Threaded Openings. Pipe connections if threaded shall be tapped into material having a minimum thickness as specified in Table H-9.

TABLE H-9 MINIMUM THICKNESS OF MATERIAL FOR THREADED CONNECTIONS TO BOILERS

Size of Pipe Connection, In.	Minimum Thickness of Material Required, In.
¾ and under	$\frac{5}{16}$
1 to 2½ inclusive	$\frac{7}{16}$
3 to 3½ inclusive	⅝
4 to 5 inclusive	⅞
6 to 8 inclusive	1
9 to 12 inclusive	1¼

Installation

H-90. Provisions shall be made for the expansion and contraction of steam mains connected to boilers by providing substantial anchorage at suitable points, so that there shall be no undue strain transmitted to the boiler.

H-91. When feed or make-up water is introduced from a pressure line, it shall be connected to the piping system and not directly to the boiler.

H-92. All hot-water heating systems shall be so installed that there will be no opportunity for the fluid-relief column to freeze or to be accidentally shut off.

H-93. If valves are used in the supply and return mains, they shall be locked and sealed open and bear tags stating that provision must be made to prevent pressure from building up in boiler whenever the valves are closed. It is recommended that no valves be placed in the supply and return mains of a single boiler installation. Provision shall be made for cleaning the interior of the return main at or near the boiler.

H-94. When a valve is placed in the top connection from a hot-water supply boiler to a storage tank, an additional connection without valve shall be made between the boiler and top of storage tank.

Appendix B

MAXIMUM ALLOWABLE STRESS VALUES FOR FERROUS MATERIALS, ksi (MPa)
(Multiply by 1000 to Obtain psi)

Spec. No.	Grade	Nominal Composition	P-No.	Group No.	External Pressure Chart	Spec. Min. Tensile Strength, ksi (MPa)	Spec. Min. Yield Strength, ksi (MPa)	Note(s)	Max. Allow. Stress Value, ksi (MPa)
Plate Steels									
Carbon Steels									
SA-36	. . .	Carbon steel	1	1	CS-2	58.0 (400)	36.0 (248)	(7)(19)	11.6 (80)
SA-285	A	Carbon steel	1	1	CS-1	45.0 (311)	24.0 (166)	. . .	9.0 (62)
SA-285	B	Carbon steel	1	1	CS-1	50.0 (345)	27.0 (186)	. . .	10.0 (69)
SA-285	C	Carbon steel	1	1	CS-2	55.0 (379)	30.0 (207)	. . .	11.0 (75)
SA-455	. . .	Carbon steel	1	2	CS-2	75.0 (517)	38.0 (262)	(8)	15.0 (103)
SA-455	. . .	Carbon steel	1	2	CS-2	73.0 (503)	37.0 (255)	(9)	14.6 (100.5)
SA-455	. . .	Carbon steel	1	2	CS-2	70.0 (483)	35.0 (242)	(10)	14.0 (97)
SA-515	60	Carbon steel	1	1	CS-2	60.0 (414)	32.0 (220)	. . .	12.0 (82)
SA-515	65	Carbon steel	1	1	CS-2	65.0 (448)	36.0 (248)	. . .	13.0 (89)
SA-515	70	Carbon steel	1	2	CS-2	70.0 (483)	38.0 (262)	. . .	14.0 (97)
SA-516	55	Carbon steel	1	1	CS-2	55.0 (379)	30.0 (207)	. . .	11.0 (75)
SA-516	60	Carbon steel	1	1	CS-2	60.0 (414)	32.0 (220)	. . .	12.0 (82)
SA-516	65	Carbon steel	1	1	CS-2	65.0 (448)	36.0 (248)	. . .	13.0 (89)
SA-516	70	Carbon steel	1	2	CS-2	70.0 (483)	38.0 (262)	. . .	14.0 (97)
Sheet Steels									
Carbon Steels									
SA-414	A	Carbon steel	1	1	CS-1	45.0 (311)	25.0 (173)	. . .	9.0 (62)
SA-414	B	Carbon steel	1	1	CS-2	50.0 (345)	30.0 (207)	. . .	10.0 (69)
SA-414	C	Carbon steel	1	1	CS-2	55.0 (379)	33.0 (228)	. . .	11.0 (75)
SA-414	D	Carbon steel	1	1	CS-2	60.0 (414)	35.0 (242)	. . .	12.0 (82)
SA-414	E	Carbon steel	1	1	CS-2	65.0 (448)	38.0 (262)	. . .	13.0 (89)
SA-414	F	Carbon steel	1	2	CS-3	70.0 (483)	42.0 (290)	. . .	14.0 (97)
SA-414	G	Carbon steel	1	2	CS-3	75.0 (517)	45.0 (311)	. . .	15.0 (103)
Pipe and Tubes									
Seamless Carbon Steel									
SA-53	A	Carbon steel	1	1	CS-2	48.0 (331)	30.0 (207)	. . .	9.6 (66)
SA-53	B	Carbon steel	1	1	CS-2	60.0 (414)	35.0 (242)	. . .	12.0 (82)
SA-106	A	Carbon steel	1	1	CS-2	48.0 (331)	30.0 (207)	. . .	9.6 (66)
SA-106	B	Carbon steel	1	1	CS-2	60.0 (414)	35.0 (242)	. . .	12.0 (82)
SA-106	C	Carbon steel	1	2	CS-3	70.0 (483)	40.0 (276)	. . .	14.0 (97)
SA-192	. . .	Carbon steel	1	1	CS-1	(47.0)(324)	26.0 (180)	(1)	9.4 (65)
SA-210	A-1	Carbon steel	1	1	CS-2	60.0 (414)	27.0 (186)	. . .	12.0 (82)

MAXIMUM ALLOWABLE STRESS VALUES FOR FERROUS MATERIALS, ksi (MPa) (CONT'D)
(Multiply by 1000 to Obtain psi)

Spec. No.	Grade	Nominal Composition	P-No.	Group No.	External Pressure Chart	Spec. Min. Tensile Strength, ksi (MPa)	Spec. Min. Yield Strength, ksi (MPa)	Note(s)	Max. Allow. Stress Value, ksi (MPa)
Pipe and Tubes (Cont'd)									
Electric Resistance Welded Carbon Steel									
SA-53	A	Carbon steel	1	1	CS-2	48.0 (331)	30.0 (207)	(2)	8.2 (56)
SA-53	B	Carbon steel	1	1	CS-2	60.0 (414)	35.0 (242)	(2)	10.2 (70)
SA-135	A	Carbon steel	1	1	CS-2	48.0 (331)	30.0 (207)	(2)	8.2 (56)
SA-135	B	Carbon steel	1	1	CS-2	60.0 (414)	25.0 (173)	(2)	10.2 (70)
SA-178	A	Carbon steel	1	1	CS-1	(47.0) ((324)	26.0 (180)	(1)(2)	8.0 (55)
SA-178	C	Carbon steel	1	1	CS-2	60.0 (414)	37.0 (255)	(2)	10.2 (70)
Butt Welded									
SA-53	F	Carbon steel	1	1	...	48.0 (331)	30.0 (207)	(4)	5.8 (40)
Forgings									
Carbon Steels									
SA-105	...	Carbon steel	1	2	...	70.0 (483)	36.0 (248)	...	14.0 (97)
SA-181	Class 60	Carbon steel	1	1	CS-2	60.0 (414)	30.0 (207)	...	12.0 (82)
SA-181	Class 70	Carbon steel	1	2	CS-2	70.0 (483)	36.0 (248)	...	14.0 (97)
SA-266	1	Carbon steel	1	1	CS-2	60.0 (414)	30.0 (207)	...	12.0 (82)
SA-266	2	Carbon steel	1	2	CS-2	70.0 (483)	36.0 (248)	...	14.0 (97)
SA-266	3	Carbon steel	1	2	CS-2	75.0 (517)	37.5 (259)	...	15.0 (103)
Castings									
Carbon Steels									
SA-216	WCA	Carbon steel	1	1	CS-2	60.0 (414)	30.0 (207)	(5)	9.6(66)
SA-216	WCB	Carbon steel	1	2	CS-2	70.0 (483)	36.0 (248)	(5)	11.2 (77)
Bolting									
Carbon Steels									
SA-307	B	Carbon steel	...	...	...	60.0 (414)	...	(6)	7.0 (48)
SA-193	B5	5Cr–½Mo	...	...	...	100.0 (689)	...	(6)	25.0 (172)
SA-193	B7	1Cr–0.2Mo	...	...	...	100.0 (689)	...	(6)	25.0 (172)
SA-311	1018, Class A	Carbon steel	...	...	...	...	...	(20) (20a)	14.0 (97)
		...	...	...	...	...	...	(20b)	13.0 (89)
SA-311	1035, Class A	Carbon steel	...	...	...	...	...	(20) (20a)	17.0 (117)
		...	...	...	...	...	...	(20b)	16.0 (110)
SA-311	1045, Class A	Carbon steel	...	...	...	...	...	(20) (20a)	19.0 (130)
		...	...	...	...	...	...	(20b)	18.0 (124)
SA-311	1045, Class B	Carbon steel	...	...	...	...	...	(20) (20a)	23.0 (158)
SA-311	1050, Class A	Carbon steel	...	...	...	...	...	(20) (20a)	20.0 (137)
		...	...	...	...	...	...	(20b)	19.0 (130)
SA-320	L7	1Cr–0.2Mo	...	...	...	125.0 (862)	...	(6)	25.0 (172)
SA-320	L43	1¾Ni–¾Cr–¼Mo	...	...	...	125.0 (862)	...	(6)	25.0 (172)
SA-325	1	Carbon steel	...	...	...	...	...	(6)	7.0 (48)
SA-354	BC	Carbon steel	...	...	...	...	...	(6)	25.0 (172)
SA-354	BD	Carbon steel	...	...	...	...	...	(6)	25.0 (172)

MAXIMUM ALLOWABLE STRESS VALUES FOR FERROUS MATERIALS, ksi (MPa) (CONT'D)
(Multiply by 1000 to Obtain psi)

Spec. No.	Grade	Nominal Composition	P-No.	Group No.	External Pressure Chart	Spec. Min. Tensile Strength, ksi (MPa)	Spec. Min. Yield Strength, ksi (MPa)	Note(s)	Max. Allow. Stress Value, ksi (MPa)
Bars and Stays									
Carbon Steels									
SA-36	...	Carbon steel	1	1	...	58.0 (400)	...	(7)	11.6 (80)
SA-675	45	Carbon steel	1	1	...	45.0 (311)	...	...	9.0 (62)
SA-675	50	Carbon steel	1	1	...	50.0 (345)	...	...	10.0 (69)
SA-675	55	Carbon steel	1	1	...	55.0 (379)	...	...	11.0 (75)
SA-675	60	Carbon steel	1	1	...	60.0 (413)	...	...	12.0 (82)
SA-675	65	Carbon steel	1	1	...	65.0 (448)	...	...	13.0 (89)
SA-675	70 (483)	Carbon steel	1	2	...	70.0 (483)	...	...	14.0 (97)
Plate									
Alloy Steel									
SA-240	304	18Cr–8Ni	8	1	HA-1	75.0 (517)	30.0 (207)	(16)	15.0 (103)
SA-240	304L	18Cr–8Ni	8	1	HA-3	70.0 (483)	25.0 (173)	(16)	14.0 (97)
SA-240	316	16Cr–12Ni–2Mo	8	1	HA-2	75.0 (517)	30.0 (207)	(16)	15.0 (103)
SA-240	316L	16Cr–12Ni–2Mo	8	1	HA-4	70.0 (483)	25.0 (173)	(16)	14.0 (97)
SA-240	316Ti	16Cr–12Ni–2Mo–Ti	8	1	HA-2	75.0 (517)	30.0 (207)	(16)	15.0 (103)
SA-240	439	18Cr–Ti	7	2	...	65.0 (448)	30.0 (207)	(11)(12)(13)	13.0 (89)
SA-240	S44400	18Cr–2Mo	7	2	CS-2	60.0 (414)	40.0 (276)	(11)(18)	12.0 (82)
Tube									
Alloy Steel									
SA-213	TP304	Smls. 18Cr–8Ni	8	1	HA-1	75.0 ((517)	30.0 (207)	(14)(15)(16)	15.0 (103)
SA-213	TP304L	Smls. 18Cr–8Ni	8	1	HA-3	70.0 (483)	25.0 (173)	(14)(15)(16)	14.0 (97)
SA-213	TP316	Smls. 16Cr–12Ni–2Mo	8	1	HA-2	75.0 (517)	30.0 (207)	(14)(15)(16)	15.0 (103)
SA-213	TP316L	Smls. 16Cr–12Ni–2Mo	8	1	HA-4	70.0 (483)	25.0 (173)	(14)(15)(16)	14.0 (97)
SA-249	TP304	Wld. 18Cr–8Ni	8	1	HA-1	75.0 (517)	30.0 (207)	(2)(14)(15)(16)	12.8 (88)
SA-249	TP304L	Wld. 18Cr–8Ni	8	1	HA-3	70.0 (483)	25.0 (173)	(2)(14)(15)(16)	11.9 (82)
SA-249	TP316	Wld. 16Cr–12Ni–2Mo	8	1	HA-2	75.0 (517)	30.0 (207)	(2)(14)(15)(16)	12.8 (88)
SA-249	TP316L	Wld. 16Cr–12Ni–2Mo	8	1	HA-4	70.0 (483)	25.0 (173)	(2)(14)(15)(16)	11.9 (82)
SA-268	S44400	18Cr–2Mo	7	2	CS-2	60.0 (414)	40.0 (276)	(11)(18)	12.0 (82)
SA-268	TP439	18Cr–Ti	7	2	CS-2	60.0 (414)	40.0 (276)	(11)(12)(13)	12.0 (82)
SA-268	S44735	Smls. 29Cr–4Mo	10J	1	CS-2	75.0 (517)	60.0 (414)	(22)	15.0 (103)
SA-268	S44735	Wld. 29Cr–4Mo	10J	1	CS-2	75.0 (517)	60.0 (414)	(2)(22)	12.7 (87)

MAXIMUM ALLOWABLE STRESS VALUES FOR FERROUS MATERIALS, ksi (MPa) (CONT'D)
(Multiply by 1000 to Obtain psi)

Spec. No.	Grade	Nominal Composition	P-No.	Group No.	External Pressure Chart	Spec. Min. Tensile Strength, ksi (MPa)	Spec. Min. Yield Strength, ksi (MPa)	Note(s)	Max. Allow. Stress Value, ksi (MPa)
Pipe Alloy Steel									
SA-312	TP304	Smls. 18Cr–8Ni	8	1	HA-1	75.0 (517)	30.0 (207)	(15)(16)	15.0 (103)
SA-312	TP304L	Smls. 18Cr–8Ni	8	1	HA-3	70.0 (483)	25.0 (173)	(15)(16)	14.0 (97)
SA-312	TP316	Smls. 16Cr–12Ni– 2Mo	8	1	HA-2	75.0 (517)	30.0 (207)	(15)(16)	15.0 (103)
SA-312	TP316L	Smls. 16Cr–12Ni– 2Mo	8	1	HA-4	70.0 (483)	25.0 (173)	(15)(16)	14.0 (97)
SA-312	TP304	Wld. 18Cr–8Ni	8	1	HA-1	75.0 (517)	30.0 (207)	(2)(15)(16)	12.8 (88)
SA-312	TP304L	Wld. 18Cr–8Ni	8	1	HA-3	70.0 (483)	25.0 (173)	(2)(15)(16)	11.9 (82)
SA-312	TP316	Wld. 16Cr–12Ni–2Mo	8	1	HA-2	75.0 (517)	30.0 (207)	(2)(15)(16)	12.8 (88)
SA-312	TP316L	Wld. 16Cr–12Ni–2Mo	8	1	HA-4	70.0 (483)	25.0 (173)	(2)(15)(16)	11.9 (82)
Bar Alloy Steel									
SA-479	S44400	18Cr–2Mo	7	2	...	60.0 (413)	...	(11)(18)	12.0 (82)
SA-479	439	18Cr–Ti	7	2	HA-27	70.0 (483)	40.0 (276)	(11)(12)(13)	14.0 (97)
SA-479	304L	18Cr–8Ni	8	1	HA-3	70.0 (483)	25.0 (173)	(14)(16)	14.0 (97)
SA-479	316L	16Cr–12Ni–2Mo	8	1	HA-4	70.0 (483)	25.0 (173)	(14)(16)	14.0 (97)
Forgings Alloy Steel									
SA-182	F304	18Cr–8Ni	8	1	HA-1	75.0 (517)	30.0 (207)	(15)(16)(21)	15.0 (103)
SA-182	F304L	18Cr–8Ni	8	1	HA-3	70.0 (483)	25.0 (173)	(15)(16)(21)	14.0 (97)
SA-182	F316	16Cr–12Ni–2Mo	8	1	HA-2	75.0 (517)	30.0 (207)	(15)(16)(21)	15.0 (103)
SA-182	F316L	16Cr–12Ni–2Mo	8	1	HA-4	70.0 (483)	25.0 (173)	(15)(16)(21)	14.0 (97)
Castings Alloy Steel									
SA-351	CF8C	18Cr–10Ni–Cb	8	1	...	70.0 (483)	...	(5)(16)(17)	11.2 (77)
SA-351	CF3M	16Cr–12Ni–2Mo	8	1	...	70.0 (483)	...	(5)(16)(17)	11.2 (77)

NOTES: See next page.

NOTES:

(1) Tensile value in parentheses is expected minimum.
(2) The stress value includes a joint factor of 0.85.
(3) The stress value includes a joint factor of 0.80.
(4) The stress value includes a joint factor of 0.60.
(5) The stress value includes a casting quality factor of 0.80. Increased casting quality factors as a result of material examination beyond the requirements of the material specifications shall not be permitted.
(6) The stress value is established from a consideration of strength only and will be satisfactory for average service. For bolted joints, where freedom from leakage over a long period of time without retightening is required, lower stress values may be necessary as determined from the relative flexibility of the flange and bolts, and corresponding relaxation properties.
(7) These allowable stress values apply also to structural shapes.
(8) For thicknesses up to $\frac{3}{8}$ in. (9.52 mm), inclusive.
(9) For thicknesses over $\frac{3}{8}$ in. to 0.580 in. (9.5 mm to 14.7 mm), inclusive.
(10) For thicknesses over 0.580 to 0.750 in. (14.7 mm to 19.0 mm), inclusive.
(11) The maximum thickness of material covered by this Table is $\frac{3}{8}$ in. (9.5 mm)
(12) The service temperature shall not exceed 200°F (93°C).
(13) Filler metal shall be Type 430 with a nominal titanium content of approximately 1.25%. The 300 series of chromium–nickel–iron filler metals shall not be used in welding vessels conforming to the requirements of Section IV.
(14) Tubing material shall be fully annealed.
(15) Limitations of HF-204.2 also apply.
(16) The water temperature shall not exceed 210°F (98°C).
(17) The minimum thickness for header material is 0.10 in. (2.6 mm)
(18) Filler metal shall be Type 430 with a nominal molybdenum content of approximately 2%. The 300 series of chromium–nickel–iron filler metals shall not be used in welding vessels conforming to the requirements of Section IV.
(19) SA/CSA-G40.21 as specified in Section IIA, grade 38W or 44W, may be used in lieu of SA-36 for plates and bars not exceeding $\frac{3}{4}$ in. (20 mm). for use at the same maximum allowable stress values as SA-36.
(20) For tie-rods and draw bolts on cast-iron sectional boilers subject to system pressure. Welding is not permitted.
   (a) To $\frac{7}{8}$ in. (22 mm) diam. incl.
   (b) Over $\frac{7}{8}$ in. to $1\frac{1}{4}$ in. incl. (22 mm to 32 mm).
   (c) To 3 in. (76 mm) incl.
(21) These allowable stresses apply only to material 5 in. (127 mm) and under in thickness.
(22) Heat treatment after forming or fabrication is neither required nor prohibited.

Appendix C

MAXIMUM ALLOWABLE STRESS VALUES FOR NONFERROUS MATERIALS, ksi (MPa)
(Multiply by 1000 to Obtain psi)

Spec. No.	Copper or Copper Alloy No.	Form	Condition	P-No.	External Pressure Chart	Spec. Min. Tensile Stress, ksi (MPa)	Spec. Min. Yield Stress, ksi (MPa)	Note(s)	Max. Allow. Stress Value, ksi (MPa)
Copper									
SB-42	Copper	Pipe	Annealed/061	31	NFC-1	30.0(207)	9.0(62)	(2)	6.0(41)
SB-42	Copper	Pipe	Hard Drawn/H80 — $\frac{1}{8}$ in. to 2 in. NPS, incl.	31	NFC-4	45.0(311)	40.0(276)	(1)(2)	9.0(62)
SB-42	Copper	Pipe	Light Drawn/H55 — $2\frac{1}{2}$ in. to 12 in. NPS, incl.	31	NFC-3	36.0(248)	30.0(207)	(1)(2)	7.2(49)
SB-75	Copper	Smls. Tubes	Annealed/050/060	31	NFC-1	30.0(207)	9.0(62)	(2)	6.0(41)
SB-75	Copper	Smls. Tubes	Light Drawn/H55	31	NFC-3	36.0(248)	30.0(207)	(1)(2)	7.2(49)
SB-75	Copper	Smls. Tubes	Hard Drawn/H80	31	NFC-4	45.0(311)	40.0(276)	(1)(2)	9.0(62)
SB-111	Copper	Smls. Condenser Tubes	Light Drawn/H55	31	NFC-6	36.0(248)	30.0(207)	(2)	7.2(49)
SB-111	Copper	Smls. Condenser Tubes	Hard Drawn/H80	31	NFC-4	45.0(311)	40.0(276)	(2)	9.0(62)
SB-152	Copper	Plate, Sheet, Strip, & Bar	Hot Rolled/025, Annealed	31	NFC-1	30.0(207)	10.0(69)	(2)	6.0(41)
SB-283	C37700	Forging Brass	As Forged/M10/M11	...	...	...	...	(8)(9)	5.8(40)
SB-395	Copper	Smls. Tubes	Light Drawn/H55	31	NFC-6	36.0(248)	30.0(207)	(1)(2)	7.2(49)
Copper–Silicon									
SB-96	C66500	Plate & Sheet	Annealed/061	33	NFC-2	50.0(345)	18.0(124)	(5)	10.0(64)
SB-98	C65500	Rods	Soft Anneal/060	33	...	52.0(360)	15.0(103)	(5)	10.1(70)
SB-98	C65500	Rods	Quarter Hard/H01	33	...	55.0(379)	24.0(166)	(5)	10.0(69)
SB-98	C65100	Rods	Soft Anneal/060	33	...	40.0(276)	12.0(82)	(5)	8.0(55)
SB-98	C65100	Rods	Half Hard/H02	33	...	55.0(379)	20.0(138)	(5)	11.0(75)
SB-315	C65500	Pipe & Tube	Annealed/030/061	33	NFC-2	50.0(345)	15.0(103)	(5)	10.0(69)
Red Brass									
SB-43	C23000	Smls. Pipe	Annealed/061	32	NFC-2	40.0(276)	12.0(82)	...	8.0(55)
SB-111	C23000	Smls. Condenser Tubes	Annealed/061	32	NFC-2	40.0(276)	12.0(82)	...	8.0(55)
SB-395	C23000	Smls. Condenser Tubes	Annealed/061	32	NFC-2	40.0(276)	12.0(82)	...	8.0(55)

MAXIMUM ALLOWABLE STRESS VALUES FOR NONFERROUS MATERIALS, ksi (MPa) (CONT'D)
(Multiply by 1000 to Obtain psi)

Spec. No.	Copper or Copper Alloy No.	Form	Condition	P-No.	External Pressure Chart	Spec. Min. Tensile Stress, ksi (MPa)	Spec. Min. Yield Stress, ksi (MPa)	Note(s)	Max. Allow. Stress Value, ksi (MPa)
Admiralty									
SB-395	C44300	Smls. Condenser Tubes	Annealed/O61	32	NFC-2	45.0(311)	15.0(103)	...	9.0(62)
SB-395	C44400	Smls. Condenser Tubes	Annealed/O61	32	NFC-2	45.0(311)	15.0(103)	...	9.0(62)
SB-395	C44500	Smls. Condenser Tubes	Annealed/O61	32	NFC-2	45.0(311)	15.0(103)	...	9.0(62)
SB-171	C44300	Plates, ≤ 4 in.	...	32	NFC-2	45.0(311)	15.0(103)	...	9.0(62)
SB-171	C44400	Plates, ≤ 4 in.	...	32	NFC-2	45.0(311)	15.0(103)	...	9.0(62)
SB-171	C44500	Plates, ≤ 4 in.	...	32	NFC-2	45.0(311)	15.0(103)	...	9.0(62)
Naval Brass									
SB-171	C46400	Plates, ≤ 3 in.	...	32	NFC-2	50.0(345)	20.0(138)	...	10.0(69)
Copper–Nickel									
SB-111	C70600	Smls. Condenser Tubes	Annealed/O61	34	NFC-3	40.0(276)	15.0(103)	...	8.0(55)
SB-111	C71000	Smls. Condenser Tubes	Annealed/O61	34	NFC-3	45.0(311)	16.0(110)	...	9.0(62)
SB-111	C71500	Smls. Condenser Tubes	Annealed/O61	34	NFC-4	52.0(360)	18.0(124)	...	10.4(71)
SB-171	C70600	Plates, ≤ 5 in.	...	34	NFC-3	40.0(276)	15.0(103)	...	8.0(55)
SB-171	C71500	Plates, ≤ 2½ in.	...	34	NFC-4	50.0(345)	20.0(138)	...	10.0(69)
SB-395	C70600	Smls. Condenser Tubes	Annealed/O61	34	NFC-3	40.0(276)	15.0(103)	...	8.0(55)
SB-395	C71000	Smls. Condenser Tubes	Annealed/O61	34	NFC-3	45.0(311)	16.0(110)	...	9.0(62)
SB-395	C71500	Smls. Condenser Tubes	Annealed/O61	34	NFC-4	52.0(360)	18.0(124)	...	10.4(71)
SB-466	C70600	Pipe & Tube	Annealed	34	NFC-3	38.0(262)	13.0(90)	...	7.6(52)
SB-466	C71000	Pipe & Tube	Annealed	34	NFC-3	45.0(311)	16.0(110)	...	9.0(62)
SB-466	C71500	Pipe & Tube	Annealed	34	NFC-4	50.0(345)	18.0(124)	...	10.0(69)
Nickel–Copper									
SB-164	N04400	Bar	Hot or Cold Worked, Annealed	42	NFN-3	70.0(483)	25.0(173)	(7)	14.0(97)
SB-164	N04400	Rounds	Hot Worked (As Worked or Stress Relieved)	42	NFN-3	80.0(552)	40.0(275)	(7)	16.0(110)

MAXIMUM ALLOWABLE STRESS VALUES FOR NONFERROUS MATERIALS, ksi (MPa) (CONT'D)
(Multiply by 1000 to Obtain psi)

Spec. No.	Copper or Copper Alloy No.	Form	Condition	P-No.	External Pressure Chart	Spec. Min. Tensile Stress, ksi (MPa)	Spec. Min. Yield Stress, ksi (MPa)	Note(s)	Max. Allow. Stress Value, ksi (MPa)
Nickel–Copper (Cont'd)									
SB-165	N04400	Smls. Pipe and Tube, 5 in. O.D. max.	Annealed	42	NFN-3	70.0(483)	28.0(193)	(7)	14.0(97)
SB-165	N04400	Smls. Pipe and Tube, Over 5 in. O.D.	Annealed	42	NFN-3	70.0(483)	25.0(173)	(7)	14.0(97)
SB-165	N04400	Smls. Pipe and Tube, All Sizes	Stress Relieved	42	NFN-3	85.0(586)	55.0(379)	(7)	17.0(117)
Integrally Finned Tubes									
SB-359	...	...	...	...	...	...	...	(6)	...
Castings, Bronze and Brass									
SB-61	C92200	Steam and Valve Bronze	...	...	NFN-1	34.0(234)	16.0(110)	(3)	5.4(37)
SB-62	C83600	85-5-5-5 Composition Bronze	...	...	NFC-1	30.0(207)	14.0(97)	(3)	4.8(33)
SB-584	C84400	81-3-7-9 Composition Semi-Red Brass	...	...	...	29.0(200)	13.0(90)	(3)	4.6(31.5)
SB-584	C90300	88-8-0-9 Tin Bronze	...	...	...	40.0(276)	18.0(124)	(3)	5.1(35)
Nickel–Iron–Chromium									
SB-409	N08810	Plate	Sol Annealed	45	NFN-9	65.0(448)	25.0(173)	(10)	13.0(89)
SB-409	N08800	Plate	Annealed	45	NFN-8	75.0(517)	30.0(207)	(10)	15.0(103)

NOTES:

(1) When nonferrous materials conforming to specifications given in Section II are used in welded or brazed construction, the maximum allowable working stresses shall not exceed the values given herein for the material in the annealed condition.

(2) When material is to be welded, the phosphorus deoxided types should be specified.

(3) The stress value includes a casting quality factor of 0.80. Increased casting quality factors as a result of material examination beyond the requirement of the material specification shall not be permitted. This is not intended to apply to valves and fittings made to recognized standards.

(4) DELETED

(5) Copper–silicon alloys are not always suitable when exposed to certain median and high temperatures, particularly steam above 212°F (100°C). Therefore, this material is limited to the construction of hot water boilers to be operated at a temperature not to exceed 200°F (93°C).

(6) Use in accordance with HF-204 and HF-204.1.

(7) To be used for HLW connections only.

(8) For use in HG-307.2(b) Eq. (2), the maximum allowable stress at room temperature [100°F (38°C), max.] shall be 10.0 [through $1\frac{1}{2}$ in. (38 mm) thickness] and 9.2 [over $1\frac{1}{2}$ in. (38 mm) thickness].

(9) No welding or brazing permitted.

(10) The maximum water temperature shall not exceed 210°F (98°C).

Appendix D

MAXIMUM ALLOWABLE STRESS VALUES IN TENSION FOR LINED MATERIALS, ksi (MPa)

Spec. No.	Grade	Nominal Composition	P-No.	Group No.	External Pressure Chart	Spec. Min. Tensile Strength, ksi (MPa)	Spec. Min. Yield Strength, ksi (MPa)	Note(s)	Max. Allowable Design Stress, ksi (MPa)
Plate									
SA-36	...	Carbon steel	1	1	...	58 (400)	...	(8)	14.5 (100.0)
SA-285	A	Carbon steel	1	1	...	45 (311)	...	...	11.3 (77.9)
	B	Carbon steel	1	1	...	50 (345)	...	...	12.5 (86.2)
	C	Carbon steel	1	1	...	55 (380)	...	...	13.8 (95.1)
SA-455		Carbon steel	1	2	...	75 (517)	...	(5)	18.8 (129.6)
SA-455		Carbon steel	1	2	...	73 (504)	...	(6)	18.3(129.6)
SA-455		Carbon steel	1	2	...	70 (483)	...	(7)	17.5 (120.7)
SA-285 Modified to Chem.								(3)	
AISI C-1012		0.10C to 0.15C	1	1	...	45 (311)	...	...	11.3 (77.9)
AISI C-1015		0.13C to 0.18C	1	1	...	50(345)	...	...	12.5 (86.2)
AISI C-1023		0.20C to 0.25C	1	1	...	55 (380)	...	...	13.8 (95.1)
SA-515	60	Carbon steel	1	1	CS-2	60 (414)	32 (221)	...	15.0 (103.4)
	65	Carbon steel	1	1	CS-2	65 (449)	36 (249)	...	16.3 (112.4)
	70	Carbon steel	1	2	CS-2	70 (483)	38 (262)	...	17.5 (120.7)
SA-516	55	Carbon steel	1	1	CS-2	55 (380)	30 (207)	...	13.8 (95.1)
	60	Carbon steel	1	1	CS-2	60 (414)	32 (221)	...	15.0 (103.4)
	65	Carbon steel	1	1	CS-2	65 (449)	36 (249)	...	16.3 (112.4)
	70	Carbon steel	1	2	CS-2	70 (483)	38 (262)	...	17.5 (120.7)
Sheet									
SA-414	A	Carbon steel	1	1	...	45 (311)	...	...	11.3 (77.9)
	B	Carbon steel	1	1	...	50 (345)	...	...	12.5 (86.2)
	C	Carbon steel	1	1	...	55 (380)	...	...	13.8 (95.1)
	D	Carbon steel	1	1	...	60 (414)	...	...	15.0 (103.4)
	E	Carbon steel	1	1	...	65 (449)	...	...	16.3 (112.4)
	F	Carbon steel	1	2	...	70 (483)	...	...	17.5 (120.7)
	G	Carbon steel	1	2	...	75 (517)	...	...	18.8 (129.6)
SA-414 Modified to Chem.								(3)	
AISI C-1012		0.10C to 0.15C	1	1	...	45 (311)	...	...	11.3 (77.9)
AISI C-1015		0.13C to 0.18C	1	1	...	50 (345)	...	...	12.5 (86.2)
AISI C-1023		0.20C to 0.25C	1	1	...	55 (380)	...	...	13.8 (95.1)
Bars and Forging									
SA-36	...	Carbon steel	1	1	...	58 (400)	...	(1)(8)	14.5 (100.0)
SA-105		Forging carbon steel	1	2	...	70 (483)	...	...	17.5 (120.7)
SA-181	Class 60	Forging carbon steel	1	1	...	60 (414)	...	...	15.0 (103.4)
Pipe									
SA-53	A	Smls. carbon steel	1	1	...	48 (331)	...	...	12.0 (82.7)
SA-53	B	Smls. carbon steel	1	1	...	60 (414)	...	...	15.0 (103.4)
SA-106	A	Smls. carbon steel	1	1	...	48 (331)	...	...	12.0 (82.7)

MAXIMUM ALLOWABLE STRESS VALUES IN TENSION FOR LINED MATERIALS, ksi (CONT'D)

Spec. No.	Grade	Nominal Composition	P-No.	Group No.	External Pressure Chart	Spec. Min. Tensile Strength, ksi (MPa)	Spec. Min. Yield Strength, ksi (MPa)	Note(s)	Max. Allowable Design Stress, ksi (MPa)
SA-106	B	Smls. carbon steel	1	1	...	60 (414)	...	...	15.0 (103.4)
SA-106	C	Smls. carbon steel	1	2	...	70 (483)	...	...	17.5 (120.7)
SA-53	A	ERW carbon steel	1	1	...	48 (331)	...	(2)	10.2 (70.3)
SA-53	B	ERW carbon steel	1	1	...	60 (414)	...	(2)	12.8 (88.3)
SA-135	A	ERW carbon steel	1	1	...	48 (331)	...	(2)	10.2 (70.3)
SA-135	B	ERW carbon steel	1	1	...	60 (414)	...	(2)	12.8 (88.3)
Tube									
SA-178	A	ERW carbon steel	1	1	...	(47) (325)	...	(2)(4)	10.0 (68.9)
SA-178	C	ERW carbon steel	1	1	...	60 (414)	...	(2)	12.8 (88.3)
SA-513	1008	ERW carbon steel	1	1	...	(42) (290)	...	(2)(4)(9)	8.9 (61.4)
SA-513	1010	ERW carbon steel	1	1	...	(45) (311)	...	(2)(4)(9)	9.6 (66.2)
SA-513	1015	ERW carbon steel	1	1	...	(48) (331)	...	(2)(4)(9)	10.2 (70.3)
Bars									
SA-675	45	Carbon steel	1	1	...	45 (311)	...	...	11.3 (77.9)
	50	Carbon steel	1	1	...	50 (345)	...	...	12.5 (86.2)
	55	Carbon steel	1	1	...	55 (380)	...	...	13.8 (95.1)
	60	Carbon steel	1	1	...	60 (414)	...	...	15.0 (103.4)
	65	Carbon steel	1	1	...	65 (449)	...	...	16.3 (112.4)
	70	Carbon steel	1	2	...	70 (483)	...	...	17.5 (120.7)

GENERAL NOTES:

(a) Nonferrous material, if utilized for connections, etc., shall be in accordance with Table HF-300.2.

(b) To convert from ksi to MPa, multiply by 6.895.

NOTES:

(1) These allowable stress values apply also to structural shapes.

(2) The stress value includes a joint factor of 0.85.

(3) For use only as shell plates, heads, tubesheets, or other surfaces to be glass lined.

(4) Tensile value in parentheses is expected minimum.

(5) For thickness up to $\frac{3}{8}$ in. (10 mm), incl.

(6) For thickness over $\frac{3}{8}$ in. to 0.580 in. (10 mm to 15 mm), incl.

(7) For thickness over 0.580 in. to 0.750 in. (15 mm to 19 mm), incl.

(8) SA/CSA-G40.21, as specified in Section II, Part A, grade 38W or 44W may be used in lieu of SA-36 for plates and bars not exceeding $\frac{3}{4}$ in. (19 mm) for use at the same maximum allowable stress values as SA-36.

(9) This tube is restricted to use in glass lined water heaters.

Appendix E

MAXIMUM ALLOWABLE STRESS VALUES FOR MATERIALS IN TENSION FOR UNLINED WATER HEATERS, ksi (MPa)

Spec. No.	Grade	Nominal Composition	P-No.	Group No.	External Pressure Chart	Spec. Min. Tensile Strength, ksi (MPa)	Spec. Min. Yield Strength, ksi (MPa)	Note(s)	Max. Allowable Design Stress, ksi (See HLW-303) Standard	Alternative
Plate										
Alloy Steel										
SA-240	304	18Cr–8Ni	8	1	HA-1	75 (517)	30 (207)	(2)(3)(5)	16.7 (115.1)	17.8(7)
	304L	18Cr–8Ni	8	1	HA-3	70 (483)	25 (173)	(2)(3)(5)	14.3 (98.6)	16.5(7)
	316	16Cr–12Ni–2Mo	8	1	HA-2	75 (517)	30 (207)	(2)(3)(5)	17.7 (122.0)	18.8(7)
	316L	16Cr–12Ni–2Mo	8	1	HA-4	70 (483)	25 (173)	(2)(3)(5)	14.1 (97.2)	16.7(7)
	439	18Cr–Ti	7	2	...	65 (449)	30 (207)	(1)(3)(4)	16.3 (112.4)	...
	S44400	18Cr–2Mo	7	2	CS-2	60 (414)	40 (276)	(1)(2)(4)	15.0 (103.4)	...
Tube										
Alloy Steel										
SA-213	TP304	Smls. 18Cr–8Ni	8	1	HA-1	75 (517)	30 (207)	(2)(3)(5)	16.7 (115.1)	17.8(7)
	TP304L	Smls. 18Cr–8Ni	8	1	HA-3	70 (483)	25 (173)	(2)(3)(5)	14.3 (98.6)	16.5(7)
	TP316	Smls. 16Cr–12Ni–2Mo	8	1	HA-2	75 (517)	30 (207)	(2)(3)(5)	17.7 (122.0)	18.8(7)
	TP316L	Smls. 16Cr–12Ni–2Mo	8	1	HA-4	70 (483)	25 (173)	(2)(3)(5)	14.1 (97.2)	16.7(7)
SA-249	TP304	Wld. 18Cr–8Ni	8	1	HA-1	75 (517)	30 (207)	(2)(3)(5)(7)	14.2 (97.9)	15.1(7)
	TP304L	Wld. 18Cr–8Ni	8	1	HA-3	70 (483)	25 (173)	(2)(3)(5)(7)	11.9 (82.0)	14.0(7)
	TP316	Wld. 16Cr–12Ni–2Mo	8	1	HA-2	75 (517)	30 (207)	(2)(3)(5)(7)	15.0 (103.4)	16.0(7)
	TP316L	Wld. 16Cr–12Ni–2Mo	8	1	HA-4	70 (483)	25 (173)	(2)(3)(5)(7)	12.0 (82.7)	14.2(7)
SA-268	TP439	Smls. 18Cr–Ti	7	2	CS-2	60 (414)	40 (276)	(1)(3)(4)	15.0 (103.4)	...
	S44400	Smls. 18Cr–2Mo	7	2	CS-2	60 (414)	40 (276)	(1)(2)(4)	15.0 (103.4)	...
	S44400	Wld. 18Cr–2Mo	7	2	CS-2	60 (414)	40 (276)	(1)(2)(4)(5)(7)	12.8 (88.3)	...
	S44735	Smls. 29Cr–4Mo	10J	1	CS-2	75 (517)	60 (414)	(8)	18.4 (126.9)	...
	S44735	Wld. 29Cr–4Mo	10J	1	CS-2	75 (517)	60 (414)	(7)(8)	15.5 (106.9)	...
Bar										
Alloy Steel										
SA-479	304	18Cr–8Ni	8	1	...	75 (517)	...	(2)(3)(5)	16.7 (115.1)	17.8(7)
	304L	18Cr–8Ni	8	1	...	70 (483)	...	(2)(3)(5)	14.3 (98.6)	16.5(7)
	316	16Cr–12Ni–2Mo	8	1	...	75 (517)	...	(2)(3)(5)	17.7 (122.0)	18.8(7)
	316L	16Cr–12Ni–2Mo	8	1	...	70 (483)	...	(2)(3)(5)	14.1 (97.2)	16.7(7)
	439	18Cr–Ti	7	2	...	70 (483)	...	(1)(3)(4)	16.6 (114.5)	...
	S44400	18Cr–2Mo	7	2	...	60 (414)	...	(1)(2)(4)	15.0 (103.4)	...
Pipe										
Alloy Steel										
SA-312	TP304	Smls. 18Cr–8Ni	8	1	HA-1	75 (517)	30 (207)	(2)(3)(5)	16.7 (115.1)	17.8(7)
	TP304	Wld. 18Cr–8Ni	8	1	HA-1	75 (517)	30 (207)	(2)(3)(5)(7)	14.2 (97.9)	15.1(7)
	TP304L	Smls. 18Cr–8Ni	8	1	HA-3	70 (483)	25 (173)	(2)(3)(5)	14.3 (98.6)	16.5(7)
	TP304L	Wld. 18Cr–8Ni	8	1	HA-3	70 (483)	25 (173)	(2)(3)(5)(7)	12.2 (84.1)	14.0(7)

MAXIMUM ALLOWABLE STRESS VALUES FOR MATERIALS IN TENSION FOR UNLINED WATER HEATERS, ksi (MPa) (CONT'D)

Spec. No.	Grade	Nominal Composition	P-No.	Group No.	External Pressure Chart	Spec. Min. Tensile Strength, ksi (MPa)	Spec. Min. Yield Strength, ksi (MPa)	Note(s)	Max. Allowable Design Stress, ksi (MPa) (See HLW-303) Standard	Alternative
Pipe										
Alloy Steel										
	TP316	Smls. 16Cr–12Ni–2Mo	8	1	HA-2	75 (517)	30 (207)	(2)(3)(5)	17.7 (122.0)	18.8(7)
	TP316	Wld. 16Cr–12Ni–2Mo	8	1	HA-2	75 (517)	30 (207)	(2)(3)(5)(7)	15.0 (103.4)	16.0(7)
	TP316L	Smls. 16Cr–12Ni–2Mo	8	1	HA-4	70 (483)	25 (173)	(2)(3)(5)	14.1 (97.2)	16.7(7)
	TP316L	Wld. 16Cr–12Ni–2Mo	8	1	HA-4	70 (483)	25 (173)	(2)(3)(5)(7)	12.0 (82.7)	14.2(7)
Forgings										
Alloy Steel										
SA-182	F304	18Cr–8Ni	8	1	HA-1	75 (517)	30 (207)	(2)(3)(5)	16.7 (115.1)	17.8(7)
	F304L	18Cr–8Ni	8	1	HA-3	70 (483)	25 (173)	(2)(3)(5)	14.3 (98.6)	16.5(7)
	F316	16Cr–12Ni–2Mo	8	1	HA-2	75 (517)	30 (207)	(2)(3)(5)	17.7 (122.0)	18.8(7)
	F316L	16Cr–12Ni–2Mo	8	1	HA-4	70 (483)	25 (173)	(2)(3)(5)	14.1 (97.2)	16.7(7)

GENERAL NOTE: To convert from ksi to MPa, multiply by 6.895.

NOTES:

(1) The maximum is $^3/_8$ in. (10 mm)

(2) The maximum thickness is $^1/_2$ in. (12.7 mm)

(3) The service temperature shall not exceed 210°F (99°C).

(4) Filler metal shall be Type 430 with a nominal molybdenum content of approximately 2%. The 300 series of chromium–nickel–iron filler metals shall not be used in welding vessels conforming to the requirements of Section IV.

(5) Water heaters using this material are to be operated only on deionized water having a minimum specific resistivity of 1.0 MΩ/cm.

(6) Due to the relatively low yield strength of the austenitic stainless steel materials, these higher stress values were established at temperatures at which the short time tensile properties govern to permit the use of these alloys where slightly greater deformation is acceptable. These higher stress values exceed two-thirds but do not exceed 90% of the yield strength at temperature. Use of these stress values may result in dimensional changes due to permanent strain. These stress values are not recommended for flanges of gasketed joints or other applications where slight amounts of distortion can cause leakage or malfunction.

(7) The stress value includes a joint factor of 0.85.

(8) Heat treatment after forming or fabrication is neither required nor prohibited.

Appendix F

DIVISION OF STATE FIRE MARSHAL
BUREAU OF FIRE PREVENTION
BOILER SAFETY PROGRAM

Boiler or Pressure Vessel Data Report

FIRST INSPECTION REPORT

This inspection is intended for your safety and the safety of the citizens of Florida. Your cooperation is greatly appreciated.

1	DATE INSPECTED MO DAY YR	CERT EXP DATE MO YR	CERTIFICATE POSTED ❑ YES ❑ NO	OWNER NO.	JURISDICTION NUMBER	NAT'L BD NO. ❑ / OTHER NO. ❑
2	OWNER	NATURE OF BUSINESS	KIND OF INSPECTION INT ❑ EXT ❑	CERTIFICATE INSPECTION YES ❑ NO ❑		
	OWNER STEET ADDRESS NUMBER	OWNERS CITY	STATE	ZIP		
3	USER NAME - OBJECT LOCATION	SPECIFIC LOCATION IN PLANT	OBJECT LOCATION - COUNTY			
	USERS STEET ADDRESS NUMBER	USERS CITY	STATE	ZIP		
4	TYPE ❑ FT ❑ WT ❑ CI ❑ AIR TANK ❑ WATER TANK ❑ OTHER	YEAR BUILT	MANUFACTURER			
5	USE ❑ POWER ❑ PROCESS ❑ STEAM HTG. ❑ HWH ❑ HWS ❑ STORAGE ❑ HEAT EXCHANCE ❑ OTHER	FUEL (BOILER)	METHOD OF FIRING (BOILER)	PRESSURE GAGE TESTED YES ❑ NO ❑		
6	PRESSURE THIS INSPECTION PREV INSPECTION	SAFETY - RELIEF VALES SET AT	EXPLAIN IF PRESSURE CHANGED			
7	IS CONDITION OF OBJECT SUCH THAT A CERTIFICATE MAY BE ISSUED? YES ❑ NO ❑ (If no explain fully on back of form listing code violations)	HYDRO TEST ❑ YES PSI DATE ❑ NO				
8	SHELL No.	DIAMETER in. ❑ ID ❑ OD	OVERALL LENGTH ft. in.	THICKNESS in.	TOTAL HTG SURFACE (BOILER) Sq Ft	MATERIAL ASME Spec Nos
9	ALLOWABLE STRESS psi	BUTT STRAP Thks in. ❑ Single ❑ Double	HEADERS - W T BOILERS Thickness in	TYPE ❑ Box ❑ Sinuous ❑ Wtr Wall ❑ Other		
10	TYPE LONGITUDINAL SEAM ❑ Lap ❑ Butt ❑ Welded ❑ Brazed ❑ Riveted	REVETED Dia Hole in	PITCH in. X in. X	SEAM EFF %		
11	HEAD THICKNESS in.	HEAD TYPE ❑ Plus ❑ Minus ❑ Fixed ❑ Flat ❑ Movable ❑ Quick Opening	RADIUS DISH in.	ELLIP RATIO	BOLTING No. Dia. in. Material	
12	TUBE SHEET THICKNESS in	TUBES No. Dia. in Length ft. in.	PITCH (W T BLRS) in. X in.	LIGAMENT EFF %		
13	FIETUBE BOILERS	DISTANCE UPPER TUBES TO SHELL Front in. Rear in.	STAYED AREA Front Head Above Tubes Below Tubes	Rear Head Above Tubes Below Tubes		
	STAYS ABOVE TUBES Front No. Rear No	TYPE ❑ Head to Head ❑ Diagonal ❑ Welded ❑ Weldiess	AREA OF STAYS Front Rear			
	STAYS BELOW TUBES Front No Rear No.	TYPE ❑ Head to Head ❑ Diagonal ❑ Welded ❑ Weldless	AREA OF STAYS Front Rear			
14	FURNACE TYPE Adamson (No. Sect.) ❑ Corrugated ❑ Plain ❑ Other	THICKNESS in.	TOTAL LENGTH ft in.	TYPE LONG SEAM ❑ Welded ❑ Riveted ❑ Seamless		
15	STAYBOLTS - TYPE ❑ Threaded ❑ Welded ❑ Hollow ❑ Drilled (Size Hole in.)	DIAMETER in	PITCH in. X in.	NET AREA sq. in.		
16	SAFETY RELIEF VALUES No. Size	TOTAL CAPACITY Lb/Hr Cfm Btu/hr	OUTLETS No. Size	PROPERLY DRAINED ❑ Yes ❑ No (If "No" explain on the back of form)		
17	STOP VALVES	ON STEAM LINE ❑ Yes ❑ No	ON RETURN LINE ❑ Yes ❑ No	OTHER CONNECTIONS ❑ Yes ❑ No	STEAM LINES PROPERLY DRAINED ❑ Yes ❑ No (If "No" explain on the back of form)	
18	FEED PIPE Size in.	FEED APPLIANCE No.	TYPE DRIVE ❑ Steam ❑ Motor	CHECK VALVES	FEED LINE ❑ Yes ❑ No	RETURN LINE ❑ Yes ❑ No
19	WATER GAGE GLASS No.	TRY COCKS No.	BLOW OFF PIPE Size in. Location	INSPECTION OPENINGS COMPLY WITH CODE ❑ Yes ❑ No (If "No" explain on back of form)		
20	CAST IRON BOILERS Length in. Width in. Height in	SECTIONS No.	DOES WELDING OF STEAM, FEED, BLOWOFF & OTHER PIPING COMPLY WITH CODE ❑ Yes ❑ No (If "No" explain on the back of form)			
21	SHOW ALL CODES STAMPING ON BACK OF FORM Give details (use sketch) for special objects NOT covered above - such as Double wall vessel, etc	DOES ALL MATERIAL OTHER THAN AS INDICATED ABOVE COMPLY WITH CODE ❑ Yes ❑ No (If "No" explain on back of form)				
22	NAME AND TITLE OF PERSON TO WHOM REQUIREMENTS WERE EXPLAINED:					
23	I HEREBY CERTIFYTHIS IS A TRUE REPORT OF MY INSPECTION Signature of Inspector:	IDENT. NO	EMPLOYED BY	IDENT. NO.		

DI4-380
10-01-2000

OTHER CONDITIONS AND REQUIREMENTS

CODE STAMPING

BACK, DI4-380

Appendix G

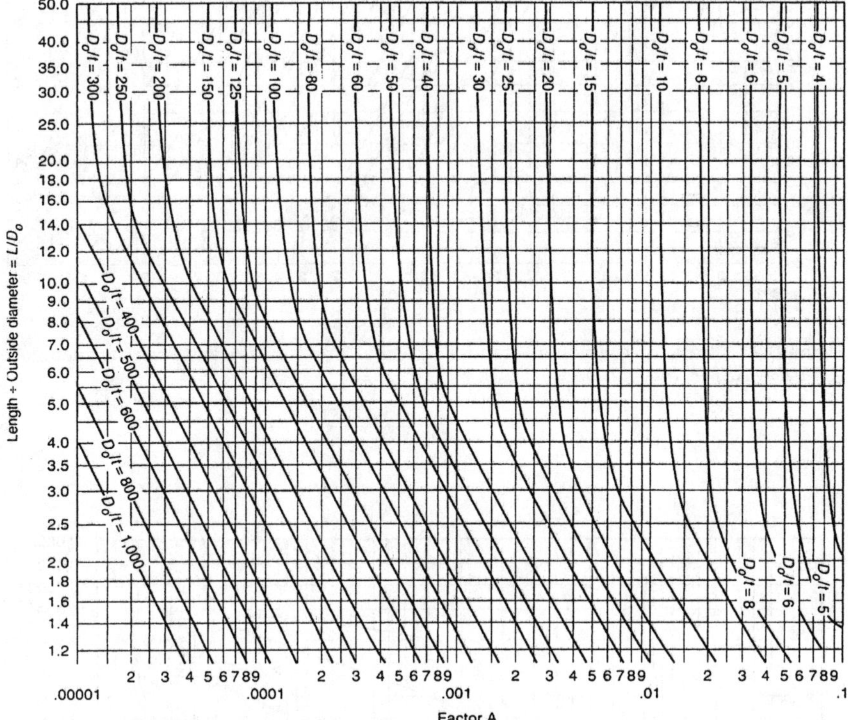

Figure G.1 Chart for components under external pressure for determining factor A. (*Courtesy ASME International*)

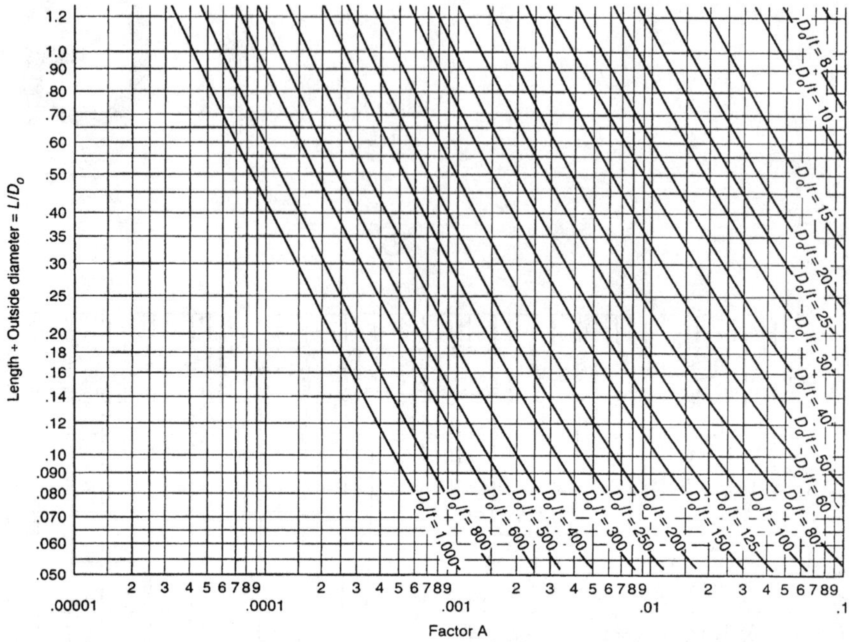

Figure G.2 Chart for components under external pressure for determining factor A. (*Courtesy ASME International*)

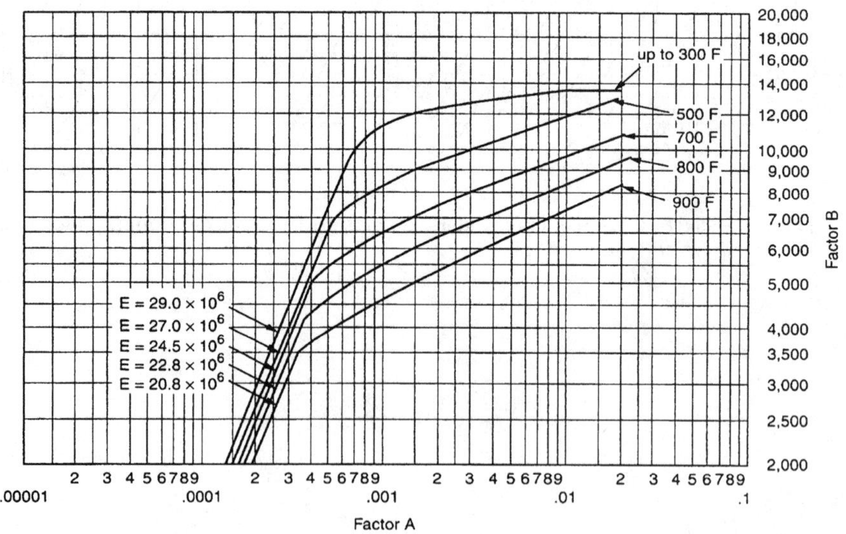

Figure G.3 Chart for determining factor B. (*Courtesy ASME International*)

Appendix H

FORM H-2 MANUFACTURER'S DATA REPORT FOR ALL TYPES OF BOILERS EXCEPT WATERTUBE AND THOSE MADE OF CAST IRON
As Required by the Provisions of the ASME Code Rules, Section IV

1. Manufactured and certified by: ___ (1) ___ (name and address of manufacturer)

2. Manufactured for: ___ (2) ___ (name and address of purchaser)

3. Location of installation: ___ (3) ___ (name and address)

4. Unit identification: ___ (4) ___ (complete boiler, superheater, waterwall, economizer, etc.) ___ (5) ___ (mfr.'s serial no.) ______ (CRN) ______ (drawing no.) ______ (Nat'l Bd. no.) ___ (6) ___ (year built)

5. The chemical and physical properties of all parts meet the requirements of material specifications of the ASME BOILER AND PRESSURE VESSEL CODE. The design, construction, and workmanship conform to ASME Code, Section IV, ___ (7) ___ (year) ___ (8) ___ (addenda (date)) ___ (9) ___ (Code Case no.)

Manufacturer's Partial Data Reports properly identified and signed by Commissioned Inspectors have been furnished for the following items of this report: ___ (30) (56) ___

______ (name of part, item number, mfr.'s name and identifying stamp)

6. Shells or drums: ___ (10) ___ (no.) ___ (11) ___ (mat'l spec., gr.) ___ (42) ___ (thickness (in.)) ___ (10) ___ (inside diameter) ______ (length (overall)) ______ (inside diameter) ______ (length (overall))

7. Joints: ___ (12) ___ (long. (seamless, welded)) ___ (13) ___ (eff. (compared to seamless)) ___ (12) ___ (girth (seamless, welded)) ______ (no. of shell courses)

8. Tubesheet: ___ (11) ___ (mat'l spec., grade) ______ (thickness) Tube holes: ______ (number and diameter)

9. Tubes: No. ___ (11) ___ (mat'l spec., grade) ______ (straight or bent) Diameter ______ Length ______ (if various, give max. and min.) Gauge ______ (or thickness)

10. Heads: ___ (11) ___ (mat'l spec. no.) ______ (thickness) ______ (flat, dished, ellipsoidal) ______ (radius of dish)

11. Furnace: ___ (11) ___ (mat'l spec., gr.) ______ (thickness) ___ (14) ___ (no.) ______ (size (O.D. or W x H)) ______ (length (each section)) ___ (15) ___ (total) ___ (16) ___ (type (plain, corrugated, etc.)) Seams: ______ (type (seamless, welded))

12. Staybolts: ______ (no.) ___ (17) ___ (size (diameter)) ___ (11) ___ (mat'l spec., gr.) ______ (size) ______ (telltale) ___ (18) ___ (net area) ______ (pitch (horiz. and vert.)) ___ (19) ___ (MAWP (psi))

13. Stays or braces:

Location	Material Spec.	Type	Number and Size	Pitch	Total Net Area	Fig. HG-343 L/l	Dist. Tubes to Shell	MAWP, psi
(a) F.H. above tubes	(11)	(20)			(21)	(22)		
(b) R.H. above tubes								
(c) F.H. below tubes								
(d) R.H. below tubes								
(e) Through stays								

14. Other parts: 1. ___ (23) ___ 2. ______ 3. ______ (brief description, i.e., dome, boiler piping)

1. ___ (24) ___ (11) ___

2. ______

3. ______ (mat'l spec., grade, size, material thickness, MAWP)

15. Nozzles, inspection and safety valve openings: (25)

Purpose (inlet, outlet, drain, etc.)	No.	Diameter or Size	Type	How Attached	Material	Nominal Thickness	Reinforcement Material	Location
Handhole up to 3 x 4 in.					NA		NA	
Manhole								
					(11)			

FORM H-2 (Back)

16. Boiler supports: ______ (no.) ______ (type [saddles, legs, lugs]) ______ (attachment [bolted or welded])

17. MAWP: ___(26)___ (psi) Based on ___(27)___ (Code par. and/or formula) Heating surface ___(28)___ [sq ft or kW (total)] Shop hydro. test ___(29)___ [psi (complete boiler)]

18. Maximum water temperature ___(74)___ °F

19. Remarks: ______

CERTIFICATE OF SHOP COMPLIANCE

We certify that the statements made in this data report are correct and that all details of design, material, construction, and workmanship of this boiler conform to Section IV of the ASME BOILER AND PRESSURE VESSEL CODE.

"H" Certificate of Authorization no. ___(31)___ expires ______ , ______ .

Date ___(32)___ Signed ______ (by representative) Name ______ (manufacturer that constructed and certified boiler)

(33) CERTIFICATE OF SHOP INSPECTION

Boiler constructed by ______ at ______ .

I, the undersigned, holding a valid commission issued by the National Board of Boiler and Pressure Vessel Inspectors and/or the state or province of ___(34)___ and employed by ______ have inspected parts of this boiler referred to as data items ___(35)___ and have examined Manufacturers' Partial Data Reports for items ___(36)___ and state that, to the best of my knowledge and belief, the manufacturer has constructed this boiler in accordance with the applicable sections of the ASME BOILER AND PRESSURE VESSEL CODE.

By signing this certificate neither the Inspector nor his employer makes any warranty, expressed or implied, concerning the boiler described in this Manufacturer's Data Report. Furthermore, neither the Inspector nor his employer shall be liable in any manner for any personal injury or property damage or a loss of any kind arising from or connected with this inspection.

Date ______ Signed ______ (Authorized Inspector) Commissions ___(37)___ [Nat'l. Bd. (incl. endorsements), state, prov., and no.]

CERTIFICATE OF FIELD ASSEMBLY COMPLIANCE

We certify that the field assembly construction of all parts of this boiler conforms with the requirements of SECTION IV of the ASME BOILER AND PRESSURE VESSEL CODE.

"H" Certificate of Authorization no. ___(38)___ expires ______ , ______ .

Date ___(39)___ Signed ______ (by representative) Name ______ (assembler that certified and constructed field assembly)

(40) CERTIFICATE OF FIELD ASSEMBLY INSPECTION

I, the undersigned, holding a valid commission issued by the National Board of Boiler and Pressure Vessel Inspectors and/or the state or province of ___(34)___ and employed by ______ have compared the statements in this Manufacturer's Data Report with the described boiler and state that the parts referred to as data items ___(41)___ , not included in the certificate of shop inspection, have been inspected by me and that to the best of my knowledge and belief the manufacturer and/or the assembler has constructed and assembled this boiler in accordance with the applicable sections of the ASME BOILER AND PRESSURE VESSEL CODE. The described boiler was inspected and subjected to a hydrostatic test of ______ psi.

By signing this certificate neither the Inspector nor his employer makes any warranty, expressed or implied, concerning the boiler described in this Manufacturer's Data Report. Furthermore, neither the Inspector nor his employer shall be liable in any manner for any personal injury or property damage or a loss of any kind arising from or connected with this inspection.

Date ______ Signed ______ (Authorized Inspector) Commissions ___(37)___ [Nat'l. Bd. (incl. endorsements) state, prov., and no.]

FORM H-3 MANUFACTURERS' DATA REPORT FOR WATERTUBE BOILERS
As Required by the Provisions of the ASME Code Rules, Section IV

HW-520/610/670

1. Manufactured and certified by A.O. Smith Water Products Co. – 5 25589 Highway # 1 North, McBee, South Carolina 29101
(name and address of manufacturer)

2. Manufactured for Stock
(name and address of purchaser)

3. Location of installation Not Known
(name and address)

4. Unit Identification HW-520 HW-610 HW-670 Hot Water Boiler (complete boiler, superheater, waterwall, etc.) | D04 02676- D04 02682 (mfr's serial no) | (CRN) | 191183-000 (drawing no.) | C 02676- C 02682 (Nat'l Bd. no.) | 2004 (year built)

CRN# K3517.518643YT2097

5. The chemical and physical properties of all parts meet the requirements of material specifications of the ASME BOILER AND PRESSURE VESSEL CODE. The design, construction and workmanship conform to ASME Code, Section IV, 2001 (year) 2003 (addenda (date)) N/A (Code Case no.)

6. (a) Drums:

No	Inside Diameter In.	Inside Length Ft.	Inside Length In.	Shell Plates Mat'l. Spec. Grade	Shell Plates Thickness In.	Shell Plates Inside Radius In.	Tube Sheets Thickness In.	Tube Sheets Inside Radius In.	Tube Hole Ligament Efficiency, % Longitudinal	Tube Hole Ligament Efficiency, % Circum-Ferential
1										
2										

No	Longitudinal Joints No. & Type	Longitudinal Joints Efficiency	Circum. Joints No. & Type	Circum. Joints Efficiency	Heads Mat'l Spec. Grade	Heads Thickness In.	Heads Type**	Heads Radius of Dish	Hydrostatic Test psi
1									
2									

*Indicate if (1) seamless, (2) fusion welded

**Indicate if (1) flat, (2) dished, (3) ellipsoidal, (4) hemispherical

6. (b) Boiler tubes:

Diameter	Thickness	Mat'l. Spec.No., Grade	No.	How Attached
.875"	.045"	SB-75 C12200	Coil	Brazed
.875"	.056"	SB-359 C12200	16	Brazed
1.00"	.057"	SB-359 C12200	1	Brazed

6. (c) Headers no. ______
(box or sinuous or round mat'l spec. no., thickness)

Heads or ends Box SB-62 .125 Hydro test, psi 240
(Shape mat'l, spec. no., thickness)

6. (d) Stay bolts: ______
(mat'l spec no., diameter, size, telltale, net area)

Pitch ______ in. Net area ______ sq. in. Design pressure ______ psi
(supported by one bolt)

6. (e) Mud drum: ______ Heads or ends ______ Hydro test, psi ______
(for sect. header boilers, state size, shape, mat'l, spec. no., thickness) (shape, mat'l spec. no, thickness)

7. Waterwall headers:

No.	Size and Shape	Material Spec. No., Gr.	Thickness In.	Shape	Thickness In.	Material Spec. No., Gr.	Hydro Test. Psi	Diameter In	Thickness In.	Material Spec. No., Gr.
1										
2										
3										

8. (a) Other parts (1)______(2)______(3)______ 8. (b) Tubes for other parts: ______

1										
2										
3										

9. Nozzles, inspection and safety valve openings:

Purpose (inlet, outlet, drain, etc.)	No.	Dia. or size	Type	How Attached	Mat'l	Nom. Thickness	Reinforcement Mat'l	Location
Handhole up to 3" x 4"					NA			
Manhole								
Inlet	1	2.00"	Thr'd	Brazed	SB-62	.125"	N/A	Bottom Manifold
Outlet	1	2.00"	Thr'd	Brazed	SB-62	.125"	N/A	Top Manifold
Temp. Probe	1	.750"	Thr'd	Brazed	SB-62	.125"	N/A	Top Manifold
Relief	1	1.00"	Thr'd	Brazed	SB-62	.125"	N/A	Top Manifold
Auxiliary	1	.500	Thr'd	Brazed	SB-62	.125"	N/A	Top Manifold

AOS QC-118

FORM H-3 (back)

		MAWP	Maximum Water Temp.	Shop Hydro Test Psi	Heating Surface sq. ft.
a	Boiler	160 P.S.I.	250 Deg. F	240	61.1
b	Waterwall				
c	Superheater				
d	Other parts				

Heating surface to be stamped on drum heads.

This heating surface not to be used for determining minimum safety valve capacity.

11. Field Hydro Test psi

12. Manufacturers' Partial Data Reports properly identified and signed by Commissioned Inspectors have been furnished for the following items of this report

N/A

"Wet 20.67 SQ. FT." (name of part, item number, mfr's name, and identifying stamp)

13. Remarks: Constructed under the provision of HG 515.4(b)

CERTIFICATE OF SHOP COMPLIANCE

We certify that the statements made in this data report are correct and that all details of design, material, construction, and workmanship of this boiler conform to Section IV of the ASME BOILER AND PRESSURE VESSEL CODE.
"H" Certificate of Authorization No. 34,077 expires August 13, 2006.

Date 4-23-04 Signed [signature] (by representative) G. Jackson M. Wait Name A. O. Smith Water Products Company – 5 (manufacturer that constructed and certified boiler)

CERTIFICATE OF SHOP INSPECTION

Boiler constructed by A.O. Smith Water Products Co.- 5 at 25589 Highway # 1 North, McBee, South Carolina 29101
I, the undersigned, holding a valid commission issued by The National Board of Boiler and Pressure Vessel Inspectors and the state or province of NC and employed by "HSB CT" have inspected parts of this boiler referred to as data items 5, 6b, 6c, 9, 10a and have examined Manufacturer's Partial Data Re- ports for items N/A and state that, to the best of my knowledge and belief, the manufacturer has constructed this boiler in accordance with Section IV of the ASME BOILER AND PRESSURE VESSELCODE.
By signing this certificate neither the inspector nor his employer makes any warranty, expressed or implied, concerning the boiler described in this Manufacturers' Data Report. Furthermore, neither the inspector nor his employer shall be liable in any manner for any personal injury or property damage or a loss of any kind arising from or connected with this inspection.

Date 4-23-04 Signed [signature] (Authorized Inspector) R.R. Williamson Commissions Nat'l. Bd. No. 7745A (Nat'l Bd.(incl. Endorsements) state, prov. and no.)

CERTIFICATE OF FIELD ASSEMBLY COMPLIANCE

We certify that the field assembly construction of all parts of this boiler conforms with the requirements of SECTION IV of the ASME BOILER AND PRESSURE VESSEL CODE.
"H" Certificate of Authorization no. ______ expires ______, 20 ______

Date ______ Signed ______ (by representative) Name ______ (assembler that certified and constructed field assembly)

CERTIFICATE OF FIELD ASSEMBLY INSPECTION

I, the undersigned, holding a valid commission issued by The National Board of Boiler and Pressure Vessel Inspectors and the state or province of ______ and employed by ______ have compared statements in this Manufacturers' Data Report with the described boiler and state that the parts referred to as data items ______, not included in the certificate of shop inspection, have been inspected by me and that to the best of my knowledge and belief, the manufacturer and/or the assembler has constructed and assembled this boiler in accordance with Section IV of the ASME BOILER AND PRESSURE VESSEL CODE. The described boiler was inspected and subjected to a hydrostatic test of 240 psi.
By signing this certificate neither the inspector nor his employer makes any warranty, expressed or implied, concerning the boiler described in this Manufacturers' Data Report. Furthermore, neither the inspector nor his employer shall be liable in any manner for any personal injury or property damage or a loss of any kind arising from or connected with this inspection.

Date ______ Signed ______ (Authorized Inspector) Commissions ______ (Nat'l Bd.(incl. endorsements) state, prov. and no.)

AOS QC-118

Appendix

J

Manufacturer's Master Data Report for Boilers Constructed from Cast Iron

FORM H-5 MANUFACTURER'S MASTER DATA REPORT FOR BOILERS CONSTRUCTED FROM CAST IRON

As Required by the Provisions of the ASME Code Rules, Section IV

1. Manufactured and certified by ______ (1) ______
(name and address of manufacturer) (foundry identification)

______

2. Boiler type or model no. ______ (4) ______

3. Boiler section data: (57)

Section Designation (list each individual section in boiler assembly)	Pattern and/or Part No	Metal Thickness According to Drawing
(a)		
(b)		
(c)		
(d)		
(e)		

4. Boiler section bursting data: (58)

Bursting Pressure, Metal Thickness Measured at *Break or Fracture, and Weight of Section*

Section Designation	Test No. 1	Test No. 2	Test No. 3
(a)	psi	psi	psi
	in.	in.	in.
	lb.	lb.	lb.
(b)	psi	psi	psi
	in	in.	in.
	lb	lb.	lb.
(c)	psi	psi	psi
	in	in.	in.
	lb	lb.	lb.
(d)	psi	psi	psi
	in	in	in.
	lb	lb	lb.
(e)	psi	psi	psi
	in	in.	in.
	lb.	lb	lb.

5. Minimum specified tensile strength ______ (59) ______ psi.

6. Tensile strength of associated test bars (60)

Section Designation	Bar for Test No 1	Bar for Test No 2	Bar for Test No. 3
(a)	psi	psi	psi
(b)	psi	psi	psi
(c)	psi	psi	psi
(d)	psi	psi	psi
(e)	psi	psi	psi

7. Specification no and class of gray iron ______

______

8 Maximum Allowable working pressure of boiler* ______ (61) ______ psi

*Determined by using formula in HC-402 and by selection of lowest values of bursting pressure from tests recorded in item 4 of this report.

FORM H-5 (Back)

9. Sketch of section with lowest bursting pressure

Show location of
failure and indicate whether
principally in bending or tension.

10. Examination data: (62)
 (a) Test engineer ____________________
 (name)
 (b) Witness of test(s) ____________________
 (name)
 (c) Date(s) of destruction tests ____________________
 ____
 ____

11. Manufacturer's certification: (63)
 Date ________, ____
 Certified to be true record
 ____
 (name and title)

Note: Signature of designated responsible engineering head of the manufacturer must be notarized.

12. ASME Certificate of Authorization no. ________ (31) ________ to use the "H" symbol (cast iron).
 Certificate expires ________, ____

Appendix K

BTP-540A-650A-740A
Model

Model

FORM HLW-6 MANUFACTURER'S DATA REPORT FOR WATER HEATERS OR STORAGE TANKS
As Required by the Provisions of the ASME Code Rules

1. Manufactured and certified by A. O. Smith Water Products Company-5, 25589 Highway 1 North, McBee, South Carolina, 29101, USA
(Name and address of manufacturer)

2. Manufactured for STOCK
(Name and address of purchaser)

3. Location of Installation Not Known
(Name and address)

4. Identification 86542-86556 (mfr's serial no.) L5131-5C (CRN) 195303 (drawing no.) 86542-86556 (Nat'l Bd. no.) 2004 (year built)

5. The chemical and physical properties of all parts meet the requirements of material specifications of the ASME BOILER AND PRESSURE VESSEL CODE. The design, construction, and workmanship conform to Part HLW, Section IV, 2001 (year) 03 (addenda (date)) (Code Case no.)

6. Shell: 1 (no.) SA414G (mat'l. spec., gr.) .125 (thickness (in.)) GLASS (lining) 25' 5.00"ID (dia. (in.)) 4' 4.00" (length (ft. & in.) (overall))

7. Joints: Welded (long. (seamless, welded)) 85% (eff. (as compared to seamless)) Welded (girth (seamless, welded)) 1 (no. of shell courses)

8. Heads:

Location	Mat'l Spec. Gr. Thickness	Crown Radius	Knuckle Radius	Elliptical Ratio	Hemispherical Radius	Flat Diameter	Side to Pressure (convex or concave)
TOP	SA414C .250"	N/A	1.06"		N/A	28.1875"	
BOTTOM	SA414C .250"	N/A	1.06"		N/A	28.1875"	

9. Tubesheet: SA414C (mat'l spec., grade) Tubes 24 (no.) 2.5" (size (in.)) 4' 2.5" (length (ft. & in.) (overall)) ASTM 513 (mat'l spec. gr.) 0.75" (thickness (in.)) Welded (rolled or welded)

10. Nozzles, inspection and safety valve openings:

Purpose (inlet, outlet, drain, etc.)	No.	Dia. or Size	Type	How Attached	Mat'l	Nom. Thickness	Reinforcement Mat'l.	Location
Handhole up to 3" x 4"	1	2.75X3.75	NA	Welded	NA	.688"	NA	Shell
Inlet and Outlet	2	1.50"	Thd'd	Welded	SA181-70	.375"	None	Head
Relief	1	1.00"	Thd'd	Welded	SA181-70	.300"	None	Shell
Anode	4	0.75"	Thd'd	Welded	SA181-70	.300"	None	Head
Thermostat	2	0.75"	Thd'd	Welded	SA181-70	.300"	None	Shell
Drain	1	0.75"	Thd'd	Welded	SA181-70	.300"	None	Shell
Inlet and Outlet	4	1.50"	Thd'd	Welded	SA181-70	.375"	None	Shell

11. MAWP: 160 (psi) MAX Input 740,000 BTU (BTU/hr or kW) MAX temp 210 (°F) Hydro test 160 (psi)

12. Manufacturer's Partial Data Reports properly identified and signed by Commissioned Inspectors,have been furnished for the following items for this report:

MAWP established by proof test HLW502.1 (d) (2) (b) and HLW503.1
Constructed under the provisions of HG515.4 (b)
(name of part, item no., mfr's name, identification stamps)

13. Remarks:
N/A

HLW-6 (BACK) NB Number 86542-86556

CERTIFICATE OF SHOP COMPLIANCE

We certify that the statements made in this report are correct and that all details of design, material, construction, and workmanship of the water heater or storage tank conform to Section IV of the ASME BOILER AND PRESSURE VESSEL CODE.

"HLW" Certificate of Authorization No. 21779 expires: 8/13/2006

date: 6/10/2004 Name A. O. Smith Water Products Company-5 (manufacturer that constructed the certified water heater or storage tank) signed (by representative)

CERTIFICATE OF SHOP INSPECTION

Constructed by A. O. Smith Water Products Company-5 at 25589 Highway 1 North, McBee, South Carolina, 29101, USA

I, the undersigned, holding a valid commission issued by The National Board of Boiler and Pressure Vessel Inspectors and the State or Province of NC and employed by HSB CT of Hartford, CT have inspected parts referred to as data items 4 thru 11 and have examined Manufacturer's Partial Data Report for items NA and state that, to the best of my knowledge and belief, the manufacturer has constructed the water heater or storage tank in accordance with Section IV of the ASME BOILER AND PRESSURE VESSEL CODE.

By signing this certificate, neither the Inspector, nor his employer makes any warranty, expressed or implied, concerning the water heater or storage tank described in this Manufacturer's Data Report. Furthermore, neither the Inspector nor his employer shall be liable in any manner for any personal injury or property damage or loss of any kind arising from or connected with this inspection.

Date: 6/10/2004 Signed (Authorized Inspector) Commissions: 7745A, NC1193 (Nat'l Board (incl. endorsements), State, Prov., and No.)

254665 exe: v3.5.10 NB-30 Rev. 3

Appendix L

MANUFACTURER'S/INSTALLING CONTRACTOR'S REPORT FOR ASME CSD-1

Certification and Reporting (CG-500) for Controls and Safety Devices

Unit Manufacturer

Name ____________________

Address ____________________ Zip ____________________

Telephone ____________________ Fax ____________________

Unit Identification (Boiler)

Manufacturer's Model # ____________________ Year Built ____________________

ASME # ____________________ Nat. Bd. # ____________________

UL # ____________________ AGA # ____________________

Jurisdiction ____________________

Steam

Max. W.P. ____________________ psig

Min. Safety Valve Cap. ____________________ PPH

Hot Water

Max. W.P. ____________________ psig

Max. Temp. ____________________ °F

Min. Safety Relief Valve Cap. ____________________ PPH or Btu

Boiler Unit Description (Type) ____________________

If Modular (No. of Modules) ____________________

Boiler Unit Capacity (Output) ____________________

Burner

Manufacturer ____________________ Model ____________________

UL or AGA # ____________________ Serial # ____________________

Fuels (as shipped) ____________________

	Indicate Units [1]
Gas Manifold Pressure	____________________
Oil Nozzle/Delivery Pressure (at max input)	____________________
High Gas Pressure Switch Setting	____________________
Low Oil Pressure Switch Setting	____________________

[1] Where not applicable, indicate "N/A."

Installation Location (if Known)

Customer Name ____________________

Address ____________________

City ____________________ State ____________________ Zip ____________________

Telephone ____________________ Fax ____________________

(*continued*)

Control/Device	Manufacturer	Model #	Operational Test Performed, Date
Operating Controls			
Low-Water Fuel Cutoff CW-120(a), CW-140			
Forced Circulation CW-210(a)			
Steam Pressure CW-310(b)			
Water Temperature CW-410(b)			
Safety Controls			
Low-Water Fuel Cutoff CW-120(a), CW-120(b) CW-130, CW-140			
Forced Circulation CW-210(b)			
High Steam Pressure Limit CW-310(c)			
High Water Temperature Limit CW-410(b)			
Fuel Safety Shutoff Valve, Main CF-180(b)(2), CF-180(b)(3)			
Pilot Safety Shutoff Valve CF-180(c)			
Atomizing Medium Switch CF-450(b)			
Combustion Air Switch CF-220			
High Gas Pressure CF-162			
Low Gas Pressure CF-162			
Low Oil Pressure CF-450(a)			
High Oil Temperature CF-450(c)			
Low Oil Temperature CF-450(d)			
Purge Air Flow CF-210			
Flame Safeguard (Primary) CF-310, CF-320			
Flame Detector CF-310, CF-320			
Low Fire Start			
Low-Fire Start Switch CF-610			
Safety or Safety Relief Valve(s) CW-510, CW-520			

Manufacturer ____________ Operational Test Performed, Date ____/____/____

Model ____________

Size ____________

Capacity ____________ PPH/Btu/hr

Representing Equipment Manufacturer, Name ____________

Signature ____________ Date ____________

Representing Installing Contractor, Name ____________

Signature ____________ Date ____________

Appendix M

Properties of Water

Temperature of Water (°F)	Saturation Pressure (Pounds per Square Inch Absolute)	Weight (Pounds per Gallon)	Specific Gravity 60°F/60°F	Conversion Factor,[1] Lbs./Hr. to GPM
32	0.0885	8.345	1.0013	0.00199
40	0.1217	8.345	1.0013	0.00199
50	0.1781	8.340	1.0007	0.00199
60	0.2653	8.334	1.0000	0.00199
70	0.3631	8.325	0.9989	0.00200
80	0.5069	8.314	0.9976	0.00200
90	0.6982	8.303	0.9963	0.00200
100	0.9492	8.289	0.9946	0.00201
110	1.2748	8.267	0.9919	0.00201
120	1.6924	8.253	0.9901	0.00201
130	2.2225	8.227	0.9872	0.00202
140	2.8886	8.207	0.9848	0.00203
150	3.718	8.182	0.9818	0.00203
160	4.741	8.156	0.9786	0.00204
170	5.992	8.127	0.9752	0.00205
180	7.510	8.098	0.9717	0.00205
190	9.339	8.068	0.9681	0.00206
200	11.526	8.039	0.9646	0.00207
210	14.123	8.005	0.9605	0.00208
212	14.696	7.996	0.9594	0.00208
220	17.186	7.972	0.9566	0.00209
240	24.969	7.901	0.9480	0.00210
260	35.429	7.822	0.9386	0.00211
280	49.203	7.746	0.9294	0.00215
300	67.013	7.662	0.9194	0.00217
350	134.63	7.432	0.8918	0.00224
400	247.31	7.172	0.8606	0.00232
450	422.6	6.892	0.8270	0.00241
500	680.8	6.553	0.7863	0.00254
550	1045.2	6.132	0.7358	0.00271
600	1542.9	5.664	0.6796	0.00294
700	3093.7	3.623	0.4347	0.00460

1. Multiply flow in pounds per hour by the factor to get equivalent flow in gallons per minute. Weight per gallon is based on 7.48 gallons per cubic foot.

Appendix N

Properties of Saturated Steam						
			Heat Content			
Gauge Pressure in. Hg Vac.	Absolute Pressure psia	Temperature °F	Sensible h_f Btu/lb	Latent h_{fg} Btu/lb	Total h_g Btu/lb	Specific Volume Steam V_g cu ft/lb
27.96	1	101.7	69.5	1032.9	1102.4	333.0
25.91	2	126.1	93.9	1019.7	1113.6	173.5
23.87	3	141.5	109.3	1011.3	1120.6	118.6
21.83	4	153.0	120.8	1004.9	1125.7	90.52
19.79	5	162.3	130.1	999.7	1129.8	73.42
17.75	6	170.1	137.8	995.4	1133.2	61.89
15.70	7	176.9	144.6	991.5	1136.1	53.57
13.66	8	182.9	150.7	987.9	1.138.6	47.26
11.62	9	188.3	156.2	984.7	1140.9	42.32
9.58	10	193.2	161.1	981.9	1143.0	38.37
7.54	11	197.8	165.7	979.2	1144.9	35.09
5.49	12	202.0	169.9	976.7	1146.6	32.35
3.45	13	205.9	173.9	974.3	1148.2	30.01
1.41	14	209.6	177.6	972.2	1149.8	28.00
Gauge Pressure psig						
0	14.7	212.0	180.2	970.6	1150.8	26.80
1	15.7	215.4	183.6	968.4	1152.0	25.20
2	16.7	218.5	168.8	966.4	1153.2	23.80
3	17.7	221.5	189.8	964.5	1154.3	22.50
4	18.7	224.5	192.7	962.6	1155.3	21.40
5	19.7	227.4	195.5	960.8	1156.3	20.40
6	20.7	230.0	198.1	959.2	1157.3	19.40
7	21.7	232.4	200.6	957.6	1158.2	18.60
8	22.7	234.8	203.1	956.0	1159.1	17.90
9	23.7	237.1	205.5	954.5	1160.0	17.20
10	24.7	239.4	207.9	952.9	1160.8	16.50
11	25.7	241.6	210.1	951.5	1161.6	15.90
12	26.7	243.7	212.3	950.1	1162.3	15.30
13	27.7	245.8	214.4	948.6	1163.0	14.80
14	28.7	247.9	216.4	947.3	1163.7	14.30
15	29.7	249.8	218.4	946.0	1164.4	13.90
16	30.7	251.7	220.3	944.8	1165.1	13.40
17	31.7	253.6	222.2	943.5	1165.7	13.00
18	32.7	255.4	224.0	942.4	1166.4	12.70
19	33.7	257.2	225.8	941.2	1167.0	12.30
20	34.7	258.8	227.5	940.1	1167.6	12.00
22	36.7	262.3	230.9	937.8	1168.7	11.40
24	38.7	265.3	234.2	935.8	1170.0	10.80
26	40.7	268.3	237.3	933.5	1170.8	10.30
28	42.7	271.4	240.2	931.6	1171.8	9.87
30	44.7	274.0	243.0	929.7	1172.7	9.46

continued

Properties of Saturated Steam (*Continued*)

Gauge Pressure psig	Absolute Pressure psia	Temperature °F	Heat Content: Sensible h_f Btu/lb	Heat Content: Latent h_{fg} Btu/lb	Heat Content: Total h_g Btu/lb	Specific Volume Steam V_g cu ft/lb
32	46.7	276.7	245.9	927.6	1173.5	9.08
34	48.7	279.4	248.5	925.8	1174.3	8.73
36	50.7	281.9	251.1	924.0	1175.1	8.40
38	52.7	284.4	253.7	922.1	1175.8	8.11
40	54.7	286.7	256.1	920.4	1176.5	7.83
42	56.7	289.0	258.5	918.6	1177.1	7.57
44	58.7	291.3	260.8	917.0	1177.8	7.33
46	60.7	293.5	263.0	915.4	1178.4	7.10
48	62.7	295.6	265.2	913.8	1179.0	6.89
50	64.7	297.7	267.4	912.2	1179.6	6.68
52	66.7	299.7	269.4	910.7	1180.1	6.50
54	68.7	310.7	271.5	909.2	1180.7	6.32
56	70.7	303.6	273.5	907.8	1181.3	6.16
58	72.7	305.5	275.3	906.5	1181.8	6.00
60	74.7	307.4	277.1	905.3	1182.4	5.84
62	76.7	309.2	279.0	904.0	1183.0	5.70
64	78.7	310.9	280.9	902.6	1183.5	5.56
66	80.7	312.7	282.8	901.2	1184.0	5.43
68	82.7	314.3	284.5	900.0	1184.5	5.31
70	84.7	316.0	286.2	898.8	1185.0	5.19
72	86.7	317.7	288.0	897.5	1185.5	5.08
74	88.7	319.3	289.4	896.5	1185.9	4.97
76	90.7	320.9	291.2	895.1	1186.3	4.87
78	92.7	322.4	292.9	893.9	1186.8	4.77
80	94.7	323.9	294.5	892.7	1187.2	4.67
82	96.7	325.5	296.1	891.5	1187.6	4.58
84	98.7	326.9	297.6	890.3	1187.9	4.49
86	100.7	328.4	299.1	889.2	1188.3	4.41
88	102.7	329.9	300.6	888.1	1188.7	4.33
90	104.7	331.2	302.1	887.0	1189.1	4.25
92	106.7	332.6	303.5	885.8	1189.3	4.17
94	108.7	333.9	304.9	884.8	1189.7	4.10
96	110.7	335.3	306.3	883.7	1190.0	4.03
98	112.7	336.6	307.7	882.6	1190.3	3.96
100	114.7	337.9	309.0	881.6	1190.6	3.90
102	116.7	339.2	310.3	880.6	1190.9	3.83
104	118.7	340.5	311.6	879.6	1191.2	3.77
106	120.7	341.7	313.0	878.5	1191.5	3.71
108	122.7	343.0	314.3	877.5	1191.8	3.65
110	124.7	344.2	315.5	876.5	1192.0	3.60
112	126.7	345.4	316.8	875.5	1192.3	3.54
114	128.7	346.5	318.0	874.5	1192.5	3.49

continued

Properties of Saturated Steam (*Continued*)

Gauge Pressure psig	Absolute Pressure psia	Temperature °F	Heat Content: Sensible h_f Btu/lb	Heat Content: Latent h_{fg} Btu/lb	Heat Content: Total h_g Btu/lb	Specific Volume Steam V_g cu ft/lb
116	130.7	347.7	319.3	873.5	1192.8	3.44
118	132.7	348.9	320.5	872.5	1193.0	3.39
120	134.7	350.1	321.8	871.5	1193.3	3.34
125	139.7	352.8	324.7	869.3	1194.0	3.23
130	144.7	355.6	327.6	866.9	1194.5	3.12
135	149.7	358.3	330.6	864.5	1195.1	3.02
140	154.7	360.9	333.2	862.5	1195.7	2.93
145	159.7	363.5	335.9	860.3	1196.2	2.84
150	164.7	365.9	338.6	858.0	1196.6	2.76
155	169.7	368.3	341.1	856.0	1197.1	2.68
160	174.7	370.7	343.6	853.9	1197.5	2.61
165	179.7	372.9	346.1	851.8	1197.9	2.54
170	184.7	375.2	348.5	849.8	1198.3	2.48
175	189.7	377.5	350.9	847.9	1198.8	2.41
180	194.7	379.6	353.2	845.9	1199.1	2.35
185	199.7	381.6	355.4	844.1	1199.5	2.30
190	204.7	383.7	357.6	842.2	1199.8	2.24
195	209.7	385.7	359.9	840.2	1200.1	2.18
200	214.7	387.7	362.0	838.4	1200.4	2.14
210	224.7	391.7	366.2	834.8	1201.0	2.04
220	234.7	395.5	370.3	831.2	1201.5	1.96
230	244.7	399.1	374.2	827.8	1202.0	1.88
240	254.7	402.7	378.0	824.5	1202.5	1.81
250	264.7	406.1	381.7	821.2	1202.9	1.74
260	274.7	409.3	385.3	817.9	1203.2	1.68
270	284.7	412.5	388.8	814.8	1203.6	1.62
280	294.7	415.8	392.3	811.6	1203.9	1.57
290	304.7	418.8	395.7	808.5	1204.2	1.52
300	314.7	421.7	398.9	805.5	1204.4	1.47
310	324.7	424.7	402.1	802.6	1204.7	1.43
320	334.7	427.5	405.2	799.7	1204.9	1.39
330	344.7	430.3	408.3	796.7	1205.0	1.35
340	354.7	433.0	411.3	793.8	1205.1	1.31
350	364.7	435.7	414.3	791.0	1205.3	1.27
360	374.7	438.3	417.2	788.2	1205.4	1.24
370	384.7	440.8	420.0	785.4	1205.4	1.21
380	394.7	443.3	422.8	782.7	1205.5	1.18
390	404.7	445.7	425.6	779.9	1205.5	1.15
400	414.7	448.1	428.2	777.4	1205.6	1.12
420	434.7	452.8	433.4	772.2	1205.6	1.07
440	454.7	457.3	438.5	767.1	1205.6	1.02
460	474.7	461.7	443.4	762.1	1205.5	.98

continued

Properties of Saturated Steam (*Continued*)						
			Heat Content			Specific Volume
Gauge Pressure psig	Absolute Pressure psia	Temperature °F	Sensible h_f Btu/lb	Latent h_{fg} Btu/lb	Total h_g Btu/lb	Steam V_g cu ft/lb
480	494.7	465.9	448.3	757.1	1205.4	.94
500	514.7	470.0	453.0	752.3	1205.3	.902
520	534.7	474.0	457.6	747.5	1205.1	.868
540	554.7	477.8	462.0	742.8	1204.8	.835
560	574.7	481.6	466.4	738.1	1204.5	.805
580	594.7	485.2	470.7	733.5	1204.2	.776
600	614.7	488.8	474.8	729.1	1203.9	.750
620	634.7	492.3	479.0	724.5	1203.5	.726
640	654.7	495.7	483.0	720.1	1203.1	.703
660	674.7	499.0	486.9	715.8	1202.7	.681
680	694.7	502.2	490.7	711.5	1202.2	.660
700	714.7	505.4	494.4	707.4	1201.8	.641
720	734.7	508.5	498.2	703.1	1201.3	.623
740	754.7	511.5	601.9	698.9	1200.8	.605
760	774.7	514.5	505.5	694.7	1200.2	.588
780	794.7	517.5	509.0	690.7	1199.7	.572
800	814.7	520.3	512.5	686.6	1199.1	.557

Appendix

Unit Conversions

O.1 Flow Rate

1 ft^3/hr	= 0.02832 m^3/hr
1 ft^3/min	= 0.472 L/sec
1 ft^3/sec	= 448.8 gals/min
1 m^3/hr	= 35.315 ft^3/hr
	= 264.2 U.S. gals/hr
	= 4.403 gals/min
1 Imperial gal/hr	= 1.2009 U.S. gals/hr
1 lb/hr	= 0.4536 kg/hr
1 kg/min	= 132.3 lb/hr
1 gal/min	= 0.06309 L/sec
	= 3.785 L/min
	= 0.2271 m^3/hr
	= 0.002228 ft^3/sec
1 L/hr	= 0.004403 gals/min
1 L/sec	= 15.8529 gals/min

O.2 Power and Heat

1 British thermal unit (Btu)	= 778 foot·pounds (ft·lb)
	= 0.252 kilocalorie (kcal)
	= 107.6 meter kilogram (mkg)
	= 1.055×10^3 joules (j)
1 therm·hour	= 100,000 Btu/hour (Btu/h)
	= 39.3082 brake horsepower (hp)
	= 2.9873 boiler horsepower (bhp)
1 calorie (cal)	= 3088 (ft·lb)
	= 3.968 Btu
	= 4.186 j
1 joule (j)	= 10^7 ergs
	= 9486×10^{-4} Btu
	= 0.7736 ft·lb
	= 1 watt second (W·s)
1 kilowatt (kW)	= 3414 Btu/h
	= 1000 W
	= 738 (ft·lb)/s
	= 1.341 horsepower (hp)
	= 102 mkg/s
1 kilowatt·hour (kWh)	= 3414 Btu
	= 860 kcal
	= 2.665×10^6 (ft·lb)
	= 3.6×10^6 j
1 megawatt (MW)	= 1000 kilowatts (kW)
1 horsepower (hp)	= 33,000 foot·pounds per minute
	= 550 ft·lb/s
	= 0.746 kW
	= 746 W
	= 76.0 mkg/s
1 horsepower·hour (hp·h)	= 2545 Btu
	= 1.98×10^6 ft·lb
	= 64.17 kcal

1 boiler horsepower (bhp)	= 33,475 Btu/h
	= 34.5 pounds per hour of evaporation from and at 212°F
	= 9.803 kW
	= 10 square feet of boiler heating surface (watertube boiler)
	= 12 square feet of boiler heating surface (firetube boiler)

O.3 Pressure

1 pound per square inch (psi)	= 0.06804 atmosphere (atm)
	= 6.895×10^3 pascals (pa)
	= 0.0703 kg/cm^2
1 kg/cm^2	= 0.9678 atm
	= 14.22 psi
	= 32.81 ft of water
	= 28.96 in of mercury
1 bar	= 0.9869 atm
	= 14.5 psi
	= 1×10^6 dynes/cm^2
1 atmosphere	= 14.696 psi
	= 29.92 inches (in) of mercury
	= 1.0333 kg/cm^2
1 in of water at 62°F	= 0.0361 psi
	= 0.07355 in of mercury
	= 5.20 pounds per square foot (psf)
	= 0.00254 kg/cm^2
1 ft head of water at 62°F	= 0.433 psi
	= 0.0295 atm
	= 0.8826 in of mercury
	= 0.03048 kg/cm^2
1 in of mercury	= 0.491 psi
	= 1.133 ft of water
	= 0.03453 kg/cm^2

O.4 Temperature

°F = (1.8 × °C) + 32

°C = (°F − 32)/1.8

°R = °F + 460

°K = °C + 273

O.5 Density of Water (at 62°F)

1 cubic foot (ft^3)	= 62.5 lb
	= 7.48 gallons (gal)
1 pound (lb)	= 0.01604 cu ft^3
	= 0.1198 gal
1 gallon (gal)	= 8.33 lb
	= 277.3 cubic inches (in^3)
1 long ton of water	= 36 ft^3

O.6 Length

1 inch (in)	= 2.54 centimeters (cm)
	= 25.4 millimeters (mm)
1 foot (ft)	= 12 in
	= 30.48 cm
1 yard (yd)	= 3 ft
	= 0.914 meter (m)
1 mile (mi)	= 5280 ft
	= 1760 yd
	= 1.609 kilometers (km)
1 meter (m)	= 100 cm
	= 1000 mm
	= 1.094 yd
	= 3.28 ft
	= 39.37 in
1 kilometer (km)	= 1000 m
	= 0.621 mile (mi)

O.7 Area

1 square inch (in^2)	= 6.45 cm^2
	= .006944 ft^2
1 square foot (ft^2)	= 144 in^2
	= 0.0929 m^2
1 square yard (yd^2)	= 9 ft^2
	= 0.836 m^2
1 square mile (mi^2)	= 640 acres
	= 2.590 km^2
	= 3.098×10^6 yd^2
	= 2.590×10^6 m^2
1 acre	= 43,560 ft^2
	= 4840 yd^2
	= 4.047×10^3 m^2
1 square meter (m^2)	= 10,000 cm^2
	= 11.196 yd^2
	= 10.76 ft^2
1 square centimeter (cm^2)	= 100 mm^2
	= 0.155 in^2
1 nautical mile	= 6080 ft
	= 1.853 km

O.8 Volume

1 cubic inch (in^3)	= 16.39 cm^3
	= 0.0005787 ft^3
1 cubic foot (ft^3)	= 1728 in^3
	= 28.32 liters (L)
	= 0.02832 m^3
1 cubic yard (yd^3)	= 27 ft^3
	= 0.765 m^3
1 cubic meter (m^3)	= 1000 L
	= 1.308 yd^3
	= 35.31 ft^3
1 Imperial gallon	= 277.4 in^3
	= 4.55 L
1 U.S. gallon	= 0.833 Imperial gal
	= 3.785 L
	= 231 in^3
1 liter (l)	= 1000 cm^3
	= 0.22 Imperial gal
	= 0.2642 U.S. gal
	= 61 in^3

O.9 Weight

1 pound (lb)	= 16 ounces (oz)
	= 700 grains (gr)
	= 454 grams (g)
	= 0.454 kg
1 grain (gr)	= 64.8 mg
	= 0.0648 g
	= 0.0023 oz
1 gram (g)	= 1000 mg
	= 0.03527 oz
	= 15.43 gr
1 kilogram (kg)	= 1000 g
	= 2205 lb
1 U.S. short ton	= 2000 lb
	= 907 kg
1 U.S. long ton	= 2240 lb
	= 1016 kg
1 metric ton	= 1000 kg
	= 0.948 U.S. long ton
	= 1.102 U.S. short ton
	= 2205 lb

Bibliography

Codes and Standards

ASME Boiler and Pressure Vessel Code Section IV—Rules for Construction of Heating Boilers, 2004 Edition, ASME International, NY.

ASME Boiler and Pressure Vessel Code Section II, Materials Part A, Part B, Part C, & Part D, 2004 Edition, ASME International, NY.

ASME Boiler and Pressure Vessel Code Section V—Nondestructive Examination, 2004 Edition, ASME International, NY.

ASME Boiler and Pressure Vessel Code Section VI—Recommended Rules for the Care and Operation of Heating Boilers, 2004 Edition, ASME International, NY.

ASME Boiler and Pressure Vessel Code Section IX—Welding and Brazing Qualifications, 2004 Edition, ASME International, NY.

ASME/ANSI B16.34, *Valves—Flanged, Threaded and Welding Ends,* 1998 Edition, ASME International, NY.

ASME CSD-1, *Control and Safety Devices for Automatically Fired Boilers,* 2002 Edition, ASME International, NY.

ASME PTC 25, *Power Test Code for Pressure Relief Valves,* 2001 Edition, ASME International, NY.

NFPA 31, Standard for the *Installation of Oil-Burning Equipment,* 2001 Edition, National Fire Protection Association, Quincy, MA.

NFPA 54, *National Fuel Gas Code,* 2002 Edition, National Fire Protection Association, Quincy, MA.

UL 726, *Standard for Oil-fired Boiler Assemblies,* 2001 Edition, Underwriters Laboratories, Northbrook, IL.

UL 795, *Standard for Commercial—Industrial Gas-Fired Heating Equipment,* 1999 Edition, Underwriters Laboratories, Northbrook, IL.

UL 834, *Standard for Electric Boilers,* 2004 Edition, Underwriters Laboratories, Northbrook, IL.

ANSI Z21.13, *Gas-Fired Low-Pressure Steam and Hot Water Boilers,* American National Standards Institute, Washington, DC.

Books

Chen, Elliot, and Swanekamp, *Standard Handbook of Power Plant Engineering,* Second Edition, New York: McGraw-Hill.

Improving Steam System Performance—A Sourcebook for Industry, DOE/GO-102002-1557, June 2002, Office of Industrial Technologies, U.S. Department of Energy.

Malek, M. A., and G. M. Halley, *Chapter 19—ASME Section VI: Recommended Rules for the Care and Operation of the Heating Boilers, Companion Guide to the ASME Boiler and Pressure Vessel Code,* 2002, New York: ASME International.

Malek, M. A., and J. I. Woodworth, *Chapter 18—ASME Section IV: Rules for the Construction of Heating Boilers, Companion Guide to the ASME Boiler and Pressure Vessel Code,* 2002, New York: ASME International.

Spring, H. M., and A. L. Kohan, *Boiler Operator's Guide,* Second Edition, New York: McGraw-Hill.

Woodruff, E. B., H. B. Lammers, and T. F. Lammers, *Steam Plant Operation,* Sixth Edition, New York: McGraw-Hill.

Magazines

Malek, M. A., Special Edition 1998, "Fundamentals of Boiler Scale," *National Board Bulletin,* an Official Publication of the National Board of Boiler and Pressure Vessel Inspectors.

Malek, M. A., "Inspection of Water Heater," *National Engineer*, The Journal of the National Association of Power Engineers, July 1999.

Malek, M. A., "Rules for De-rating a Boiler," *National Engineer,* The Journal of the National Association of Power Engineers, November/December 2001.

Manufacturer's Catalogs

"Steam and Hot Water. High and Low Pressure." Sellers Engineering Company.

"Product Overview Catalog," Triad Boiler Systems, Inc.

"Water Heater Catalog," RBI Water Heaters, Canada.

"Efficiency at Work," Weben—Jarco, Inc.

"Copper Finned Gas-Fired Mighty Therm," Teledyne Laars.

"Hydronic Fitters Guide," Slant/Fin Corporation.

"Multiple Boiler Systems," Weil-McLain

"Rite Boilers," Rite Engineering & Manufacturing Corporation

"Industrial/Commercial Cast Iron Boilers," Peerless Industries, Inc.

"Commercial Gas Water Heaters," Lochinvar Corporation

"Cast Iron Gas Fired Steam Boilers," Dunkirk Boilers

"Precision Cooking Systems," Vulcan-Hart

"The Market Forge Sterilmatic Steam Sterilizer," Market Forge Industries, Inc.

"Steam Separators and Heat Recovery Systems for Low and High Pressure Boilers," Penn Separator Corporation.

"Plumbing and Heating Products," Conbraco Industries International.

"Boiler Tube Expanders and Accessories," Thomas C. Wilson, Inc

"Firetube Boiler Cleaning Equipment," Goodway Technologies Corporation.

"Maintenance Free Circulators," Grundfos

"General Catalog," McDonnell & Miller

"Turbo-Ring Forced Draft Burners," Gordon-Piatt Energy Group, Inc.

"General Purpose Flow Switch," Johnson Controls, Inc.

Index